Industrial and Applied Mathematics

The Industrial and Applied Mathematics series publishes high-quality research-level monographs, lecture notes and contributed volumes focusing on areas where mathematics is used in a fundamental way, such as industrial mathematics, bio-mathematics, financial mathematics, applied statistics, operations research and computer science.

More information about this series at http://www.springer.com/series/13577

Pammy Manchanda · René Lozi
Abul Hasan Siddiqi
Editors

Industrial Mathematics and Complex Systems

Emerging Mathematical Models, Methods and Algorithms

Springer

Editors
Pammy Manchanda
Department of Mathematics
Guru Nanak Dev University
Amritsar, Punjab
India

Abul Hasan Siddiqi
School of Basic Sciences and Research
Sharda University
Greater Noida, Uttar Pradesh
India

René Lozi
CNRS, Dieudonné Center of Mathematics
University Côte d'Azur
Nice
France

ISSN 2364-6837 ISSN 2364-6845 (electronic)
Industrial and Applied Mathematics
ISBN 978-981-10-3757-3 ISBN 978-981-10-3758-0 (eBook)
DOI 10.1007/978-981-10-3758-0

Library of Congress Control Number: 2017946022

Printed on acid-free paper

This Springer imprint is published by Springer Nature
The registered company is Springer Nature Singapore Pte Ltd.
The registered company address is: 152 Beach Road, #21-01/04 Gateway East, Singapore 189721, Singapore

Preface

The present volume is based on the selected invited and contributory talks during the international conference held at Sharda University, Greater Noida, India, during January 29–31, 2016, on the occasion of the Silver Jubilee of the Indian Society of Industrial and Applied Mathematics. The conference was inaugurated by Mr. Rajnath Singh, the Minister for Home Affairs, Government of India. Professor U.P. Singh, former Vice-Chancellor of Purvanchal University, and former President of the Indian Mathematical Society, chaired the inaugural session. Professor KR. Sreenivasan, former Director of Abdus Salam ICTP, Trieste, Italy (UNESCO organization), and currently a senior functionary of New York University, was the guest of honor who delivered the keynote address. There were 20 invited speakers and more than 300 participants from different parts of India and abroad. A good number of participants presented their research work as contributory talks. Abdus Salam International Centre for Theoretical Physics (ICTP) and International Mathematical Union (IMU) provided financial support for the participants from Malaysia, Turkey, and Uzbekistan. Several functionaries of the International Council for Industrial and Applied Mathematics (ICIAM) and eminent industrial and applied mathematicians such as Prof. Barbara Lee Keyfitz (immediate past President of ICIAM), Prof. Maria J. Esteban (President, ICIAM), Prof. Alistair Fitt (Vice-Chancellor, Oxford Brooke University, and former secretary, ICIAM), Prof. Leon O. Chua (University of Berkeley), Prof. Guenter Leugering (Vice-President, International Affairs, Friedrich Alexander University, Erlangen-Nurenberg, Germany), and Prof. Maria Skopina (Euler Institute of Mathematical Sciences, Saint Petersburg State University, Russia) participated and delivered lectures. Three well-known Indian centers of applied and industrial mathematics: Tata Institute of Fundamental Research—Centre For Applicable Mathematics (TIFR CAM), Bengaluru; Indian Institute of Science (IISc), Bengaluru; and Indian Institute of Technology (IIT) Bombay, were represented by Prof. G.D.V. Gowda, Prof. A. Adimurthi, Prof. Mythly Ramaswamy, Dr. Venkateswaran P. Krishnan, Prof. G. Rangarajan, and Prof. A.K. Pani.

The Silver Jubilee Committee of ISIAM honored Prof. Gowda, Prof. Pani, and Prof. Rangarajan for their valuable contributions in the field of industrial and applied mathematics. Professor A. Adimurthi (TIFR CAM, Bengaluru) and Prof. Mushahid Husain (Vice-Chancellor, Mahatama Jyotiba Phule (MJP) Rohilkhand University, Bareilly, Uttar Pradesh) were given the Dr. Zakir Husain award instituted by the Duty Society, Aligarh Muslim University (AMU). Messages by the then President of India, Shri Pranab Mukherjee, and the Minister for Science and Technology, Dr. Harsh Vardhan, wished for a successful event.

The invited and contributory talks cover various areas of applied mathematics and represent the latest advances in the interdisciplinary fields such as mathematics, environmental science, medical sciences, oil exploration and production, dynamical systems, and biological sciences.

In Chap. 1, Barbara Lee Keyfitz surveys the current development in linear and nonlinear waves in gas dynamics. She has discussed many open research problems in this field. Maria J. Esteban presents an account of nonlinear flows and optimality for functional inequalities in Chap. 2, based on her joint work with Jean Dolbeault and Michael Loss. It is mainly related to rigidity results for nonnegative solutions of the semi-linear elliptic equation on infinite cylinder-like domains or in the Euclidean space and as a consequence, about optimal symmetry properties for the optimizers of the Caffarelli–Kohn–Nirenberg inequalities.

Chapter 3 by David Walnut deals with theory and applications of frames. He discusses some situations in which frames have proven an especially useful tool, namely noise reduction, robust communications, compressive sensing, and phaseless recovery. In Chap. 4, challenging problems of industrial applications of multicore-implemented nonlinear mappings are discussed by Rene Lozi, Jean-Pierre Lozi, and Oleg Garasym.

In Chap. 5, Guenter Leugering, Falk M. Hante, Alexander Martin, Lars Schewe, and Martin Schmidt discuss the challenges in optimal control problems for gas and fluid flow in networks of pipes and canals. Chapter 6 is devoted to the recent work of Majaz Moonis jointly with Ahmedul Kabir, Carolina Ruiz, and Sergio A. Alvarez on "comparison of conventional regression in the machine learning methods for stroke outcome prediction," which will be published in full in the next issue of the *Indian Journal of Industrial and Applied Mathematics*. Professor Moonis in his talk also discussed the role of imaging in medical sciences, particularly in acute ischemic stroke. Besides stroke outcome prediction, the chapter is also devoted to the controversy of CAT versus MRI in stroke management, which is an ongoing hot debate.

In Chap. 7, B.I. Golubov and S.S. Volosivets present their new results on "Fourier transforms of multiplicative convolutions." Chapter 8 by Maria Skopina studies tight wavelet frames with matrix dilations, which can be used in many practical situations. Chapter 9 by Akhtar Khan jointly with M. Cho, B. Jadamba, R. Kahler, and M. Sama is devoted to the development of a computational framework for the inverse problem of identifying variable parameters appearing nonlinearly in a variational problem.

In Chap. 10, M. Brokate and M. Yu Rasulova present the solution of the hierarchy of quantum kinetics equations with delta potential. In Chap. 11, Yeliz Karaca, Zafer Aslan, and A.H. Siddiqi present the applications and comparison of 1-DWT transform and partial correlation multiple sclerosis and subgroup diagnostic classification. In Chap. 12, V. Gowda discusses his recent results in the domain of finite volume method for nonlinear system of hyperbolic conservation laws arising in oil reservoir simulation. In Chap. 13, Meenakshi and P. Manchanda study certain properties of Haar–Vilenkin wavelets for solving differential equations. It may be mentioned that Haar–Vilenkin wavelet was introduced by them along with A.H. Siddiqi in 2008. In Chap. 14, Rohit Khokher and Ram Chandra Singh study footprint-based personal recognition using the dactyloscopy technique.

In Chap.15, M. Dilshad, A.H. Siddiqi, Rais Ahmad, and Faizan A. Khan present their latest results on a class of variational inequalities. In Chap. 16, Sudip Chakraborty, Sonia Chowdhury, Joydeep Pal, and Priti Kumar Roy discuss the impact of vaccination to control HPV dynamics.

In Chap. 17, Chhavi Mangla, Harsh Bhasin, Musheer Ahmad, and Moin Uddin present their work on the innovative solution of nonlinear equations using genetic algorithm. Rakesh Kumar and Bhupender Kumar Som describe their results on an M/M/c/N feedback queuing model with reverse balking and reneging in Chap. 18. In Chap. 19, U.M. Pirzada and D.C. Vakaskar study solution of fuzzy heat equation under fuzzified thermal diffusivity. Saureesh Das and Rashmi Bhardwaj present their work on chaos in nanofluidic convection of CuO nanofluid in Chap. 20.

Bhanumati Panda, Anumeha Dube, and Sushil Kumar study the dynamics of the seasonal variability of plankton and forage fish in Chilika Lagoon using npzf model in Chap. 21. Chapter 22 by Fahad Al Basir, Sushil Kumar, and Priti Kumar Roy deals with the effect of glycerol kinetics and mass transfer during enzymatic biodiesel production from Jatropha oil.

Chapter 23 by Jahangir Chowdhury, Sourav Rana, Sabyasachi Bhattacharya, and Priti Kumar Roy is devoted to the work "role of bio-pest control on theta logistic populations: a case study on Jatropha curcus cultivation system." In Chap. 24, Sudipa Chauhan, Sumit Kaur Bhatia, and Nidhi Purohit present their work on the dynamics of SIRS model with single time delay.

All invited and contributory talks could not find a place in this volume due to one reason or the other. Therefore, a summary of some of the presented talks, compiled by Pooja, is presented in Chapter 25. Chapter 25 contains a summary of some of the invited and contributory talks by L.O. Chua, G. Pfander, G. Rangarajan, A. Adimurthi, G. Fairweather, Venky Krishnan, Samares Pal, L.M. Saha, Vikram, Pooja, Mamta Rani, Abdullah, Renu Chugh and Mandeep Kumari, M.K. Ahmad and Santosh Kumar, Deepti Gupta, Puneet Kaur, Mazibar Rahman, Javed Miya and M.A. Ansari, A.K. Sahoo and G.S. Mishra, Shelly Arora and Amandeep Kaur, S. Prabhakaran and L. Jones T. Doss, Vivek Kumar and Bhola Ishwar, Noor e Zahra, Ruchira Aneja and A.H. Siddiqi, Mijanur Rehman, Nitendra and Khursheed, Nagma Irfan and A.H. Siddiqi.

Amritsar, India Pammy Manchanda
Nice, France René Lozi
Greater Noida, India Abul Hasan Siddiqi

Acknowledgements

Editors and organizers gratefully acknowledge the financial support of the National Board for Higher Mathematics (NBHM), Department of Science and Technology, New Delhi; Abdus Salam International Centre of Theoretical Physics (ICTP), Trieste, Italy; International Mathematics Union (IMU), Berlin, Germany; Indian National Science Academy (INSA); and Duty Society, Aligarh Muslim University (AMU). The organizing committee and editors also express their gratitude to the Chancellor of Sharda University, Mr. P.K. Gupta, for generous support without which the conference could not have been organized at this large scale. We also take this opportunity to thank Dr. Thomas Hempfling of Springer Basel, Switzerland, who spared his valuable time to grace the inaugural function. Cooperation of Mr. Shamim Ahmad of Springer India is highly appreciated.

Pammy Manchanda
René Lozi
Abul Hasan Siddiqi

Contents

About the Editors

Pammy Manchanda is a senior professor of mathematics at the Guru Nanak Dev University, Amritsar, India. She has published 44 research papers in several reputed international journals, edited two proceedings for international conferences of the Indian Society of Industrial and Applied Mathematics (ISIAM), and co-authored three books. She has attended, delivered talks, and chaired sessions at reputed academic conferences and workshops across the world, including ICIAM (1999–2015) and the International Congress of Mathematicians (ICM) since 2002. She is the Managing Editor of the *Indian Journal of Industrial and Applied Mathematics* and a member of the editorial board of the Springer book series *Industrial and Applied Mathematics*.

René Lozi is a professor at the Dieudonné Center of Mathematics, University of Nice, France. In 1991, he became a full professor at the University of Nice and the Institute for Teacher Trainees (IUFM), France. Earlier, he served as the Director of this institute (2001–2006) and as the Vice Chairman of the French Board of Directors of the IUFM (2004–2006). He received his French State Thesis (on chaotic dynamical systems) under the supervision of Prof. René Thom (a Fields Medalist) in 1983. In 1977, he discovered a particular mapping of the plane having a strange attractor (now classically known as the "Lozi map"). Today, his research areas include complexity and emergence theory, dynamical systems, bifurcation, control of chaos, cryptography-based chaos, and recently memristor (a physical device for neuro-computing).

Abul Hasan Siddiqi is a professor emeritus at the School of Basic Sciences and Research, Sharda University, Greater Noida, India. He is also the elected President of the Indian Society of Industrial and Applied Mathematics (ISIAM). He has jointly published more than 100 research papers and five books with his research collaborators, and edited the proceedings of nine international conferences. He is the Founder Secretary of the ISIAM, which celebrated its Silver Jubilee in January 2016. He is Editor in Chief of the *Indian Journal of Industrial and Applied Mathematics*, published by ISIAM and *Industrial and Applied Mathematics*, a book series with Springer.

Chapter 1
Linear and Nonlinear Waves in Gas Dynamics

Barbara Lee Keyfitz

Abstract Although systems of hyperbolic conservation laws form an important model for many phenomena in fluid dynamics, including compressible flow, surface waves in shallow water, reacting fluids, magnetohydrodynamics, and multiphase flow, the underlying theory of quasilinear hyperbolic systems in more than one space variable is poorly developed. This survey outlines a few reasons for the absence of a comprehensive theory and examines some current research on multidimensional problems. When one examines the structure of the characteristics of the gas dynamics equations, it is noteworthy that they fall into two distinct types, which could be called "nonlinear" and "linear". Each type governs some aspects of a solution, and the two types interact in complicated ways. The study of examples gives many suggestions for further research, although we are still far from a theory.

Keywords Hyperbolic conservation laws · Multidimensional conservation laws Wave interactions · Linear and nonlinear characteristics

1.1 Introduction and Background

We begin with a review of the terminology of "characteristics". As a motivating example, the first-order equation

$$\frac{\partial u}{\partial t} + a\frac{\partial u}{\partial x} = 0 \tag{1.1}$$

has the general solution

$$u(x, t) = f(x - at),$$

where f is a function of a single variable. When f is sufficiently smooth, say $f \in \mathscr{C}^1$, we speak of a *classical* solution; but the usual mechanism—multiply by a test function

B.L. Keyfitz (✉)
The Ohio State University, Columbus, OH, USA
e-mail: keyfitz.2@osu.edu

P. Manchanda et al. (eds.), *Industrial Mathematics and Complex Systems*,
Industrial and Applied Mathematics, DOI 10.1007/978-981-10-3758-0_1

and integrate by parts—allows us to consider any distribution, $f \in \mathscr{D}'$, as a *weak* solution. (For quasilinear hyperbolic systems, of which the equations of gas dynamics are an example, additional admissibility conditions are generally required in order to distinguish between physically meaningful weak solutions and spurious ones. That is not the focus of the paper, so I will not go into detail on this point. See [15] for background on admissibility conditions, and [16], mentioned in Sect. 1.4.1, for an example of how the current knowledge of admissibility criteria appears to be incomplete.)

In this example, the *characteristic curves* are the lines $x - at$ constant. Their significance is well known:

- They identify the space-time paths for propagation of signals
- They separate regions of smooth flow (to see this, consider isolated singularities in f)
- They are unsuitable for prescribing data: data on a line $x - at = x_0$ cannot be given consistently and would not serve to determine f.

Characteristics also have a geometric significance:

- In physical space, $\mathbb{R}^2 = (x, t)$, the vector $\mathbf{t} = (a, 1)$ (constructed from the coefficients of the equation) is tangent to each characteristic at every point.
- In a dual space $\mathbb{R}^2 = (\xi, \tau)$, the *characteristic normals* $\boldsymbol{\nu} = (1, -a)$ are normal to those tangents: $\boldsymbol{\nu} \cdot \mathbf{t} = 0$.

The characteristic normals, which are the basic objects, are the solutions of the equation $\tau + a\xi = 0$ obtained from the *principal symbol*, $\tau + a\xi$, of the linear differential operator $\partial_t + a\partial_x$.

This generalizes to a first-order system of n equations,

$$\sum_0^d A_j \frac{\partial \mathbf{u}}{\partial x_j} + \mathbf{b} = 0\,,$$

where now the coefficients A_j are $n \times n$ matrices, and the state variable $\mathbf{u}$ and lower order term $\mathbf{b}$ are n-vectors. The principal symbol is now a matrix, $L_0 = \sum_0^d A_j \xi_j$, and the characteristic normals are the solutions $\boldsymbol{\nu} = (\xi_0, \ldots, \xi_d)$ of the determinantal equation $\det(\sum A_j \xi_j) = 0$. Now the surfaces in $\mathbb{R}^{d+1} = \{(x_0, \ldots, x_d)\}$ whose normals are characteristic are the characteristic surfaces. From here the story plays out differently depending on whether the matrices A_j are constant, depend on x ($A_j = A_j(x)$, giving a linear system) or depend on x and $\mathbf{u}$ ($A_j = A_j(x, \mathbf{u})$, a quasilinear system). In the case of a quasilinear or a fully nonlinear system, the characteristic normals are defined with respect to linearization of the equation around a state $\mathbf{u}_0$.

The theory of partial differential equations began with the Cauchy–Kovalevsky theorem, see [20] for example, which proved the existence of a unique analytic solution locally for any system, provided that the system was defined entirely by analytic functions, and that analytic data were given on an analytic, noncharacteristic surface. Characteristics appear again in the classification of partial differential equations by type. A system is said to be *elliptic* if there are no real characteristics, and *hyperbolic* if it has a maximal set of real characteristics (in a sense that we will explain). Clearly, these two categories are not exhaustive. In addition, while characteristics and their detailed properties play an important—practically a defining—role in the theory of quasilinear hyperbolic partial differential equations (conservation laws), that is a reflection of the dominant physical phenomenon—wave motion—modeled by conservation laws. For many applications, including processes in which reaction, diffusion, or dispersion plays the decisive role, an analysis that focuses on characteristics misses the point.

We turn our attention now to hyperbolic systems. For completeness, we note that much classical theory of hyperbolic equations treats second- (or higher-)order equations rather than first-order systems. The two approaches can be unified, see [39]; we will not attempt to do so here, but we will draw on familiar examples of second-order equations for reference.

The notion of a "maximal set of characteristics" is not well defined without the identification of a distinguished variable. For example, the characteristic equation of the two-dimensional wave equation, $u_{tt} = c^2(u_{xx} + u_{yy})$, is $\tau^2 = c^2(\xi^2 + \eta^2)$. The roots of this equation are always real if we are solving the equation for τ but not if we are solving for ξ or η. In examples that come from a physical problem, it is usually (though not always) clear which variable should be distinguished as *timelike*. We will now assume that has been done, and will rename the first variable, x_0, as t. Then the quasilinear system of n equations,

$$A_0 \frac{\partial \mathbf{u}}{\partial t} + \sum_1^d A_j \frac{\partial \mathbf{u}}{\partial x_j} + \mathbf{b} = 0\,,$$

with $A_j = A_j(x, t, \mathbf{u})$ and $\mathbf{b} = b(x, t, \mathbf{u})$ allowed to depend on $\mathbf{u}$, is *hyperbolic* at any states $(x, t, \mathbf{u})$ where the equation

$$\det(A_0\tau + \sum A_j\xi_j) = 0$$

has, for all $\xi = (\xi_1, \ldots, \xi_d)$, n real roots, $\tau_i(x, t, \mathbf{u}; \xi)$, the characteristic normals. A system is said to be *symmetrizable hyperbolic* if it can be put in this form with A_0 symmetric and positive definite, and the remaining A_j symmetric for all j. It is an important, now classic, result that the Cauchy problem for *linear* symmetrizable hyperbolic systems is well-posed in the Sobolev space H^s, see [24, 44] for example.

1.1.1 An Example

To fix ideas, consider the linear wave equation in two space dimensions, $u_{tt} = c^2(u_{xx} + u_{yy})$. We can write it as a system, defining

$$u_1 = u_y\,, \quad u_2 = u_t - cu_x\,.$$

Then the wave equation becomes

$$\partial_t \begin{pmatrix} u_1 \\ u_2 \end{pmatrix} = c \begin{pmatrix} 1 & 0 \\ 0 & -1 \end{pmatrix} \partial_x \begin{pmatrix} u_1 \\ u_2 \end{pmatrix} + c \begin{pmatrix} 0 & 1 \\ 1 & 0 \end{pmatrix} \partial_y \begin{pmatrix} u_1 \\ u_2 \end{pmatrix}.$$

The characteristic normals are $\boldsymbol{\nu} = (\xi, \eta, \pm c\sqrt{\xi^2 + \eta^2})$; we can also define characteristic variables,

$$\mathbf{v} = \begin{pmatrix} \eta \\ \xi \mp \sqrt{\xi^2 + \eta^2} \end{pmatrix};$$

these are the eigenvectors $\{\mathbf{v} = \mathbf{v}(\boldsymbol{\nu}) \in \mathbb{R}^n \mid L_0\mathbf{v} = 0\}$ of the principal symbol. The characteristic normals can be written in the form $\tau^2 = c^2(\xi^2 + \eta^2)$; they determine the *characteristic cone*. For each generator of the cone, there is a characteristic surface, the plane $\xi x + \eta y \pm c\sqrt{\xi^2 + \eta^2}t = 0$. The envelope of the characteristic surfaces through $(0, 0, 0)$ forms the *wave cone*, $x^2 + y^2 = c^2t^2$. The wave cone exemplifies several important features of solutions of hyperbolic systems:

- It forms the boundary of the domain of influence of the origin;
- It bounds the support of the fundamental solution of the wave equation (the support includes the interior of the cone);
- The *singular support* of a solution is the boundary of the cone;
- A typical solution to a Cauchy problem, say with data $u(x, y, 0) = 0, u_t(x, y, 0) = u_0(x, y)$, takes the form

$$u(t, x, y) = \frac{1}{4\pi c} \int_B \frac{u_0(\xi, \eta)}{\sqrt{c^2t^2 - (x - \xi)^2 - (y - \eta)^2}} \, d\xi \, d\eta \equiv K_W * u_0,$$

  where $B = \{(\xi, \eta) \mid (x - \xi)^2 + (y - \eta)^2 \le c^2t^2\}$;
- The convolution kernel, $K_W(\cdot, t)$ lies in $H^{-d/2+1-\varepsilon}$ for any $\varepsilon > 0$.

As a reminder, the Sobolev spaces H^s or $W^{s,2}$ consist of functions whose s^{th} order derivatives are L_2 integrable. For negative and fractional values of s, $W^{s,p}$ is most conveniently defined by means of Fourier transforms, see [1].

1.1.2 The Problem

Now we get to the point: The analysis of quasilinear hyperbolic systems is faced with what appear to be incompatible restrictions. On the one hand, the linear theory of hyperbolic systems extends to local (meaning for a limited time) existence for quasilinear symmetrizable hyperbolic systems in H^s, for $s > d/2 + 1$. (While linear systems have solutions in H^s when the data is in H^s (even negative values), quasilinear systems are constrained by a natural limitation that s must be large enough for the coefficient matrices $A_j = A_j(x, t, \mathbf{u})$ to be defined.) The life span of H^s solutions depends on $\|\mathbf{u}_0\|_s$ (and on the geometry of the problem, expressed through the flux matrices A_j).

However, solutions of quasilinear systems do not remain in H^s for all time. Burgers equation provides a simple, but compelling, example: The equation $u_t + uu_x = 0$ with data $u(x, 0) = u_0$ is well known to have a solution given in implicit form by $u(x + u_0(x)t, t) = u_0(x)$, and a simple application of the implicit function theorem makes clear that this solution breaks down at $t = \min(-1/u_0'(x))$. For larger values of t, the problem has only a weak solution, with a discontinuity along a shock line, $x = s(t)$. The well-developed theory of conservation laws in a single space dimension gives global-in-time existence (in fact well-posedness) in BV, the space of functions of bounded variation. (There are some additional restrictions in this theory, which we will not detail here.) However, BV is not compatible with the Sobolev spaces H^s, which are based on L_2, and it is not clear whether, even in a single space dimension, systems of conservation laws are well-posed in L_2.

On the other hand, besides the local result mentioned above, there are results for linear and quasilinear systems, which we will discuss below, that suggest that no multidimensional hyperbolic systems are well-posed except in the spaces H^s.

Some 25 years ago, I, along with a group of other people, began to explore this quandary by looking at a number of examples, and by focusing on self-similar solutions to conservation laws in two space dimensions, where the self-similar reduction gives one a system in two independent variables. I would like to begin by acknowledging the efforts and contributions of my coauthors, some of whom continue to work on these problems. Their names appear in the appropriate sections, and in the references. Other researchers also entered this field at about the same time, and I have tried to include references, although my list is far from complete, to their work. Besides Chen, Feldman, and their associates [10–12], note the work of Elling [17], Elling, and Liu [18, 19]. Slemrod, Wang, and their coauthors [9, 43] have noted a relationship between multidimensional problems coming from conservation law theory and the geometric problem of isometric embedding.

1.2 Compressible Gas Dynamics in Two Space Dimensions

The equations of compressible ideal gas dynamics in two space dimensions form a system of four equations. This system is sometimes called the "full Euler system", a terminology which is easily confused with the incompressible Euler equations, about which more later. Compressible, or high-speed, flow is an important and well-studied topic, both in engineering and computational science (computational fluid dynamics, or CFD). These equations are an appealing target for analysis. In addition to the possibility of comparison with computational results, many simplified models are available. We emphasize, though, that the theory for this example is just as incomplete as for general quasilinear hyperbolic systems.

The compressible gas dynamics equations take the form

$$\begin{aligned} \rho_t + (\rho u)_x + (\rho v)_y &= 0 \\ (\rho u)_t + (\rho u^2 + p)_x + (\rho u v)_y &= 0 \\ (\rho v)_t + (\rho u v)_x + (\rho v^2 + p)_y &= 0 \\ (\rho E)_t + (\rho u H)_x + (\rho v H)_y &= 0\,, \end{aligned} \tag{1.2}$$

expressing conservation of mass, momentum, and energy. The state variables are ρ (density), (u, v) (velocity), and p (pressure); the nonlinear functions appearing in the equations are the energy and enthalpy, defined by

$$E = \frac{1}{\gamma - 1}\frac{p}{\rho} + \tfrac{1}{2}(u^2 + v^2), \quad \text{and} \quad H = \frac{\gamma}{\gamma - 1}\frac{p}{\rho} + \tfrac{1}{2}(u^2 + v^2)\,. \tag{1.3}$$

These involve a parameter γ, the ratio of specific heats, a constant determined by the chemistry of the gas (for air, it has the value 1.4).

When we linearize these equations at a constant state $\mathbf{u} = (\rho, u, v, p)$, we find that the characteristic equation $\det L_0 = 0$ has four real roots:

$$\bar{\tau} = 0, \;\; 0, \;\; \pm\sqrt{\gamma p/\rho}\sqrt{\xi^2 + \eta^2}\,, \;\; \text{where } \bar{\tau} = \tau - (\xi u + \eta v)\,.$$

1.2.1 Characteristic Normals, Familiar and Unfamiliar

The normals are sketched in Fig. 1.1, for $u = v = 0$. The four families of roots fall into two classes:

1. The pair $\tau = (\xi u + \eta v)$. In this case, all normals lie in a plane normal to $(u, v, -1)$. The corresponding characteristic surfaces are planes whose envelope is the direction $(u, v, -1)$. The dynamic behavior of a characteristic variable associated with this characteristic is that of a solution to a scalar equation with such

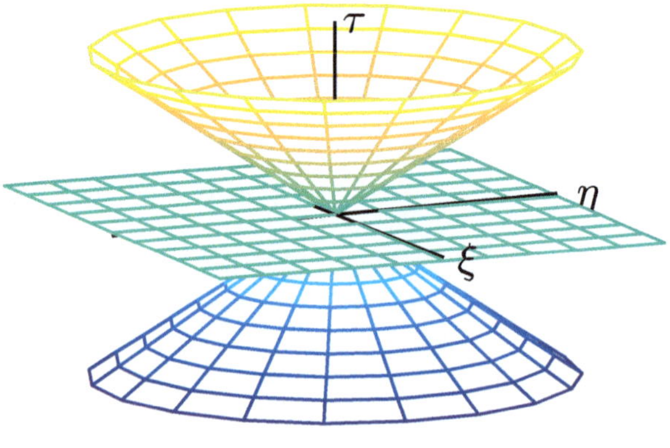

Fig. 1.1 Characteristic normals for the compressible Euler equations

a characteristic normal. That is, a characteristic variable corresponding to this family is modeled by a transport equation, $w_t + uw_x + vw_y = 0$, similar to (1.1). The domain of influence of the point (0, 0, 0) is the line $(x, y, t) = (-ut, -vt, t)$. Since this is a double characteristic, one might also anticipate flow features seen in hyperbolic equations with multiple characteristics. Mizohata [37] and Lax [33] discuss the question of multiple characteristics in some detail.
2. The pair $\tau = (\xi u + \eta v) \pm \sqrt{\gamma p/\rho}\sqrt{\xi^2 + \eta^2}$. The characteristic normals form a pair of conical surfaces (forward and backward in time) somewhat like the linear wave equation discussed in Sect. 1.1.1. This time, the wave cone (the envelope of the corresponding characteristic surfaces) is a tilted cone whose axis depends on u and v, and whose opening angle depends on p and ρ.

We can contrast the families in 1 and 2 in two ways. First, we see "transport equation" versus "wave equation" as the mode of propagation. Either a signal travels along a ray (in 1), or it spreads into a circle (in 2). Second, the waves in 2 exhibit a "nonlinear" type of propagation, where the speed of a signal depends on the states (as is the case for Burgers equation), while the waves in 1 are *linearly degenerate* in the terminology of conservation laws. (The definition here is that a characteristic family is linearly degenerate or genuinely nonlinear according as $\nabla\tau(\mathbf{u}, \xi) \cdot \mathbf{r}(\mathbf{u}, \xi) = 0$ or $\neq 0$.) In terms of the phenomena modeled by the gas dynamics system, the nondegenerate, nonlinear waves in 2 are the acoustic waves (where discontinuities are shocks), while the other two families, in 1, correspond to entropy and vorticity waves, where the discontinuities are known as contact discontinuities or slip lines.

For general quasilinear hyperbolic systems, other possible combinations may exist, but they may not correspond to anything that occurs in physical models.

1.2.2 Acoustic Waves in Two Dimensions

We began our study of two-dimensional problems by focusing on the nonlinear, acoustic waves. My coauthors in this work include Čanić, Lieberman, Kim, Jegdić, Tesdall, Popivanov, Payne, and Ying. Most of this research concerns self-similar problems, where the data and the solution are taken to be functions of

$$\xi = \frac{x}{t} \quad \text{and} \quad \eta = \frac{y}{t}.$$

Note that these reduced variables are not the same as the dual variables (ξ, η) introduced above to define characteristic normals. Note also that initial data for self-similar problems appear to be given at infinity. In practice, a self-similar problem has data $\mathbf{u}(x, y, 0)$ that are constant on rays from the origin. These data give rise, naturally, to boundary value problems in the ξ-η plane with boundary conditions at infinity. For sectorially constant data (the situation we consider), these problems can be restated as boundary value problems on a bounded domain. A general discussion of the

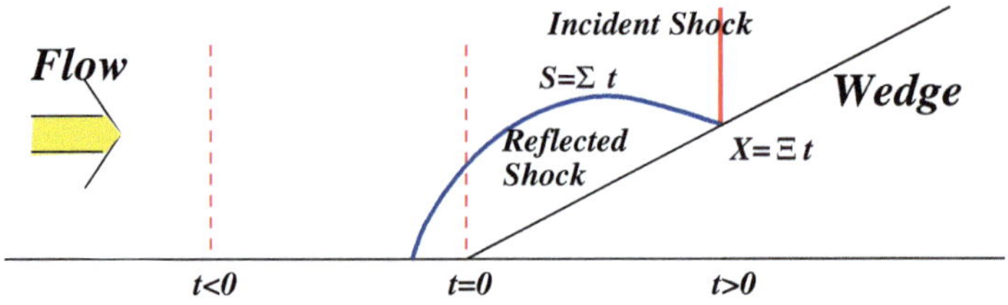

Fig. 1.2 Shock reflection by a wedge

background of self-similar problems can be found in [30]. A brief investigation of self-similar problems when that data are not piecewise constant is the subject of [49].

Such problems, also known as two-dimensional Riemann problems, are interesting for several reasons, [3, 30]. By analogy with one-dimensional problems, they allow one to focus on simple transport and wave interactions. In one space dimension, solutions to Riemann problems form the bedrock of the theory, allowing one to construct solutions to general initial value problems, as well as providing both local and asymptotic estimates. While there is little hope that they will be as instrumental in solving Cauchy problems in higher dimensions, they are interesting in themselves in two dimensions, as there are important problems that admit a self-similar formulation, such as the benchmark problem of shock reflection by a wedge, pictured in Fig. 1.2.

Our investigation began with Čanić, Lieberman, and myself [8] developing a method, using Schauder-type fixed point theorems in weighted Hölder spaces, for the steady transonic small disturbance equation, a simple model system that mimics the qualitative behavior of the nonlinear waves, and has no linear waves at all. Then this was modified, with Eun Heui Kim [4, 5], to handle first strong and then weak regular reflection for the unsteady transonic small disturbance system.

This research was further advanced by Jegdić [26, 27] to handle the nonlinear wave system, for which the behavior of the 'acoustic' waves is very similar to that for the gas dynamics equations, and which, in addition, contains a single linear wave family (with a much simpler structure than in gas dynamics, since it is possible to decouple the acoustic variables from the linear ones). Jegdić's result, like most what we have done, is local in the sense that it shows the existence of the reflected shock and downstream flow only in a neighborhood of the reflection point, introducing a cutoff function to replace the downstream boundary. For the nonlinear wave system, Čanić and I, with Kim, also solved a related problem, constructing a prototype of a Mach stem [7]. For that problem, we were able to prove existence of a solution in the entire domain, without cutoff functions.

Before describing our work, I should mention that other researchers have adopted a similar approach using a slightly different idea, originally conceived by Gui-Qiang Chen and Mikhail Feldman, and explained in their book, [13]. A principal difference between their method and ours is that they have chosen a model for gas dynamics that assumes potential flow. This allows for a stream function, and reduces the problem to computing a flow potential. It also simplifies the boundary conditions, and removes the linear waves completely. We refer to their monograph, [13], for details on their approach and further references to related work.

In the self-similar coordinates ξ and η, we obtain a reduced equation of the form

$$(-\xi I + A(\mathbf{u}))\mathbf{u}_\xi + (-\eta I + B(\mathbf{u}))\mathbf{u}_\eta = 0\,.$$

Here A and B are the Jacobian matrices of the flux functions. (The system can also be put in conservation form, with a source term.) What is significant is that the type of this equation changes. It is hyperbolic for $(\xi, \eta) >> 1$ but, at least for gas dynamics, there is a "subsonic" region near the origin. The change of type occurs in the nondegenerate, acoustic waves only, [6], and in general leads to a system of "mixed" type with some characteristics real and some complex. The change of type can be understood by noting that near the origin one is "inside" the wave cone for the interesting part of the flow, where waves are interacting.

It is reasonable to reformulate the system. In the subsonic region, we replace one of the equations in the first-order system by a second-order equation, to take advantage of the well-developed theory of second-order elliptic equations. For the purpose of this overview we take as our model the simpler, three-equation system for isentropic gas dynamics, rather than the full system (1.2). For the isentropic system, the pressure is given as a function of density and there is no equation for the pressure. We have the first three equations of (1.2) with $p = p(\rho)$ (often taken as $p = A\rho^\gamma$ with the same parameter γ). In the subsonic region, where there are two complex characteristics, the equation for ρ becomes

$$\begin{aligned} Q(\rho; U, V) = (c^2 - U^2)\rho_{\xi\xi} - 2UV\rho_{\xi\eta} + (c^2 - V^2)\rho_{\eta\eta} + 2cc'(\rho_\xi^2 + \rho_\eta^2) \\ -2\rho_\xi\left(U(1 + U_\xi + V_\eta) - c^2\tfrac{U(V_\eta+1)-VV_\xi}{U^2+V^2}\right) \\ -2\rho_\eta\left(V(1 + U_\xi + V_\eta) + c^2\tfrac{UU_\eta-V(U_\xi+1)}{U^2+V^2}\right) = 0 \end{aligned} \tag{1.4}$$

Examination of the principal part shows that the equation changes type at the sonic line,

$$(u - \xi)^2 + (v - \eta)^2 \equiv U^2 + V^2 = c^2 \equiv c^2(\rho)\,.$$

Equation (1.4) is an equation for ρ, but it is clearly coupled with the pseudovelocity components $U = u - \xi$ and $V = v - \eta$ and their derivatives. To complete the system, we also use the transport equations for U and V. These take the form

$$\begin{aligned} (U, V)\cdot\nabla U + U &= -p_\xi/\rho = -c^2\rho_\xi/\rho \equiv q_\xi(\rho) \\ (U, V)\cdot\nabla V + V &= -p_\eta/\rho = -c^2\rho_\eta/\rho \equiv q_\eta(\rho)\,; \end{aligned}$$

and in turn involve ρ.

1.2.3 The Free Boundary Problem and a Local Solution

We have solved boundary value problems for this equation, as well as for the simpler models listed in Sect. 1.2.2 where the density and velocity variables are not coupled in such a complicated way. As a first attempt, we showed existence of solutions for the unsteady transonic small disturbance system, for both strong [4] and weak [5] regular reflection. Later, these results were extended to an artificial system, the nonlinear wave system, in both the strong [27] and weak [26] case. Our most recent results are for the isentropic gas dynamics system [28]. Our method to handle strong regular reflection is described in this section. We are currently completing existence theorems for weak regular reflection for the isentropic case, and extending this construction to the full gas dynamics system. In all cases, we look at regular reflection, and in all cases we have solved a local problem, near the reflection point, with a cutoff function replacing a downstream boundary condition.

The key to this approach is to formulate a free boundary problem for the reflected shock, whose position is coupled with the subsonic flow behind the reflection point. Because Eq. (1.4) is nonlinear, we have found that classical methods for handling the free boundary problem to be the most suitable.

The complete problem reads:
Equations

$$\left.\begin{array}{l} Q(\rho; U, V) = 0 \\ (U, V)\cdot\nabla U + U = -p_\xi/\rho\,, \\ (U, V)\cdot\nabla V + V = -p_\eta/\rho \end{array}\right\} \quad \text{in} \quad \Omega\,. \tag{1.5}$$

Boundary conditions (see Fig. 1.3 for the geometry of the boundary curves)

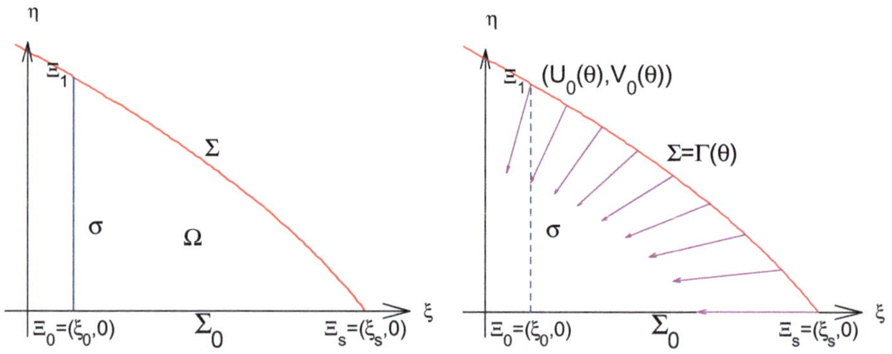

Fig. 1.3 Domain for the subsonic flow *Left*; Subsonic domain with the transport flow *Right*

$$\left.\begin{array}{lll} \rho_\eta = 0 & \text{on} & \Sigma_0 \\ N(\rho) \equiv \beta \cdot \nabla\rho = 0 & \text{on} & \Sigma \\ \rho = f & \text{on} & \sigma \\ \rho = \rho_s & \text{at} & \Xi_s \\ U|_\Sigma = U_0 = u_0 - \xi \;\; V|_\Sigma = V_0 = v_0 - \eta\,. \end{array}\right\} \tag{1.6}$$

Condition on the free boundary given by evolution of the shock

$$\frac{d\eta}{d\xi} = F(\rho, U, V). \tag{1.7}$$

The oblique derivative condition on Σ, as well as the boundary conditions for U and V there, and the equation for the evolution of the shock, are all determined from the Rankine–Hugoniot conditions, which define what it means to be a weak solution along a line of discontinuity. The condition $\rho(\Xi_s) = \rho_s$ in (1.6), where ρ_s is the density immediately behind the shock as calculated from the equations of the shock polar, is necessary, since otherwise the homogeneous boundary conditions would produce a constant solution if the cutoff function f happened to be constant. The theory of elliptic boundary value problems does not permit a one-point condition like this. However, we have the following result.

Theorem 1.1 *The boundary condition at Ξ_s cannot be prescribed but there are choices of f that give $\rho(\Xi_s) = \rho_s$.*

We summarize our approach to the free boundary problem. As applied to the isentropic gas dynamics equations, where one first encounters a serious interaction between the linear and the nonlinear waves, the approach begins by linearizing the system about states $(\rho, U, V) = (w, W, Z)$, fixing an approximate position for the free boundary and solving the linearized version of the first equation in (1.5),

$$L\rho = \sum a^{ij}(w, W, Z)\partial_{ij}\rho + \sum b^i(w, W, Z, \nabla W, \nabla Z)\partial_i\rho = 0$$

along with the first three boundary conditions in (1.6), in what is now a fixed domain Ω. The nonlinear oblique derivative condition is also linearized, as

$$M\rho = \beta(w, W, Z) \cdot \nabla\rho = 0$$

on the approximate shock curve Σ. We proved, consistent with Theorem 1.1 above,

Theorem 1.2 *For each (w, W, Z), there are choices of f that give a solution to the linearized problem for ρ with $\rho(\Xi_s) = \rho_s$.*

The proof of this theorem applies the theory of oblique derivative problems and mixed (oblique derivative and Dirichlet) problems in Lipschitz domains, developed by Lieberman. Then, from the solution of the linear equation, using compactness of the solution operator of a linear elliptic equation, one can prove, using a fixed point theorem,

Theorem 1.3 *For each (W, Z), the mapping $w \mapsto \rho$ has a fixed point.*

To handle the coupling of the nonlinear waves, as represented by ρ, and the linear family, as represented by the pseudovelocities U and V, we then proved that the complete nonlinear system consisting of the Eq. (1.5) and the boundary conditions (1.6), always with a suitable fixed approximate shock position, has a solution. The key to this is the observation that

Theorem 1.4 *The mapping $(W, Z) \mapsto (U, V)$ is a contraction.*

In order to explain this somewhat surprising result, we note the important point that the dependence of ρ on (W, Z) is very smooth. This, again, is a consequence of properties of elliptic equations. If we let $\rho[W, Z]$ stand for the solution in Theorem 1.2, then an estimate is given by

Theorem 1.5 *For $\rho_1 = \rho[W_1, Z_1]$ and $\rho_2 = \rho[W_2, Z_2]$, we have*

$$\left|\rho_1 - \rho_2\right|_{2+\varepsilon}^{(-(\gamma+1))} \leq M\left(\left|W_1 - W_2\right|_{1+\varepsilon}^{(-\gamma)} + \left|Z_1 - Z_2\right|_{1+\varepsilon}^{(-\gamma)}\right) . \tag{1.8}$$

With this in hand, we can show that the contraction property gives a fixed point in a sufficiently small domain, the hence a solution to the fixed boundary problem. Finally, we integrate the shock evolution equation (1.7). This gives a mapping $T\eta = \widetilde{\eta}$ from a given approximation η for the shock position to a new approximation, $\widetilde{\eta}$. The mapping T is also compact, and a classical fixed point theorem yields the existence of the desired solution, which we state here as

Theorem 1.6 (Jegdić, Keyfitz, Čanić and Ying) *In a small region behind the reflection point, there is a solution to the self-similar equations, and a corresponding position for the reflected shock.*

Summarizing, the components of this method involve

- the use of Hölder norms weighted at corners; the parameter γ, between 0 and 1 and not related to the ratio of specific heats in (1.3), is determined by the geometry of the domain, and allows for less smoothness of the solution at the corners
- standard elliptic estimates for $\rho_1 - \rho_2$
- obtaining good bounds near Σ, since we may use the final boundary condition in (1.6) to specify that $(W_1, Z_1) = (W_2, Z_2)$ at Σ.

The details of the proof can be found in [28].

1.2.4 Mysteries

While the technique we have outlined above may yet prove to be quite useful in solving shock reflection problems that are more complex than the cases we have studied up to this time, there are many self-similar problems for which we do not yet have a good enough understanding of the underlying dynamics to formulate solvable boundary value problems.

One example is the phenomenon of Guderley Mach reflection. This may occur for Mach numbers and wedge angles where regular reflection is not possible but simple Mach reflection is also ruled out. A calculation based on shock properties (specifically shock polar analysis, see the monograph of Courant and Friedrichs, [14], for example) shows that there are situations where regular reflection is not geometrically possible, but there is not enough energy in the linear waves to resolve the so-called triple-point paradox (that three shocks cannot meet at a single point in space). For instance, the unsteady transonic small disturbance equation, where there are no linear waves at all, and the nonlinear wave system, where energy is not transferred between the nonlinear and the linear waves, generate examples.

Guderley [21] conjectured that for initial conditions that do not permit either regular or Mach reflection, then a reflected shock that looks somewhat like a Mach stem but is fundamentally different might appear, and he suggested a structure in which a supersonic region appears behind the reflected shock. In 2002, Tesdall and Hunter exhibited this phenomenon numerically in the unsteady transonic small disturbance equations [48], and later Tesdall, Sanders and I carried out similar computations that found the same pattern in the nonlinear wave system [46] and in the gas dynamics equations [47].

The computations indicate that Guderley's theory is correct, with the modification that there is a whole series of supersonic patches and secondary shocks. The pattern was confirmed experimentally by Skews and his group [42], motivated by Tesdall's computations. Figure 1.4 shows a comparison of the computational and experimental results. The computational picture (on the left), showing a downstream flow containing at least four supersonic patches, is quite complicated, and up to this point has resisted a rigorous analysis, although Tesdall and I made an initial attempt [45], and Jegdić and Jegdić [25] did some further analysis. A related problem was studied in [32]. However, all we can be certain about, from the fact that the same phenomenon occurs in models both with and without linear waves present, is that it is behavior associated with the nonlinear, acoustic waves.

1.3 The Puzzle of L_p Solutions

In Sect. 1.1.2, we mentioned an apparent inconsistency between conservation law theory in one space dimension and what might be expected in higher dimensions. There does not appear to be a good candidate for an appropriate function space in which to locate solutions of multidimensional problems. This difficulty does not appear in the simple, self-similar examples that have been studied so far, although the complicated solutions found numerically for Guderley Mach reflection, as pictured in Fig. 1.4, hint that shock interactions might generate oscillations. But in fact there is already evidence from the theory of linear hyperbolic systems that anticipates

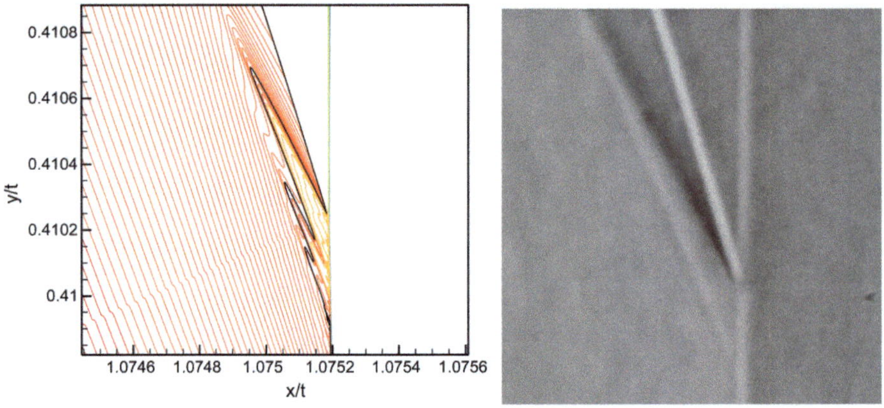

Fig. 1.4 Computed *left* and experimental *right* visualizations of Guderley Mach reflection

the possibility that some generalization of BV, the space of functions of bounded variation, which served so well in a single space dimension, is not suitable for multidimensional problems.

This story begins with a pioneering result of Walter Littman in 1963 [34]. Littman proved a theorem for the wave equation in d space dimensions,

$$\sum_1^d \frac{\partial^2 u}{\partial x_i^2} - \frac{\partial^2 u}{\partial t^2} = 0\,,$$

and a basic seminorm, the energy in L^p:

$$E_p(t) = \int \left(\sum_1^d \left| \frac{\partial u}{\partial x_i} \right|^p + \left| \frac{\partial u}{\partial t} \right|^p \right) dx\,.$$

Theorem 1.7 (Littman 1963) *An estimate $E_p(t) \le C(t) E_p(0)$ holds only if either $p = 2$ or $d = 1$.*

A short time later, Philip Brenner [2] generalized this result to any first-order, constant coefficient linear hyperbolic system,

$$\frac{\partial \mathbf{u}}{\partial t} = \sum_1^d A_j \frac{\partial \mathbf{u}}{\partial x_j} + B\mathbf{u}\,.$$

Theorem 1.8 (Brenner 1966) *Suppose $p \neq 2$, $1 \le p \le \infty$. The Cauchy problem is well-posed in L_p if and only if the matrices $A_1, \ldots, A_d$ commute.*

Based on the earlier discussion in this paper, one can penetrate the mystery in this condition. One need recall only that symmetric commuting matrices have a common set of real eigenvectors, and are simultaneously diagonalizable, whence all characteristic families are of transport type. That means that a system with acoustic waves, such as any of the examples in Sect. 1.2, cannot have commuting matrices.

Rauch [38] has shown that Brenner's result applies equally well to quasilinear systems. The title of his paper states the important conclusion: "BV Estimates Fail for Most Quasilinear Hyperbolic Systems in Dimensions Greater Than One".

1.4 Recent Results on Linear Waves

Some recent research on the equations of incompressible flow offers insight into the complex behavior of the linear waves in the gas dynamics system. Although the two systems—compressible and incompressible flow—are mathematically quite different, and are studied by different techniques, they are related. If we are looking at smooth solutions, then the system (1.2) can be written in a simpler form, because conservation is no longer important. By manipulating the equations, one obtains the system

$$\begin{aligned} \rho_t + u\rho_x + v\rho_y + \rho(u_x + v_y) &= 0 \\ u_t + uu_x + uv_y + p_x/\rho &= 0 \\ v_t + uv_x + vv_y + p_y/\rho &= 0 \\ p_t + up_x + vp_y + \gamma p(u_x + v_y) &= 0\,. \end{aligned} \tag{1.9}$$

Setting $\rho \equiv 1$ and ignoring the fourth equation, one obtains the incompressible system (often called the Euler equations),

$$\begin{aligned} u_x + v_y &= 0 \\ u_t + uu_x + uv_y + p_x &= 0 \\ v_t + uv_x + vv_y + p_y &= 0\,. \end{aligned} \tag{1.10}$$

Majda [35] and others have derived the Euler equations as the "low Mach number limit" of compressible gas dynamics. That is, one takes the limit for the acoustic speed, $c = \sqrt{dp/d\rho} \to \infty$ in the isentropic system. For the full system, (1.2), the value of c is $\sqrt{\gamma p/\rho}$. In either case, the limit is taken with respect to some scaling that relates the acoustic speed to the speed $q = |\mathbf{u}|$ of the fluid. A formal derivation of (1.10) from the first three equations in (1.9) can be made by assuming that ρ is constant and that p is not a function of ρ. This does not give a very accurate picture of the relationship, but it is consistent with the fact that as $c \to \infty$ the acoustic characteristic speeds become infinite and the system loses its hyperbolic character. The degenerate, linear characteristics are unchanged.

1.4.1 Nonuniform Dependence in Incompressible and Compressible Flow

There has been considerable interest in the well-posedness of the incompressible system (1.10). In [16], De Lellis and Székelyhidi constructed weak solutions of (1.10) that satisfy the currently accepted definitions of admissibility but are “wild” or “unphysical” in any reasonable sense, as they can be compactly supported in space and time. De Lellis and Székelyhidi's result generalizes earlier work of Scheffer [40] and of Shnirelman [41], also concerning wild and nonphysical solutions that are necessarily weak. De Lellis and Székelyhidi's construction extends to weak solutions of the compressible gas dynamics system. This may mean that the current definition of admissibility is inadequate, or it may suggest inherent problems in establishing well-posedness.

But there may also be questions even about classical (that is, smooth) solutions. Recent work of Himonas and Misiołek [22] on classical solutions of (1.10) has opened the way to results by Holmes, Tiğlay and myself. Himonas and Misiołek study dependence of solutions on initial conditions. The Cauchy problem for (1.10) poses an initial condition $\mathbf{u}(\cdot, 0) \equiv (u(\cdot, 0), v(\cdot, 0)) = \mathbf{u}_0$. Standard results for (1.10) [36] show that the pressure is determined, up to a constant, by the velocity data, so there is no initial condition for the pressure. As is the case for conservation laws, classical solutions of (1.10) have a finite life span, and are well-posed in suitably regular Sobolev spaces. The classical theorem (see [36, Chap. 3] for a history of this theorem and details of the proof) is

Theorem 1.9 (Classical) *If $s > d/2 + 1$ and $\mathbf{u}_0 \in H^s(\Omega, \mathbb{R}^d)$ with $\nabla \cdot \mathbf{u}_0 = 0$, then there is a $T > 0$ for which a unique $\mathbf{u} \in C([0, T], H^s(\Omega, \mathbb{R}^d))$ exists and depends continuously on $\mathbf{u}_0$. We have the bound*

$$\|\mathbf{u}(\cdot, t)\|_{H^s} \leq \frac{\|\mathbf{u}_0\|_{H^s}}{1 - Ct\|\mathbf{u}_0\|_{H^s}}.$$

Here the space dimension is $d \geq 2$; the domain of the data, Ω is a subset of $\mathbb{R}^d$. We are interested in periodic solutions, $\Omega = \mathbb{T}^d$, or in solutions in all of space, $\Omega = \mathbb{R}^d$. (In other cases additional boundary conditions are also needed; we do not go into detail about this.)

Since the incompressible system can be solved for the velocity alone by projecting into the space of divergence-free vector fields $\mathbf{u}$, and then the pressure recovered from a Poisson equation, [36, Chap. 1], the pressure can be ignored in constructing solutions. The dependence of pressure on velocity is smooth.

The basic nonuniformity result for periodic data, which is demonstrated by an explicit construction, is

Theorem 1.10 (Himonas–Misiołek) *Let $\Omega = \mathbb{T}^2$. The solution map, $\mathbf{u}_0 \mapsto \mathbf{u}(\cdot, t)$, is not uniformly continuous from the unit ball in $H^s(\Omega, \mathbb{R}^2)$ into $C([0, T], H^s(\Omega, \mathbb{R}^2))$.*

To prove this result, Himonas and Misiołek take two sequences of 2π-periodic initial velocities, with the property $\|\mathbf{u}_0^{1,n} - \mathbf{u}_0^{-1,n}\|_s \simeq \frac{1}{n}$ (the expression $x \simeq y$ means $Cy \leq x \leq y/C$ for a constant $C > 0$ independent of x and y):

$$\mathbf{u}_0^{1,n}(x, y) = \left(\frac{1}{n} + \frac{1}{n^s} \cos ny, \frac{1}{n} + \frac{1}{n^s} \cos nx\right)$$
$$\mathbf{u}_0^{-1,n}(x, y) = \left(-\frac{1}{n} + \frac{1}{n^s} \cos ny, -\frac{1}{n} + \frac{1}{n^s} \cos nx\right),$$

and write down the exact 2π-periodic solution:

$$\begin{aligned} \mathbf{u}^{1,n}(x, y, t) &= \left(\frac{1}{n} + \frac{1}{n^s} \cos(ny - t), \frac{1}{n} + \frac{1}{n^s} \cos(nx - t)\right) \\ \mathbf{u}^{-1,n}(x, y, t) &= \left(-\frac{1}{n} + \frac{1}{n^s} \cos(ny + t), -\frac{1}{n} + \frac{1}{n^s} \cos(nx + t)\right). \end{aligned} \tag{1.11}$$

Now an explicit calculation shows that

$$\|u^{1,n} - u^{-1,n}\|_{H^s} \simeq |\sin t| - \frac{1}{n}.$$

This means that for a given n the difference between corresponding solutions from the two families grows at a constant rate in t, even though the difference in the data tends to 0 as $n \to \infty$.

In the same paper [22], it is shown that nonuniform dependence holds also when $\Omega = \mathbb{R}^2$ or $\Omega = \mathbb{R}^3$. To obtain functions that are in $H^s(\Omega)$ when Ω is all of space, Himonas and Misiołek introduce cutoff functions to make the solutions tend to zero sufficiently rapidly at infinity.

The structure of these velocity functions is interesting. The functions consist of a low-frequency part (which is constant in the periodic case) and a high-frequency part which is highly oscillatory in space but not in time. When these functions are substituted into the Euler system, they yield an exact solution because there is some cancelation between high- and low-frequency components owing to the nonlinearity of the system. The low-frequency components tend to 0 in H^s as $n \to \infty$, but the high-frequency parts do not. The high-frequency components of the data approach each other in H^s (in fact, they are identical), but that is not the case for the solutions.

The study that Holmes, Tiğlay, and I have performed carries out a similar construction for the compressible system, (1.9). A hint for how to do this is given by the fact that the functions in (1.11) are both oscillatory in space and slow-moving in time. They can be loosely identified with the linear waves in the compressible system. As a sidenote, though, we found that although there is a single real, finite characteristic speed in the system (1.10), we needed the flexibility of the full system (1.2) or (1.9), with a pair of linear characteristics, to carry out our construction. We could not reproduce this result for the isentropic system.

Since solutions of (1.9) do not remain in H^s for $s > d/2 + 1$, our construction is also valid for only a finite time. This is not important, since the lack of uniform dependence on t appears instantaneously when $t > 0$. Our result [31] in $\mathbb{T}^2$ (two space dimensions, periodic data) for the compressible Euler system is almost identical to that in [22] for the periodic case. We construct two families of approximate solutions, for $\omega = 1$ and $\omega = -1$, of the form

$$\begin{aligned} \rho^{\omega,n} &\equiv \rho_0 > 0 \\ u^{\omega,n} &= \frac{\omega}{n} + \frac{1}{n^s}\cos(ny - \omega t) \\ v^{\omega,n} &= \frac{\omega}{n} + \frac{1}{n^s}\cos(nx - \omega t) \\ p^{\omega,n} &= \rho_0 + \frac{1}{n^{2s}}\sin(nx - \omega t)\sin(ny - \omega t)\,. \end{aligned}$$

Unlike the exact solutions (1.11) in the case of the incompressible system, these are only approximate solutions. We show they are close in H^s to classical solutions of (1.9). The life span of classical solutions depends only on the H^s norm of the data, as proved, using various methods, by Kato, Lax and Beale and Majda see [29, 35]. We state it, again using the notation $\mathbf{u} = (\rho, u, v, p)$:

Theorem 1.11 *For $s > 2$ (in our two-dimensional case, $d/2 + 1 = 2$), and $T > 0$ depending on $\|\mathbf{u}_0\|_s$, there is a unique solution $\mathbf{u}$ to (1.9) with $\mathbf{u}(\cdot, 0) = \mathbf{u}_0 \in H^s$ which maps into $C[(0, T), H^s(\Omega, \mathbb{R}^2)]$ and satisfies a bound like that in Theorem 1.9.*

Our first result is for periodic data [31],

Theorem 1.12 (Keyfitz–Tığlay) *For the given data, the exact solution(s) $\mathbf{u}_{\omega,n}$ are H^s-close to the approximate solutions $\mathbf{u}^{\omega,n}$ and so, for t sufficiently small,*

$$\|\mathbf{u}_{1,n} - \mathbf{u}_{-1,n}\|_{H^s} \simeq |\sin t| - \frac{1}{n}\,.$$

For data in $H^s(\mathbb{R}^2)$, we have the same conclusion, but the form of the approximate solutions is different, taking account of the need to keep the data and solutions in H^s. The modification is similar to that in [22]. The proof of this result is the subject of [23].

Readers who consult Kato's important paper [29] will find a claim of uniformly continuous dependence, stated as follows.

Theorem 1.13 (Kato) *If at $t = 0$, $\|\mathbf{u}_0^n - \mathbf{v}_0\|_s \le M$ and $\|\mathbf{u}_0^n - \mathbf{v}_0\|_0 \to 0$ as $n \to \infty$, then $\|\mathbf{u}^n(\cdot, t) - \mathbf{v}(\cdot, t)\|_{s-1} \to 0$ uniformly in t.*

Kato uses a subtly different notion of uniformity. Each of our sequences is bounded in H^s and tends to $v_0 \equiv 0$, and $u^{\omega,n} \to 0$ in H^{s-1} (not in H^s). In fact, what we have, as in [22], are two families of data whose difference converges to zero in H^s, but they do not separately converge. This leads to the amusing observation that even though we are tracking linear waves, we are using the nonlinearity of the system.

Acknowledgements This paper is based on a talk at the Silver Jubilee Conference of the Indian Society of Industrial and Applied Mathematics, which took place at Sharda University in Greater Noida, Uttar Pradesh, India, January 29–31, 2016. It is a pleasure to acknowledge the hospitality and support of ISIAM and Sharda University.

References

1. Adams, R.A.: Sobolev Spaces. Academic Press, New York (1975)
2. Brenner, P.: The Cauchy problem for symmetric hyperbolic systems in L_p. Math. Scand. **19**, 27–37 (1966)
3. Čanić, S., Keyfitz, B.L.: Riemann problems for the two-dimensional unsteady transonic small disturbance equation. SIAM J. Appl. Math. **58**, 636–665 (1998)
4. Čanić, S., Keyfitz, B.L., Kim, E.H.: Free boundary problems for the unsteady transonic small disturbance equation: transonic regular reflection. Methods Appl. Anal. **7**, 313–336 (2000)
5. Čanić, S., Keyfitz, B.L., Kim, E.H.: A free boundary problem for a quasilinear degenerate elliptic equation: regular reflection of weak shocks. Commun. Pure Appl. Math. **LV**, 71–92 (2002)
6. Čanić, S., Keyfitz, B.L., Kim, E.H.: Mixed hyperbolic-elliptic systems in self-similar flows. Boletim da Sociedade Brasileira de Matemática **32**, 1–23 (2002)
7. Čanić, S., Keyfitz, B.L., Kim, E.H.: Free boundary problems for nonlinear wave systems: mach stems for interacting shocks. SIAM J. Math. Anal. **37**, 1947–1977 (2005)
8. S. Čanić, B. L. Keyfitz, and G. M. Lieberman, A proof of existence of perturbed steady transonic shocks via a free boundary problem, Communications on Pure and Applied Mathematics **LIII** (2000), 1–28
9. Cao, W., Huang, F., Wang, D.: Isometric immersions of surfaces with two classes of metrics and negative Gauss curvature. Arch. Ration. Mech. Anal. **218**, 1431–1457 (2015)
10. Chen, G.-Q., Deng, X., Xiang, W.: Shock diffraction by convex cornered wedges for the nonlinear wave system. Arch. Ration. Mech. Anal. **211**, 61–112 (2014)
11. Chen, G.-Q., Feldman, M.: Multidimensional transonic shocks and free boundary problems for nonlinear equations of mixed type. J. Am. Math. Soc. **16**, 461–494 (2003)
12. Chen, G.-Q., Feldman, M.: Global solutions of shock reflection by large-angle wedges for potential flow. Ann. Math. **171**, 1067–1182 (2010)
13. Chen, G.-Q., Feldman, M.: Mathematics of shock reflection-diffraction, von Neumann's conjectures, and related analysis. University Press, Princeton (2017)
14. Courant, R., Friedrichs, K.O.: Supersonic flow and shock waves. Wiley-Interscience, New York (1948)
15. Dafermos, C.M.: Hyperbolic Conservation Laws in Continuum Physics. Springer, Berlin (2000)
16. De Lellis, C., Székelyhidi Jr., L.: On admissibility criteria for weak solutions of the Euler equations. Arch. Ration. Mech. Anal. **195**, 225–260 (2010)
17. Elling, V.: Non-existence of strong regular reflections in self-similar potential flow. J. Differ. Eq. **252**, 2085–2103 (2012)
18. Elling, V., Liu, T.-P.: The ellipticity principle for selfsimilar polytropic potential flow. J. Hyperb. Differ. Eq. **2**, 909–917 (2005)
19. Elling, V., Liu, T.-P.: Supersonic flow onto a solid wedge. Commun. Pure Appl. Math. **61**, 1347–1448 (2008)
20. Garabedian, P.R.: Partial Differential Equations, 2nd edn. Chelsea, New York (1986)
21. Guderley, K.G.: The Theory of Transonic Flow. Pergamon Press, Oxford (1962)
22. Himonas, A.A., Misiołek, G.: Non-uniform dependence on initial data of solutions to the Euler equations of hydrodynamics. Commun. Math. Phys. **296**, 285–301 (2010)
23. Holmes, J., Keyfitz, B.L., Tiğlay, F.: Nonuniform dependence on initial data for compressible gas dynamics: the Cauchy problem on $\mathbb{R}^2$. In preparation

24. Hörmander, L.: The Analysis of Linear Partial Differential Operators. Springer, New York (1983)
25. Jegdić, I., Jegdić, K.: Properties of solutions in semi-hyperbolic patches for the unsteady transonic small disturbance equation. Electron. J. Differ. Eq. **243**, 20 (2015)
26. Jegdić, K.: Weak regular reflection for the nonlinear wave system. J. Hyperbolic Differ. Eq. **5**, 399–420 (2008)
27. Jegdić, K., Keyfitz, B.L., Čanić, S.: Transonic regular reflection for the nonlinear wave system. J. Hyperbolic Differ. Eq. **3**, 443–474 (2006)
28. Jegdić, K., Keyfitz, B.L., Čanić, S., Ying, H.: A free boundary problem for the isentropic gas dynamics equations — transonic regular reflection, submitted (2015), 40 pages
29. Kato, T.: The Cauchy problem for quasi-linear symmetric hyperbolic systems. Arch. Ration. Mech. Anal. **58**, 181–205 (1975)
30. Keyfitz, B.E.: Self-similar solutions of two-dimensional conservation laws. J. Hyperbolic Differ. Eq. **1**, 445–492 (2004)
31. Keyfitz, B.L., Tiğlay, F.: Nonuniform dependence on initial data for compressible gas dynamics: the periodic Cauchy problem, J. Differ. Eq. arXiv:1611.05840
32. Keyfitz, B.L., Tesdall, A., Payne, K.R., Popivanov, N.I.: The sonic line as a free boundary. Quart. Appl. Math. **LXXI**, 119–133 (2013)
33. Lax, P.D.: Hyperbolic Partial Differential Equations. Courant Lecture Notes in Mathematics. American Mathematical Society, Providence (2006)
34. Littman, W.: The wave operator and L_p norms. J. Math. Mech. **12**, 55–68 (1963)
35. Majda, A.: Compressible Fluid Flow and Systems of Conservation Laws in Several Space Variables. Springer, New York (1984)
36. Majda, A., Bertozzi, A.L.: Vorticity and Incompressible Flow. Cambridge University Press, Cambridge (2002)
37. Mizohata, S.: On the Cauchy Problem. Academic Press, San Diego (1985)
38. Rauch, J.: BV estimates fail in most quasilinear hyperbolic systems in dimensions greater than one. Commun. Math. Phys. **106**, 481–484 (1986)
39. Renardy, M., Rogers, R.C.: An introduction to partial differential equations. Springer, New York (1993)
40. Scheffer, V.: An inviscid flow with compact support in space-time. J. Geom. Anal. **3**, 343–401 (1993)
41. Shnirelman, A.: On the nonuniqueness of weak solution of the Euler equation. Commun. Pure Appl. Math. **50**, 1261–1286 (1997)
42. Skews, B., Li, G., Paton, R.: Experiments on Guderley Mach reflection. Shock Waves **19**, 95–102 (2009)
43. Slemrod, M.: Lectures on the isometric embedding problem $(M^n, g) \mapsto \mathbb{R}^m$, $m = \frac{n}{2}(n+1)$. Differential Geometry and Continuum Mechanics, vol. 137, pp. 77–120. Springer, Berlin (2015)
44. Taylor, M.E.: Partial differential equations i: basic theory. Springer, New York (1996)
45. Tesdall, A.M., Keyfitz, B.L.: A continuous, two-way free boundary in the unsteady transonic small disturbance equations. J. Hyperbolic Differ. Eq. **7**, 317–338 (2010)
46. Tesdall, A.M., Sanders, R., Keyfitz, B.L.: The triple point paradox for the nonlinear wave system. SIAM J. Appl. Math. **67**, 321–336 (2006)
47. Tesdall, A.M., Sanders, R., Keyfitz, B.L.: Self-similar solutions for the triple point paradox in gasdynamics. SIAM J. Appl. Math. **68**, 1360–1377 (2008)
48. Tesdall, A.M., Hunter, J.K.: Self-similar solutions for weak shock reflection. SIAM J. Appl. Math. **63**, 42–61 (2002)
49. Ying, H., Keyfitz, B.L.: A two-dimensional Riemann problem for scalar conservation laws. In: Bressan, A., Chen, G.-Q., Lewicka, M., Wang, D.-H. (eds.) Nonlinear Conservation Laws and Applications. IMA, vol. 153, pp. 447–455. Springer, New York (2011)

Chapter 2
Nonlinear Flows and Optimality for Functional Inequalities: An Extended Abstract

Maria J. Esteban

Abstract The talk given on the occasion of the ISIAM 2016 conference was mainly about rigidity results for nonnegative solutions of semilinear elliptic equation on infinite cylinder-like domains or in the Euclidean espace and as a consequence, about optimal symmetry properties for the optimizers of the Caffarelli–Kohn–Nirenberg inequalities. This text contains the main results presented in that conference. All the results will be stated in the simple case of spherical cylinders, but similar, even if less precise, results can also be stated and proved for general cylinders generated by any compact smooth Riemannian manifold without a boundary. Other consequences from the results below are optimal estimates for the principal eigenvalue of Schrödinger operators on infinite cylinders. The text below is an extended abstract of that talk.

Keywords Caffarelli–Kohn–Nirenberg inequalities · Symmetry
Symmetry breaking · Optimal constants · Rigidity results · Fast diffusion equation
Carŕe du champ · Bifurcation · Instability · Emden–Fowler transformation
Cylinders · Noncompact manifolds · Laplace–Beltrami operator
Spectral estimates · Keller–Lieb–Thirring estimate · Hardy inequality

2.1 Rigidity Results

The main result presented in this talk is the following rigidity theorem, which is contained, with its proof, in [3]. Many references about previous works and related topics can be found in this article. We will not include all those references in this short text.

Work done in collaboration with J. Dolbeault and M. Loss.

M.J. Esteban (✉)
Ceremade (UMR CNRS No. 7534), PSL Research University, Université Paris-Dauphine, Place de Lattre de Tassigny, 75775 Paris 16, France
e-mail: esteban@ceremade.dauphine.fr

P. Manchanda et al. (eds.), *Industrial Mathematics and Complex Systems*,
Industrial and Applied Mathematics, DOI 10.1007/978-981-10-3758-0_2

Theorem 2.1 *For $d \ge 2$ define $2^* = 2d/(d-2)$ if $d \ge 3$, $2^* = +\infty$ if $d = 2$. And consider the cylinder $\mathscr{C}_1 := \mathbb{R} \times S^{d-1}$. For all $p \in (2, 2^*)$ and $0 < \Lambda \le \Lambda_{\rm FS} := 4\,\frac{d-1}{p^2-4}$, any positive solution $\varphi \in \mathrm{H}^1(\mathscr{C}_1)$ of*

$$-\,\partial_s^2\,\varphi - \,\Delta_\omega\,\varphi + \Lambda\,\varphi = \varphi^{p-1} \quad in \ \ \mathscr{C}_1 \tag{2.1}$$

is equal to φ_Λ, up to a translation in the s-direction, where

$$\varphi_\Lambda(s) = \left(\tfrac{p}{2}\,\Lambda\right)^{\frac{1}{p-2}} \left(\cosh\left(\tfrac{p-2}{2}\,\sqrt{\Lambda}\,s\right)\right)^{-\frac{2}{p-2}}\,. \tag{2.2}$$

By using the Emden–Fowler transformation

$$v(r,\omega) = r^{a-a_c}\,\varphi(s,\omega) \quad \text{with} \quad r = |x|\,, \quad s = -\log r \quad \text{and} \quad \omega = \frac{x}{r}\,, \tag{2.3}$$

it can be easily seen that Theorem 2.1 is equivalent to the following result

Theorem 2.2 *Assume that $d \ge 2$. If either $a \in [0, (d-2)/2)$ and $b > 0$, or $a < 0$ and $b \ge b_{\rm FS}(a)$, with*

$$b_{\rm FS}(a) := \frac{d\,(a_c - a)}{2\,\sqrt{(a_c-a)^2 + d - 1}} + a - a_c\,, \tag{2.4}$$

then any nonnegative solution v of

$$-\,\nabla\cdot\left(|x|^{-2a}\,\nabla v\right) = |x|^{-b\,p}\,|v|^{p-2}\,v \quad in \ \ \mathbb{R}^d\setminus\{0\} \tag{2.5}$$

which satisfies $\int_{\mathbb{R}^d} \frac{|v|^p}{|x|^{b\,p}},\,dx < \infty$, is equal to $v_\star$ up to a scaling, with

$$v_\star(x) = \left(1 + |x|^{(p-2)\,(a_c-a)}\right)^{-\frac{2}{p-2}} \quad \forall\,x \in \mathbb{R}^d\,.$$

Next pick n and α such that

$$n = \frac{d - b\,p}{\alpha} = \frac{d - 2\,a - 2}{\alpha} + 2 = \frac{2\,p}{p-2}\,.$$

Then, defining

$$v(r,\omega) = w(r^\alpha,\omega) \quad \forall\,(r,\omega) \in \mathbb{R}^+ \times S^{d-1} \tag{2.6}$$

it can again be easily seen that the two above theorems are equivalent to

Theorem 2.3 *Assume that $d \ge 2$. If $0 < \alpha < \alpha_{\rm FS} := \sqrt{\frac{d-1}{n-1}}$, then any nonnegative solution $w(x) = w(r,\omega)$ $(r \in \mathbb{R}_+,\ \omega \in S^{d-1})$ of*

$$-\alpha^2 w'' - \alpha^2 \frac{n-1}{r} w' - \frac{\Delta w}{r^2} = w^{p-1} \quad in \ \mathbb{R}^d \setminus \{0\}, \tag{2.7}$$

which satisfies $\int_{I\!R^d} |x|^{n-d} |w|^p \, dx < \infty$, *is equal to* $w_\star$ *up to a scaling, and multiplication by a constant, with*

$$w_\star(x) = \left(1 + |x|^2\right)^{-n} \quad \forall\, x \in \mathbb{R}^d .$$

Notice that the above definitions imply the equivalence of the above three conditions

$$0 < \Lambda \le \Lambda_{\rm FS} := 4\,\frac{d-1}{p^2-4}\,; \quad 0 < \alpha < \alpha_{\rm FS} := \sqrt{\frac{d-1}{n-1}}\,;$$

$$a < (d-2)/2) \ \text{and} \ b > 0, \ \text{ or } \ a < 0 \ \text{and} \ b \ge b_{\rm FS}(a)$$

Finally, let us remark that the above three results are optimal, since as it is proved in [2, 4], when the above conditions are not satisfied, there are nonnegative solutions of the corresponding equations that depend on $\omega \in S^{d-1}$ in a nontrivial way.

2.2 Consequence: Optimal Symmetry Result for Optimizers of the Critical Caffarelli–Kohn–Nirenberg Inequalities

The Caffarelli–Kohn–Nirenberg inequalities

$$\left(\int_{I\!R^d} \frac{|v|^p}{|x|^{b\,p}}\,dx\right)^{2/p} \le \mathsf{C}_{a,b} \int_{I\!R^d} \frac{|\nabla v|^2}{|x|^{2\,a}}\,dx \quad \forall\, v \in \mathscr{D}_{a,b} \tag{2.8}$$

have been established in [1], under the conditions that $a \le b \le a+1$ if $d \ge 3$, $a < b \le a+1$ if $d = 2$, $a + 1/2 < b \le a+1$ if $d = 1$, and $a < a_c$ where

$$a_c := \frac{d-2}{2}\,,$$

and where the exponent

$$p = \frac{2\,d}{d-2+2\,(b-a)} \tag{2.9}$$

is determined by the invariance of the inequality under scalings. Here $\mathsf{C}_{a,b}$ denotes the optimal constant in (2.8) and the space $\mathscr{D}_{a,b}$ is defined by

$$\mathscr{D}_{a,b} := \left\{ v \in \mathrm{L}^p\left(I\!R^d, |x|^{-b}\,dx\right) \,:\, |x|^{-a}\,|\nabla v| \in \mathrm{L}^2\left(I\!R^d, dx\right) \right\}.$$

Note that, up to scaling and multiplication by a constant, any optimal solution for the above inequality is a nonnegative solution of (2.5). It was proved in [4] (see also [2] for a partial result) that whenever $a < 0$ and $b < b_{\rm FS}(a)$, the optimizers of (2.5) are never radially symmetric. What Theorem 2.2 implies is that whenever $b \geq b_{\rm FS}(a)$ or $a \in [0, a_c)$, the optimizers, which can be taken as nonnegative functions, thus yielding an optimal symmetry result.

2.3 Outline of the Proof

Let us now quickly present the main ideas of the proof of the above results in the case $d \geq 3$. We will explain it for in the context of Theorem 2.3. Let us introduce some notation:

$$u^{\frac{1}{2}-\frac{1}{n}} = |w| \quad \Longleftrightarrow \quad u = |w|^p \quad \text{with} \quad p = \frac{2\,n}{n-2} \tag{2.10}$$

and notice that, up to a multiplicative constant, the r.h.s. in (2.8) is transformed into a generalized *Fisher information*

$$\mathscr{I}[u] := \int_{(0,\infty)\times S^{d-1}} u\,|\mathsf{D}\mathsf{p}|^2\,d\mu \quad \text{where} \quad \mathsf{p} = \frac{m}{1-m}\,u^{m-1} \quad \text{and} \quad m = 1 - \frac{1}{n}\,, \tag{2.11}$$

with $\mathsf{D}p = \left(\alpha\,\frac{\partial \mathsf{p}}{\partial r}\,,\,\frac{1}{r}\,\nabla_\omega \mathsf{p}\right)$, while the l.h.s. in (2.8) is now proportional to a *mass*, $\int_{(0,\infty)\times S^{d-1}} u\,d\mu$, where the measure $d\mu$ is defined as $r^{n-1}\,dr\,d\omega$ on $(0,\infty)\times S^{d-1}$. Here p is the *pressure function*, as in [5, 5.7.1 p. 98]. If we replace m by $1-\frac{1}{n}$, we get that

$$\mathsf{p} = (n-1)\,u^{-\frac{1}{n}}\,. \tag{2.12}$$

Let us next introduce the fast diffusion flow

$$\frac{du}{dt} = \text{Ł} u^m\,, \quad m = 1 - \frac{1}{n}\,, \tag{2.13}$$

with

$$\text{Ł} w := -\,\mathsf{D}^*\,\mathsf{D}\,w = \alpha^2\,w'' + \alpha^2\,\frac{n-1}{r}\,w' + \frac{\Delta\,w}{r^2}\,, \quad ' = d/dr\,,$$

and assume that it is well defined for all times. It is immediate to verify that $\frac{d}{dt}\int_{(0,\infty)\times S^{d-1}} u\,d\mu = 0$. Moreover, long calculations, the study of the regularity of the solutions of (2.1) at $\pm\infty$ and the use of the Bochner–Lichnerowicz–Weitzenböck formula

$$\tfrac{1}{2}\,\Delta_\omega\,(|\nabla_\omega f|^2) = \|\mathrm{Hess} f\|^2 + \nabla_\omega(\Delta_\omega f)\cdot\nabla_\omega f + \mathrm{Ric}(\nabla_\omega f, \nabla_\omega f)\,,$$

among others, allow us to prove the following proposition.

Proposition 2.1 *With the notations defined by* (2.12) *if* u *is a smooth minimizer of* $\mathscr{I}[u]$ *under a mass constraint, with* $\alpha \le \alpha_{\rm FS}$, *then there exists a positive constant* $\zeta_\star$ *such that*

$$\frac{d}{dt}\mathscr{I}[u(t,\cdot)] = -\,2\,(n-1)^{n-1}\int_{(0,\infty)\times S^{d-1}} \mathsf{k}[\mathsf{p}(t,\cdot)]\,\mathsf{p}(t,\cdot)^{1-n}\,d\mu\,,$$

with

$$\mathsf{k}[\mathsf{p}] = \alpha^4\left(1-\frac{1}{n}\right)\left[\mathsf{p}''-\frac{\mathsf{p}'}{r}-\frac{\Delta_\omega\,\mathsf{p}}{\alpha^2\,(n-1)\,r^2}\right]^2 + 2\,\alpha^2\,\frac{1}{r^2}\left|\nabla_\omega\mathsf{p}'-\frac{\nabla_\omega\mathsf{p}}{r}\right|^2 + \frac{1}{r^4}\,\mathsf{k}_{\mathfrak{M}}[\mathsf{p}]$$

and

$$\int_{S^{d-1}} \mathsf{k}_{\mathfrak{M}}[\mathsf{p}]\,\mathsf{p}^{1-n}\,d\omega \ge (n-2)\big(\alpha_{\rm FS}^2-\alpha^2\big)\int_{S^{d-1}}|\nabla_\omega\mathsf{p}|^2\,\mathsf{p}^{1-n}\,d\omega + \zeta_\star\,(n-d)\int_{S^{d-1}}|\nabla_\omega\mathsf{p}|^4\,\mathsf{p}^{1-n}\,d\omega\,.$$

Therefore, if $\alpha \le \alpha_{\rm FS}$, the *Fisher information* $\mathscr{I}[u]$ is nonincreasing along the flow defined by (2.13). But actually we do not need to study the flow's properties, and we only use it as a guide for a complete rigorous result of Theorem 2.3. This can be done as follows. Let u be a critical point of $\mathscr{I}[u]$ under the mass constraint. Then, by Proposition 2.1, taking $u[0]=u$, and assuming $\alpha \le \alpha_{\rm FS}$,

$$0 = \mathscr{I}'[u]\cdot \text{Ł}u^m = \frac{d}{dt}\mathscr{I}[u(t)]_{|t=0} \ge \zeta_\star\,(n-d)\int_{(0,\infty)\times S^{d-1}}|\nabla_\omega\mathsf{p}|^4\,\mathsf{p}^{1-n}\,d\mu\,,$$

and hence, if $\alpha \le \alpha_{\rm FS}$, $\nabla_\omega\mathsf{p} \equiv 0$ and therefore, u is radially symmetric, since it does not depend on the angular variables. The precise shape of u is given by

$$\mathsf{p}''-\frac{\mathsf{p}'}{r}-\frac{\Delta_\omega\,\mathsf{p}}{\alpha^2\,(n-1)\,r^2} \equiv 0\,.$$

References

1. Caffarelli, L., Kohn, R., Nirenberg, L.: First order interpolation inequalities with weights. Compos. Math. **53**(3), 259–275 (1984)
2. Catrina, F., Wang, Z.-Q.: On the Caffarelli-Kohn-Nirenberg inequalities: sharp constants, existence (and nonexistence), and symmetry of extremal functions. Commun. Pure Appl. Math. **54**(2), 229–258 (2001)
3. Dolbeault, J., Esteban, M.J., Loss, M.: Rigidity versus symmetry breaking via nonlinear flows on cylinders and Euclidean spaces. Invent. Math. **206**(2), 397–440 (2016)

4. Felli, V., Schneider, M.: Perturbation results of critical elliptic equations of Caffarelli-Kohn-Nirenberg type. J. Differ. Equs. **191**(1), 121–142 (2003)
5. Vázquez, J.L.: Asymptotic behaviour for the porous medium equation posed in the whole space. Nonlinear Evolution Equations and Related Topics, pp. 67–118. Springer, Berlin (2004)

Chapter 3
What is a Frame? Theory and Applications of Frames

David Walnut

Abstract A *frame* in a separable Hilbert space is a generalization of an orthonormal basis that can be used to provide "painless nonorthogonal expansions" of elements in that space. In some respects, frames are easier to construct and use than orthogonal or Riesz bases, but the study of frames is tied to a number of deep and interesting results and conjectures in harmonic analysis (including the recently solved Kadisson–Singer Conjecture). Because of their relative ease of construction and their overcompleteness properties, frames have found applications in numerical harmonic analysis and were the first context in which wavelet expansions were discussed. The goal of this paper is to give a brief introduction to the theory of frames and to discuss some situations in which frames have proven an especially useful tool. These include noise reduction, robust communications, compressive sensing, and phaseless recovery.

Keywords Frame · Basis · Overcomplete · Riesz basis · Kadisson–Singer Phaseless recovery · Compressive sensing · Wavelets

3.1 Introduction

The goal of this paper is to give a brief introduction to the idea of a frame and to outline some of the uses and applications of frame theory. A frame is essentially a redundant basis, that is, a spanning set for a Hilbert space that contains more elements than necessary in order to represent elements in the space as series expansions. It is interesting that for many years after frames were first defined, the advantages of redundancy were not fully explored. In this paper, we will describe some applications of frames in which redundancy is essential for the effectiveness of the theory, and demonstrate how frames have become an essential tool and effective language in which to describe and solve some difficult and long-standing problems in harmonic analysis, time–frequency analysis, communication theory and physics.

D. Walnut (✉)
George Mason University, Fairfax, VA, USA
e-mail: dwalnut@gmu.edu

P. Manchanda et al. (eds.), *Industrial Mathematics and Complex Systems*,
Industrial and Applied Mathematics, DOI 10.1007/978-981-10-3758-0_3

In Sect. 3.2, we define frames for finite-dimensional spaces and describe some of their basic properties. This exposition requires only some knowledge of linear algebra. Motivated by finite-dimensional results in Sect. 3.3, we define frames in separable Hilbert spaces of arbitrary dimension. We define the analysis, synthesis, and frame operators and prove the existence of Fourier-like expansions in terms of frames. We also introduce by comparison the notion of a Riesz basis which can be thought of as the natural generalization of a basis in finite dimensions, or alternatively as a nonredundant frame.

In Sect. 3.4, we give some historical remarks about the early development of frame theory, beginning with its first description by Duffin and Schaeffer in 1952 through the initial recognition that embracing the notion of redundancy conveyed certain distinct advantages. In Sects. 3.5–3.7, we describe several problems in pure and applied mathematics in which the notion of frames turns out to be essential for the description and understanding of the problem. In particular, we describe how frames are used in communication theory to effectively code signals sent over noisy or lossy channels, we describe how the language of frames helps in designing measurement matrices in compressive sensing, and finally show how frames are used to given insight to the problem of phaseless recovery, a problem in which there had been almost no systematic progress in a century.

Finally in Sect. 3.7, we briefly describe how frames have had a significant impact in pure mathematics by providing a simple language in which to describe several important and long-standing conjectures in operator theory and mathematical physics. These conjectures are known to be equivalent to a simple-to-state question about finite frames. This conjecture, known to frame theorists as the Feichtinger Conjecture, has now been answered in the affirmative and speaks directly to how to best understand the limits of redundancy in frames.

3.2 Frames and Linear Algebra

The easiest way to understand frames is to define frames in finite-dimensional vector spaces. This amounts to some elementary assertions from linear algebra. Let us consider the d-dimensional vector space $\mathbb{C}^d$ of complex d-tuples with inner product $\langle \cdot, \cdot \rangle$ defined by

$$\langle x, y \rangle = \sum_{j=1}^{d} x_j \overline{y_j}$$

and norm given by $\|x\|_{\ell_d^2}^2 = \sum_{j=1}^d |x_j|^2$. It is an elementary fact from linear algebra that a collection of d vectors in $\mathbb{C}^d$, $X = \{x_1, x_2, \dots, x_d\}$ is a *basis* for $\mathbb{C}^d$ if and only if X is *linearly independent*, that is, if $c_1 x_1 + c_2 x_2 + \cdots + c_d x_d = 0$ implies that $c_1 = c_2 = \cdots = c_d = 0$. In this case, every $x \in \mathbb{C}^d$ can be written uniquely as $x = a_1 x_1 + a_2 x_2 + \cdots + a_d x_d$. In addition, there exists a collection $\widetilde{X} = \{\widetilde{x}_1, \widetilde{x}_2, \dots, \widetilde{x}_d\}$ in $\mathbb{C}^d$ called the *dual basis* such that $a_j = \langle x, \widetilde{x}_j \rangle$ for all j.

These ideas can be expressed more conveniently using matrix notation as follows. Let

$$B = \begin{bmatrix} x_1 \ x_2 \cdots x_d \end{bmatrix}$$

be the $d \times d$ matrix whose columns are the vectors x_j (written with respect to the standard basis). Since X is linearly independent, B is invertible. Let

$$B^{-1} = \begin{bmatrix} \overline{\tilde{x}_1} \\ \overline{\tilde{x}_2} \\ \vdots \\ \overline{\tilde{x}_d} \end{bmatrix}$$

so that $\overline{\tilde{x}_j}$ is the j-th row of B^{-1}. Hence given $x \in \mathbb{C}^d$,

$$x = BB^{-1}x = B \begin{bmatrix} \langle x, \tilde{x}_1 \rangle \\ \langle x, \tilde{x}_2 \rangle \\ \vdots \\ \langle x, \tilde{x}_d \rangle \end{bmatrix} = \sum_{j=1}^{d} \langle x, \tilde{x}_j \rangle \, x_j.$$

Note that the vectors $\{\tilde{x}_j\}_{j=1}^{d}$ form the dual basis of X.

To introduce the idea of a frame, let us now consider a collection $Y = \{y_1, y_2, \ldots, y_n\}$, where $n > d$. Since $n > d$ this collection is necessarily not linearly independent so in order to ensure that it truly generalizes the notion of a basis, we assume that $\operatorname{span}(Y) = \mathbb{C}^d$ (equivalent to linear independence when $n = d$). We say then that Y is a *frame* for $\mathbb{C}^d$.

Because Y is not linearly independent, there are constants η_j, $j = 1, \ldots, n$ such that not all of the $\eta_i = 0$ but $\eta_1 y_1 + \eta_2 y_2 + \cdots + \eta_n y_n = 0$. Since every spanning set for $\mathbb{C}^d$ contains a basis it follows that for all $x \in \mathbb{C}^d$ there exist coefficients α_j such that $x = \alpha_1 y_1 + \alpha_2 y_2 + \cdots + \alpha_n y_n$ but that the coefficients need not be unique. In this sense, frames mimic bases but are *redundant* in the sense that not all of the frame elements need be present in order to represent elements in the vector space, and that there are multiple representations of each vector in the space.

Once again we can express these ideas conveniently in matrix notation. Let

$$F = \begin{bmatrix} y_1 \ y_2 \cdots y_n \end{bmatrix}$$

be the $d \times n$ matrix whose columns are the vectors y_j (written with respect to the standard basis). Since $n > d$, F is necessarily not invertible but if Y constitutes a frame then $\operatorname{rank}(F) = d$ so that the d rows of F are linearly independent in $\mathbb{C}^n$. Since F has full row rank, the $d \times d$ matrix FF^* is invertible and for all $x \in \mathbb{C}^d$, $x = FF^*(FF^*)^{-1}x$. Letting

$$F^*(FF^*)^{-1} = \begin{bmatrix} \overline{\widetilde{y}_1} \\ \overline{\widetilde{y}_2} \\ \vdots \\ \overline{\widetilde{y}_n} \end{bmatrix},$$

it follows that

$$x = FF^*(FF^*)^{-1}x = F \begin{bmatrix} \langle x, \widetilde{y}_1 \rangle \\ \langle x, \widetilde{y}_2 \rangle \\ \vdots \\ \langle x, \widetilde{y}_n \rangle \end{bmatrix} = \sum_{j=1}^{n} \langle x, \widetilde{y}_j \rangle \, y_j.$$

The alert reader will recognize the $n \times d$ matrix $F^*(FF^*)^{-1}$ as the *pseudoinverse* or *Moore–Penrose inverse* of F, usually denoted by $F^\dagger$. This means that $c = F^\dagger x$ is the solution to $Fc = x$ with the smallest norm. In other words, of all coefficients $\alpha_1, \alpha_2, \ldots, \alpha_n$ that satisfy $x = \sum_{j=1}^{n} \alpha_j y_j$, the one with smallest norm is given by $\alpha_j = \langle x, \widetilde{y}_j \rangle$. For this reason, the collection of vectors $\{\widetilde{y}_j\}_{j=1}^{n}$ is called the *dual frame* of Y.

3.3 The Frame Inequality

Recall that for any $k \times m$ matrix M, its *Frobenius norm*, denoted $\|M\|_{fro}$, is defined to be the square root of the sum of the squared moduli of its elements. For our purposes, the main thing to know is that for any vector $x \in \mathbb{C}^m$, $\|Mx\|_{\ell^2_k} \le \|M\|_{fro}\, \|x\|_{\ell^2_m}$. Now suppose that the columns of the $d \times n$ matrix F form a frame for $\mathbb{C}^d$. We can write for all $x \in \mathbb{C}^d$, $x = (FF^*)^{-1}(FF^*)x$ and taking norms get

$$\|x\|_{\ell^2_d} \le \|(FF^*)^{-1}F\|_{fro}\, \|F^*x\|_{\ell^2_n}.$$

Additionally, we can see that if an inequality of the form $\|x\|_{\ell^2_d} \le C\|F^*x\|_{\ell^2_n}$ holds for some $C > 0$ then it follows that $F^*x = 0$ implies that $x = 0$, that the d columns of F^* are therefore linearly independent, and hence that $\mathrm{rank}(F^*) = \mathrm{rank}(F) = d$. Finally, since always $\|F^*x\|_{\ell^2_n} \le \|F^*\|_{fro}\|x\|_{\ell^2_d}$, we have proved the following theorem.

Theorem 3.1 *Let F be a $d \times n$ matrix with $n \ge d$. Then the columns of F form a frame in $\mathbb{C}^d$ if and only if there exist constants $c, C > 0$ such that for all $x \in \mathbb{C}^d$,*

$$c\|x\|^2_{\ell^2_d} \le \|F^*x\|^2_{\ell^2_n} \le C\|x\|^2_{\ell^2_d}.$$

The above inequality is referred to as the *frame inequality*. This turns out to be the characterization of a frame in $\mathbb{C}^d$ that is the most convenient way to generalize the notion of a frame from finite to arbitrary dimensions.

Definition 3.1 A *frame* in a separable Hilbert space H is a sequence of vectors $\{x_k\}_{k\in K}$ with the property that there exist constants $c_1,\ c_2>0$, called the *frame bounds* such that for all x in the Hilbert space

$$c_1\ \|x\|^2 \le \sum_{k\in K} |\langle x, x_k\rangle|^2 \le c_2\ \|x\|^2 .$$

A frame is *tight* if $c_1 = c_2$ and is *uniform* if $\left\|x_j\right\| = \|x_k\|$ for all j and k.

Corresponding to the operator from $\mathbb{C}^d$ to $\mathbb{C}^n$ given by $x \mapsto F^*x$, we define the *analysis operator* by

$$T\colon H \to \ell^2(K);\quad x \mapsto \{\langle x, x_k\rangle\}_{k\in K}.$$

Its adjoint, corresponding to the operator from $\mathbb{C}^n$ to $\mathbb{C}^d$ given by $y \mapsto Fy$, is the *synthesis operator*

$$T^*\colon \ell^2(K) \to H;\ \ \{c_k\} \mapsto \sum_{k\in K} c_k\, x_k.$$

The *frame operator* for $\{x_k\}_{k\in K}$ is

$$S = T^*T\colon H \to H;\ \ x \mapsto \sum_{k\in K} \langle x, x_k\rangle\, x_k.$$

The upper frame bound implies that T is bounded, and the lower that it is injective and continuously invertible on its range, which is a closed linear subspace of $\ell^2(K)$. In the case where the range of T is a proper subspace, T^* has a nontrivial nullspace, that is, there exist nonzero sequences $\{c_k\} \in \ell^2$ such that

$$\sum_{k\in K} c_k\, x_k = 0.$$

This corresponds to the situation in finite dimensions in which a frame has an excess of elements, that is, more elements than a basis would and we say in this case that the frame is *overcomplete* or *redundant*.

The frame operator $S = T^*T$ is always a bounded linear isomorphism of H. This leads to the following Fourier-like expansions of elements of H in terms of frames.

Theorem 3.2 *Given a frame $\{x_k\}_{k\in K}$ for a Hilbert space H, every $x \in H$ can be written as*

$$x = \sum_{k\in K} \langle x, x_k\rangle\, S^{-1}x_k = \sum_{k\in K} \langle x, S^{-1}x_k\rangle\, x_k.$$

The sequence $\{\widetilde{x}_k\}_{k\in K} = \{S^{-1}x_k\}_{k\in K}$ is called the dual frame *of $\{x_k\}$.*

In the case of an overcomplete frame, there are many ways to represent a vector in terms of the frame. However, as in the finite-dimensional case, the representation coefficients coming from the dual frame, $\{\langle x, S^{-1}x_k\rangle\}$, have the smallest norm in ℓ^2.

Where there is no redundancy, that is, when the operator T is a surjection onto ℓ^2, then the frame $\{x_k\}_{k\in K}$ is referred to as a *Riesz basis* which is the analog of a basis in finite dimensions. A Riesz basis is necessarily also a frame but allows for unique expansions of elements of H, and in this sense is said to be an *exact sequence*. A further characterization of Riesz bases is the following.

Theorem 3.3 *A sequence of vectors $\{x_k\}_{k\in K}$ in a Hilbert space H is a Riesz basis if and only if there exist constants A, $B > 0$ such that for every finite sequence of scalars $(c_j)_{j=1}^n$,*

$$A\sum_{j=1}^{n}|c_j|^2 \le \left\|\sum_{j=1}^{n} c_j x_j\right\|_H^2 \le B\sum_{j=1}^{n}|c_j|^2.$$

There are many excellent references for information on the basic theory of frames, Riesz bases, and their uses such as [17–19, 32, 46, 47].

3.4 Some Historical Remarks

The notion of a frame was first introduced in 1952 by Duffin and Schaeffer in [26] in the context of nonharmonic Fourier series. The paper considered the properties of sequences of functions in $L^2(-1/2, 1/2)$ of the form $\{e^{2\pi i\lambda_n x}\}_{n\in\mathbf{Z}}$, where $\Lambda = \{\lambda_n\}_{n\in\mathbf{Z}}$ is a sequence of points in $I\!R$. In this context, we write $\mathcal{E}(\Lambda) = \{e^{2\pi i\lambda x}\}_{\lambda\in\Lambda}$. The fundamental result of the paper is the following.

Theorem 3.4 ([26]) *The collection $\mathcal{E}(\Lambda)$ is a frame for $L^2(-\gamma, \gamma)$ for all $0 < \gamma < 1/2$ if the set Λ has* uniform density 1, *that is, if there exist constants δ, $L > 0$ such that for all $n \in \mathbf{Z}$, $|\lambda_n - n| \le L$ and for all $n \ne m$, $|\lambda_n - \lambda_m| \ge \delta$ (that is, Λ is* uniformly discrete*).*

The notion of uniform density is a special case of the more general concept of *Beurling density* (see [34]). Given a uniformly discrete subset Λ of $I\!R$, define for $r > 0$ $n^+(r)$ and $n^-(r)$ to be respectively the largest and smallest number of points of Λ in any interval of length r and let

$$D^+(\Lambda) = \lim_{r\to\infty}\frac{n^+(r)}{r} \quad \text{and} \quad D^-(\Lambda) = \lim_{r\to\infty}\frac{n^-(r)}{r}$$

denote the upper and lower Beurling densities of Λ. If $D^+(\Lambda) = D^-(\Lambda)$ then this common value is denoted $D(\Lambda)$ and is referred to as the Beurling density of Λ.

In the landmark paper [34], H. J. Landau showed, among other things, that if for some uniformly discrete subset Λ of $I\!R$, $\mathcal{E}(\Lambda)$ is a frame for $L^2(-\gamma, \gamma)$, then $D^-(\Lambda) \ge 2\gamma$, and that if $\mathcal{E}(\Lambda)$ is a Riesz basis for $L^2(-\gamma, \gamma)$, then $D^+(\Lambda) \le 2\gamma$.

This result shows in particular that for sets Λ with uniform density 1, the sets $\mathcal{E}(\Lambda)$ must necessarily be overcomplete frames for $L^2(-\gamma, \gamma)$ whenever $0 < \gamma < 1/2$. In this sense, Theorem 3.4 describes necessarily redundant systems, i.e., frames for $L^2(-\gamma, \gamma)$.

Theorem 3.4 can also be thought of as a perturbation result in the following sense. If $L = 0$ then $\Lambda = \{e^{2\pi inx}\}_{n\in\mathbf{Z}}$ and $\mathcal{E}(\Lambda)$ is an orthonormal basis for $L^2(-1/2, 1/2)$, hence clearly also a frame for $L^2(-\gamma, \gamma)$ for all $0 < \gamma < 1/2$. The theorem can then be interpreted as saying that any bounded perturbation of $\mathbf{Z}$ that remains uniformly discrete also has this property. Such questions in frame theory were explored by R. Young in a series of papers in the mid-1970s [40–45] for nonredundant systems. He proved several perturbation results for Riesz bases of complex exponentials which, while not results in the theory of frames per se, were important papers that highlighted clearly the flexibility achieved when overcomplete and redundant systems are permitted.

In [22], Daubechies, Grossman, and Meyer connected explicitly the notion of a frame with the expansion of a function in terms of so-called *coherent states*, in particular, the transformation of a single function by a fixed collection of transformations based on the Weil-Heisenberg group and by the affine group. The former collections are now more commonly referred to as Gabor functions and the latter as wavelets. In this paper, it was observed that requiring such decompositions to be orthogonal or nonredundant can lead to undesirable features of the expansion. In the particular examples given in the paper, these undesirable characteristics are related to poor time–frequency localization of the expansions. That is, nonlocal changes in the reconstructed function can arise from local changes in the coefficients. The important insight of the paper is the embrace of redundancy and the observation that redundancy can have value. This is stated explicitly in the abstract: *It is believed, that such "quasiorthogonal expansions" will be a useful tool in many areas of theoretical physics and applied mathematics.* In this sense [22] has had enduring influence by demonstrating that by allowing redundance explicitly in the representation of a function, you gain a great deal more than you lose.

A good example of the advantages of redundancy comes in the context of Gabor frames and is known as the Balian–Low Theorem. This theorem is related to the Heisenberg uncertainty principle and states the following.

Theorem 3.5 ([4], cf. [7, 8, 21]) *If the Gabor system* $\{e^{2\pi imbx}\, g(x - na)\}_{n,m\in\mathbf{Z}}$ *forms a Riesz basis for* $L^2(I\!R)$ *then*

$$\|x\, g(x)\|_2 \|\gamma \widehat{g}(\gamma)\|_2 = \infty. \tag{3.1}$$

(Here $\widehat{g}$ *denotes the Fourier transform of* g*).*

The theorem asserts that if a Gabor system forms a Riesz basis, then the Gabor function g cannot have good localization simultaneously in time and frequency. This can be interpreted as saying that if a function is developed in a Riesz basis of Gabor functions, then local changes in the time and frequency content of the function can result in global changes in the coefficients representing that function. In particular, the result says that any Gabor frame using the Gaussian $g(x) = e^{-\pi x^2}$, which minimizes the uncertainty product (3.1), must necessarily be overcomplete. Similar no-go theorems exist for wavelet systems, see e.g., [6].

Another notable development in the use of frames in the context of atomic decompositions and coherent state expansions came in a series of papers by H. Feichtinger and K. Gröchenig, [27–29]. In these papers, the idea of frames consisting of orbits of a single vector under the action of an irreducible unitary group representation is explored. The notion of a frame (or atomic) decomposition for general Banach spaces is described and frames generated by irregular sampling of the group representation are developed. The idea of an overcomplete representation is essential to achieve the generality of these results.

The change in perspective on frames represented by these papers, from a topic in the theory of nonharmonic Fourier series to an important and powerful tool in the time–frequency analysis of functions, was then fully established in the 1980s. With the tool of frame theory now firmly in the gaze of the harmonic analysis, wavelet, and time–frequency analysis communities, fruitful and interesting applications of the theory began to flourish. In the following section, we will explore some applications that take particular advantage of redundancy.

Consider the following model for transmission of a signal over a channel. Suppose that the signal of interest is the vector $x \in \mathbb{C}^d$. We store the vector by forming its coefficients with respect to a finite frame given by the $d \times n$ matrix F by computing $y = F^*x \in \mathbb{C}^n$, then transmit y over the channel. The received signal $\widehat{y}$ will be corrupted by quantization error and by noise, that is, $\widehat{y} = y + \varepsilon$ where ε is a random vector in $I\!R^n$. Note that in the absence of noise, the original signal x can be reconstructed from its frame coefficients y. The extent to which the original signal can be reconstructed from the noisy coefficients $\widehat{y}$ is a measure of the *robustness to noise* of the coding scheme. This question was investigated by Goyal, Kovačević, and Kelner for the scheme outlined above in [31].

Theorem 3.6 ([31]) *If the transmission error ε is modeled as zero-mean uncorrelated noise, the mean square error of the reconstructed signal is minimized if and only if the frame is uniform and tight.*

The idea behind the usefulness of frames for noise reduction comes directly from the redundancy of frames. If the $d \times n$ matrix F represents a frame, then the range of the mapping $F^* \colon I\!R^d \to I\!R^n$ is only a d-dimensional subspace of $I\!R^n$, and the vector y is in this subspace, but the distorted vector $\widehat{y}$ is unlikely to be. The reconstruction scheme $\widehat{x} = (FF^*)^{-1}F\widehat{y}$ solves the least-squares problem $\min_x \|F^*x - \widehat{y}\|_2$ with the minimum-norm solution $\widehat{x}$, since $(FF^*)^{-1}F$ is the pseudoinverse or Moore–Penrose inverse of F^*. So in fact this procedure projects the noise vector ε onto the range of F^* then reconstructs $\widehat{x}$ from those frame coefficients. Since projection automatically reduces the norm of a vector, we see that the approximation $\widehat{x}$ is better than what would have been obtained if the columns of F had constituted a nonredundant frame.

Now suppose that, in addition to quantization and noise, the channel distorts the transmitted vector $\widehat{y}$ by erasing components at random. Robustness to this sort of distortion means maximizing the number of components that can be erased while still allowing reconstruction of the signal as accurately as possible from the remaining coefficients. In the absence of noise, it would be sufficient to have the frame satisfy the following definition given in [31] (the term *Spark* is coined in [25]).

Definition 3.2 A frame $\mathscr{F} = \{x_k\}_{k=1}^{n}$ in $\mathbb{C}^d$ is *maximally robust to erasures* if the removal of any $l \leq n - d$ vectors from $\mathscr{F}$ leaves a frame. The *Spark* of an $d \times n$ matrix M is the size of the smallest linearly dependent subset of columns of M. Hence a frame with frame matrix F is maximally robust to erasures if $Spark(F^*) = d + 1$.

The paper [31] constructs such frames and analyzes the maximum distortion when the reconstruction is done on noisy measurements and when the number of erasures is specified. It should be noted also that other examples of full-spark finite frames are known and in particular it is known that full-spark finite Gabor frames exist for all dimensions d [35, 36].

3.5 Compressive Sensing

Consider the following problem in signal or image recovery. Recover an unknown vector $x \in I\!R^n$ from $d < n$ linear measurements under the assumption that x is *sparse* or *compressible*, that is, sparse in some orthonormal basis. In other words, we assume that for some $s \in \mathbb{N}$, x has no more than s nonzero elements (that is, that x is s-sparse) or equivalently that there is some $n \times n$ orthogonal matrix Φ with the property that the vector Φx is s-sparse. Note that the collection of s-sparse vectors in $I\!R^n$ is not a linear subspace. In what follows, we will assume that the unknown vector x is s-sparse.

If we define a $d \times n$ matrix F to be the *measurement matrix*, then the problem becomes to recover x from $y = Fx$ under the assumption that x is s-sparse. Without the assumption of sparsity, the problem is clearly underdetermined and hence not solvable. A very simple necessary and sufficient condition on F guaranteeing that the problem at the very least has a solution is given in the following theorem.

Theorem 3.7 *The collection of s-sparse vectors in $\mathbb{R}^n$ is uniquely determined by the measurements Fx in the sense that for all s-sparse vectors x_1 and x_2, $Fx_1 = Fx_2$ implies $x_1 = x_2$ if and only if $Spark(F) > 2s$.*

If we think of the measurement matrix F as a frame matrix, that is, as a matrix whose columns form a frame for $I\!R^d$, then it is clear that in order for the problem to be meaningful, it is required that the frame be redundant.

Theorem 3.7 is an injectivity result that leaves aside the very difficult problem of actually reconstructing x from the measurements. This can be done in principle by solving a reduced system of d equations in d unknowns for every possible collection of d columns of F and looking for the solution with the fewest number of nonzero entries. In other words, we can solve the minimization problem

$$\min_{\widetilde{x}\in\mathbb{R}^n} \|\widetilde{x}\|_0 \text{ subject to } F\widetilde{x} = y,$$

where $\|x\|_0$ is the support size of the vector x. This approach quickly becomes intractable as d increases. However the problem becomes more tractable under an additional assumption on the matrix F known as the *Restricted Isometry Property*.

Definition 3.3 ([13]) For each $s \in \mathbb{N}$, define the isometry constant δ_s of a $d \times n$ matrix F as the smallest number such that

$$(1 - \delta_s) \|x\|^2_{\ell^2_n} \leq \|Fx\|^2_{\ell^2_d} \leq (1 + \delta_s) \|x\|^2_{\ell^2_n}$$

holds for all s-sparse vectors $x \in I\!R^n$.

If δ_{2s} is sufficiently small, then the ℓ^1 minimization problem

$$\min_{\widetilde{x}\in\mathbb{R}^n} \|\widetilde{x}\|_{\ell^1_n} \text{ subject to } F\widetilde{x} = y \tag{3.2}$$

yields an exact solution for s-sparse vectors $\widetilde{x}$. Indeed the following theorem holds.

Theorem 3.8 ([13]) *If $\delta_{2s} < \sqrt{2} - 1$ then the solution to the ℓ^1 minimization problem (3.2) will yield the unique s-sparse solution to the equation $Fx = y$.*

The RIP also guarantees that stable recovery of x from noisy measurements is possible. To see how this works, suppose that we possess the measurements $\widetilde{y} = Fx + z$, where z is an unknown noise vector satisfying $\|z\|_{\ell^2_d} \leq \varepsilon$. In this case, the previous theorem becomes

Theorem 3.9 ([12]) *If $\delta_{2s} < \sqrt{2} - 1$ then there exists a constant C such that for any s-sparse solution x^* to the ℓ^1 minimization problem*

$$\min_{\widetilde{x}\in\mathbb{R}^n} \|\widetilde{x}\|_{\ell^1_n} \textit{ subject to } \|\widetilde{y} - F\widetilde{x}\|_{\ell^2_d} \leq \varepsilon$$

will satisfy $\|x^* - x\|_{\ell^2_n} \leq C\varepsilon$.

More information on the basics of this theory and hints for further research can be found in the excellent expository articles [10, 11, 23, 38].

In order to draw a connection to frame theory, we collect some fundamental results in matrix theory. Let F be a $d \times n$ measurement matrix with $d \leq n$. Suppose that $F = [f_1\ f_2\ \cdots\ f_m]$ and that the columns $\{f_i\}_{i=1}^m$ of F form a uniform frame for $\mathbb{R}^n$ with unit norm. The *coherence* of $\{f_i\}_{i=1}^m$ is the quantity

$$\mu(F) = \max_{i \neq j} |\langle f_i, f_j \rangle|.$$

The coherence of any unit norm uniform frame satisfies the Welch bound [39].

Theorem 3.10 *Let F be a $d \times n$ matrix with unit norm columns. Then*

$$\mu(F) \geq \sqrt{\frac{n-d}{d(n-1)}}.$$

This lower bound is achieved when the quantities $|\langle f_i, f_j\rangle|$ with $i \neq j$ are all equal, and in this case we say that the frame is *equiangular*. The following standard result follows from a classical theorem bounding the eigenvalues of a square matrix due to Gershgorin [30], see also [5, 24, 25].

Theorem 3.11 *The $d \times n$ matrix F satisfies the RIP with $\delta_s \leq \mu(F)(s-1)$. This means that any $d \times n$ matrix F satisfies the RIP for $\delta \geq \mu(F)(s-1)$.*

From these considerations, it follows that a good measurement matrix should have maximal spark and the frame formed from its columns should have minimal coherence. Such matrices are very difficult to construct, e.g., [5, 24].

3.6 Phaseless Recovery

In this section, we examine a similar signal recovery problem, specifically the recovery of a signal from only the magnitudes of a collection of linear measurements. This problem has been around for a very long time and has its roots in applications to X-ray crystallography about a century ago. Recently, significant breakthroughs in understanding and solving the problem have come about by interpreting the problem in terms of frames.

Specifically, given a frame $\{f_i\}_{i=1}^n$ for $\mathbb{C}^d$, the problem is to recover a vector $x \in \mathbb{R}^d$ from the magnitudes of its frame coefficients, namely from

$$\{|\langle x, f_i\rangle|\}_{i=1}^n.$$

The initial breakthrough in this work is due to Balan, Casazza, and Edidin [3] in which the following definition was offered.

Definition 3.4 A frame $\mathscr{F} = \{f_i\}_{i=1}^n$ for $\mathbb{C}^d$ is called *phase retrievable* if the mapping

$$\alpha\colon \mathbb{C}^d \to \mathbb{R}^n; \quad x \mapsto \{|\langle x, f_i\rangle|\}_{i=1}^n$$

is injective up to a constant phase factor.

The problem of algorithmic recovery of x from $\alpha(x)$ was addressed first in [2]. While a great deal of work has been done on this subject, what is interesting from our point of view is that the redundancy of frames plays a crucial role in the solution to the problem as well as provides the correct perspective from which to attack the problem. While this may seem natural to some degree, what is most interesting is that a redundancy factor of about 4 turns out to be what is required in the complex case and 2 in the real case. An excellent survey of the current state of the art on this problem as of the beginning of 2016 is given in [1], and the following theorems are stated there.

Theorem 3.12 ([3]) *Assume that* $\mathscr{F} = \{f_i\}_{i=1}^n$ *is a frame for* $\mathbb{R}^d$. *Then*

- *If* $\mathscr{F}$ *is phase retrievable for* $\mathbb{R}^d$ *then* $n \geq 2d - 1$.
- *If* $n = 2d - 1$ *then* $\mathscr{F}$ *is phase retrievable if and only if the frame matrix* F *corresponding to* $\mathscr{F}$ *has full Spark.*

Theorem 3.13 *Assume that* $\mathscr{F} = \{f_i\}_{i=1}^n$ *is a frame for* $\mathbb{C}^d$. *Then*

- ([33]) *If* $\mathscr{F}$ *is phase retrievable for* $\mathbb{C}^d$ *then*

$$n \geq 4d - 2 - 2b(n) + \begin{cases} 2 \text{ if } n \text{ is odd and } b = 3 \mod 4 \\ 1 \text{ if } n \text{ is odd and } b = 2 \mod 4 \\ 0 \text{ otherwise} \end{cases}$$

  where $b(n)$ *denotes the number of* 1s *in the binary expansion of* $n - 1$.
- ([9]) *For any positive integer* d, *a phase retrievable frame* $\mathscr{F}$ *for* $\mathbb{C}^d$ *can be constructed that contains* $n = 4d - 4$ *vectors.*
- ([20]) *If* $n \geq 4d - 4$ *then for generic frames,* $\mathscr{F}$ *is phase retrievable, and if* $d = 2^k + 1$ *and* $n < 4d - 4$ *then no frame* $\mathscr{F}$ *for* $\mathbb{C}^d$ *is phase retrievable.*

3.7 The Feichtinger Conjecture (Now Theorem)

We conclude our discussion of frames and redundancy with the Feichtinger conjecture. This assertion deals with the relationship between frames in Hilbert spaces and so-called *Riesz sequences*. A collection of vectors $\{x_n\}$ in a Hilbert space H is a *Riesz sequence* if there exist constants $A,\ B > 0$ such that for all finite sequences $\{c_k\}_{k=1}^n$,

$$A \sum_{j=1}^n |c_j|^2 \leq \left\| \sum_{j=1}^n c_j x_j \right\|^2 \leq B \sum_{j=1}^n |c_j|^2.$$

In other words, a Riesz sequence forms a Riesz basis for its closed linear span. It follows then that a Riesz sequence is nonredundant in the sense that if $(c_k) \in \ell^2$ and $\sum_{j=1}^\infty c_j x_j = 0$ then $c_j = 0$ for all j. It also follows that a Riesz sequence is a frame for its closed linear span.

Feichtinger conjectured that any uniform frame can be partitioned into a finite union of Riesz sequences. What is most notable about the Feichtinger conjecture is that it is equivalent to several other conjectures in diverse areas of mathematical analysis. This observation was made by Casazza and his collaborators in several papers starting with [14] in which they prove the equivalence of the Feichtinger conjecture to three important, then-unsolved problems: The Kadisson–Singer Conjecture (1959), the Paving Conjecture (1979), and the Bourgain–Tzafriri Conjecture (1991). These were deep, long-standing unsolved problems in operator theory and have connections to graph theory, mathematical physics, and signal processing.

Very recently, the Feichtinger Conjecture has been proved by Marcus, Spielman, and Srivastava [37] by proving in the affirmative the equivalent Kadisson–Singer conjecture. This result gives important insight into the deep structure of frames. It would imply that any reasonable frame can be realized as a finite union of well-behaved, nonredundant sets.

In fact, what is quite interesting from a practical point of view is that Feichtinger's conjecture has a finite-dimensional version, which is probably the simplest statement of the problem.

Theorem 3.14 (Feichtinger Conjecture—finite version [14]) *For every $B > 0$ there exists $M \in \mathbb{N}$ and $A > 0$ such that any uniform frame $\{f_i\}_{i=1}^n$ for $\mathbb{C}^d$ with unit norm can be written as a union of M Riesz sequences with upper bound B and lower bound A.*

There are several excellent references that discuss these conjectures and show their equivalence, for example, [14–16] as well as many other papers linked from http://www.framerc.org/.

References

1. Balan, R.: Frames and phaseless reconstruction. Finite Frame Theory: A Complete Introduction to Overcompleteness. Lecture Notes for the 2015 AMS Short Course. San Antonio, TX (2016)
2. Balan, R., Bodmann, B.G., Casazza, P.G., Edidin, D.: Painless reconstruction from magnitudes of frame coefficients. J. Fourier Anal. Appl. **15**(4), 488–501 (2009)
3. Balan, R., Casazza, P., Edidin, D.: On signal reconstruction without phase. Appl. Comput. Harmon. Anal. **20**(3), 345–356 (2006)
4. Balian, R.: Un principe d'incertitude fort en théorie du signal ou en mécanique quantique. C. R. Acad. Sci. Paris Sér. II Méc. Phys. Chim. Sci. Univ. Sci. Terre **292**(20), 1357–1362 (1981)
5. Bandeira, A.S., Fickus, M., Mixon, D.G., Wong, P.: The road to deterministic matrices with the restricted isometry property. J. Fourier Anal. Appl. **19**(6), 1123–1149 (2013)
6. Battle, G.: Phase space localization theorem for ondelettes. J. Math. Phys. **30**(10), 2195–2196 (1989)
7. Benedetto, J., Heil, C., Walnut, D.: Uncertainty principles for time-frequency operators. In: Continuous and discrete Fourier transforms, extension problems and Wiener-Hopf equations, vol. 58 of Oper. Theory Adv. Appl., pp. 1–25. Birkhäuser, Basel (1992)
8. Benedetto, J.J., Walnut, D.F.: Gabor frames for L^2 and related spaces. In: Wavelets: mathematics and applications, Stud. Adv. Math., pp. 97–162. CRC, Boca Raton (1994)
9. Bodmann, B.G., Hammen, N.: Stable phase retrieval with low-redundancy frames. Adv. Comput. Math. **41**(2), 317–331 (2015)
10. Bryan, K., Leise, T.: Making do with less: an introduction to compressed sensing. SIAM Rev. **55**(3), 547–566 (2013)
11. Candes, E.J., Wakin, M.B.: An introduction to compressive sampling. IEEE Signal Process. Mag. **25**(2), 21–30 (2008). March

12. Candès, E.J.: The restricted isometry property and its implications for compressed sensing. C. R. Math. Acad. Sci. Paris **346**(9–10), 589–592 (2008)
13. Candes, E.J., Tao, T.: Decoding by linear programming. IEEE Trans. Inf. Theory **51**(12), 4203–4215 (2005)
14. Casazza, P.G., Christensen, O., Lindner, A.M., Vershynin, R.: Frames and the Feichtinger conjecture. Proc. Amer. Math. Soc., **133**(4), 1025–1033 (2005). (electronic)
15. Casazza, P.G., Fickus, M., Tremain, J.C., Weber, E.: The Kadison-Singer problem in mathematics and engineering: a detailed account. In: Operator Theory, Operator Algebras, and Applications, Contemporary Mathematics, vol. 414, pp. 299–355. American Mathematical Society, Providence, RI (2006)
16. Casazza, P.G., Tremain, J.C.: The Kadison-Singer problem in mathematics and engineering. Proc. Natl. Acad. Sci. USA, **103**(7), 2032–2039 (2006). (electronic)
17. Christensen, O.: Applied and Numerical Harmonic Analysis. An introduction to frames and Riesz bases. Birkhäuser Boston Inc, Boston (2003)
18. Christensen, O.: Frames and Bases: An Introductory Course. Applied and Numerical Harmonic Analysis. Birkhäuser Boston Inc, Boston (2008)
19. Christensen, O.: Functions, Spaces, and Expansions: Mathematical Tools in Physics and Engineering. Applied and Numerical Harmonic Analysis. Birkhäuser Boston Inc, Boston (2010)
20. Conca, A., Edidin, D., Hering, M., Vinzant, C.: An algebraic characterization of injectivity in phase retrieval. Appl. Comput. Harmon. Anal. **38**(2), 346–356 (2015)
21. Daubechies, I.: Ten lectures on wavelets. In: CBMS-NSF Regional Conference Series in Applied Mathematics, vol. 61. Society for Industrial and Applied Mathematics (SIAM), Philadelphia (1992)
22. Daubechies, I., Grossmann, A., Meyer, Y.: Painless nonorthogonal expansions. J. Math. Phys. **27**(5), 1271–1283 (1986)
23. Davenport, M.A., Duarte, M.F., Eldar, Y.C., Kutyniok, G.: Introduction to compressed sensing. In: Compressed Sensing, pp. 1–64. Cambridge University Press, Cambridge (2012)
24. DeVore, R.A.: Deterministic constructions of compressed sensing matrices. J. Complex. **23**(4–6), 918–925 (2007)
25. Donoho, D.L., Elad, M.: Optimally sparse representation in general (nonorthogonal) dictionaries via l^1 minimization. Proc. Natl. Acad. Sci. USA, **100**(5), 2197–2202 (2003). (electronic)
26. Duffin, R.J., Schaeffer, A.C.: A class of nonharmonic Fourier series. Trans. Am. Math. Soc. **72**, 341–366 (1952)
27. Feichtinger, H.G.: Atomic characterizations of modulation spaces through Gabor-type representations. Rocky Mountain J. Math., **19**(1), 113–125 (1989). Constructive Function Theory—86 Conference (Edmonton, AB, 1986)
28. Feichtinger, H.G., Gröchenig, K.H.: Banach spaces related to integrable group representations and their atomic decompositions. I. J. Funct. Anal. **86**(2), 307–340 (1989)
29. Feichtinger, H.G., Gröchenig, K.H.: Banach spaces related to integrable group representations and their atomic decompositions. II. Monatsh. Math. **108**(2–3), 129–148 (1989)
30. Geršgorin, S.: über die abgrenzung der eigenwerte einer matrix. Bulletin de l'Académie des Sciences de l'URSS. Classe des sciences mathématiques et na, **20**(6), 749–754 (1931)
31. Goyal, V.K., Kovačević, J., Kelner, J.A.: Quantized frame expansions with erasures. Appl. Comput. Harmon. Anal. **10**(3), 203–233 (2001)
32. Gröchenig, K.: Foundations of time-frequency analysis. Applied and Numerical Harmonic Analysis. Birkhäuser Boston Inc, Boston, MA (2001)
33. Heinosaari, T., Mazzarella, L., Wolf, M.M.: Quantum tomography under prior information. Comm. Math. Phys. **318**(2), 355–374 (2013)
34. Landau, H.J.: Necessary density conditions for sampling and interpolation of certain entire functions. Acta Math. **117**, 37–52 (1967)
35. Lawrence, J., Pfander, G.E., Walnut, D.: Linear independence of Gabor systems in finite dimensional vector spaces. J. Fourier Anal. Appl. **11**(6), 715–726 (2005)
36. Malikiosis, R.-D.: A note on Gabor frames in finite dimensions. Appl. Comput. Harmon. Anal. **38**(2), 318–330 (2015)

37. Marcus, A.W., Spielman, D.A., Srivastava, N.: Interlacing families II: Mixed characteristic polynomials and the Kadison-Singer problem. Ann. Math. (2), **182**(1), 327–350 (2015)
38. Romberg, J.: Imaging via compressive sampling. IEEE Signal Process. Mag. **25**(2), 14–20 (2008). March
39. Welch, L.: Lower bounds on the maximum cross correlation of signals (corresp.). IEEE Trans. Inf. Theory **20**(3), 397–399 (1974). May
40. Young, R.M.: Inequalities for a perturbation theorem of Paley and Wiener. Proc. Am. Math. Soc. **43**, 320–322 (1974)
41. Young, R.M.: Interpolation in a classical Hilbert space of entire functions. Trans. Am. Math. Soc. **192**, 97–114 (1974)
42. Young, R.M.: A note on a trigonometric moment problem. Proc. Am. Math. Soc. **49**, 411–415 (1975)
43. Young, R.M.: On perturbing bases of complex exponentials in $L^2(-\pi, \pi)$. Proc. Am. Math. Soc. **53**(1), 137–140 (1975)
44. Young, R.M.: Interpolation for entire functions of exponential type and a related trigonometric moment problem. Proc. Am. Math. Soc. **56**, 239–242 (1976)
45. Young, R.M.: Some stability theorems for nonharmonic fourier series. Proc. Am. Math. Soc., **61**(2), 315–319 (1976)
46. Robert M. Young. *An introduction to nonharmonic Fourier series*, volume 93 of *Pure and Applied Mathematics*. Academic Press, Inc. [Harcourt Brace Jovanovich, Publishers], New York-London (1980)
47. Young, R.M.: An Introduction to Nonharmonic Fourier Series, 1st edn. Academic Press Inc, San Diego (2001)

Chapter 4
The Challenging Problem of Industrial Applications of Multicore-Generated Iterates of Nonlinear Mappings

Jean-Pierre Lozi, Oleg Garasym and René Lozi

Abstract The study of nonlinear dynamics is relatively recent with respect to the long historical development of early mathematics since the Egyptian and the Greek civilization, even if one includes in this field of research the pioneer works of Gaston Julia and Pierre Fatou related to one-dimensional maps with a complex variable, nearly a century ago. In France, Igor Gumosky and Christian Mira began their mathematical researches in 1958; in Japan, the Hayashi' School (with disciples such as Yoshisuke Ueda and Hiroshi Kawakami), a few years later, was motivated by applications to electric and electronic circuits. In Ukraine, Alexander Sharkovsky found the intriguing Sharkovsky's order, giving the periods of periodic orbits of such nonlinear maps in 1962, although these results were only published in 1964. In 1983, Leon O. Chua invented a famous electronic circuit that generates chaos, built with only two capacitors, one inductor and one nonlinear negative resistance. Since then, thousands of papers have been published on the general topic of chaos. However, the pace of mathematics is slow, because any progress is based on strictly rigorous proof. Therefore, numerous problems still remain unsolved. For example, the long-term dynamics of the Hénon map, the first example of a strange attractor for mappings, remain unknown close to the classical parameter values from a strictly mathematical point of view, 40 years after its original publication. In spite of this lack of rigorous mathematical proofs, nowadays, engineers are actively working on applications of chaos for several purposes: global optimization, genetic algorithms, CPRNG (Chaotic Pseudorandom Number Generators), cryptography, and so on. They use nonlinear maps for practical applications without the need of sophisticated theorems. In this chapter, after giving some prototypical examples of the industrial

J.-P. Lozi
I3S laboratory, UMR 7271, Université Côte d'Azur, CNRS, Euclide B,
Les Algorithmes, 2000 Route des Lucioles, 06900 Sophia Antipolis, France

O. Garasym
SOC, IBM, Wroclaw, Poland

R. Lozi (✉)
J. A. Dieudonné laboratory, UMR 7351, Université Côte d'Azur, CNRS,
28 Avenue Valrose, 06108 Nice Cedex 02, France
e-mail: rlozi@unice.fr

P. Manchanda et al. (eds.), *Industrial Mathematics and Complex Systems*,
Industrial and Applied Mathematics, DOI 10.1007/978-981-10-3758-0_4

applications of iterations of nonlinear maps, we focus on the exploration of topologies of coupled nonlinear maps that have a very rich potential of complex behavior. Very long computations on modern multicore machines are used: they generate up to one hundred trillion iterates in order to assess such topologies. We show the emergence of randomness from chaos and discuss the promising future of chaos theory for cryptographic security.

Keywords Chaos · Cryptography · Mappings · Chaotic pseudorandom numbers Attractors

AMS Subject Classification 37N30 · 37D45 · 65C10 · 94A60

4.1 Introduction

The last few decades have seen the tremendous development of new IT technologies that incessantly increase the need for new and more secure cryptosystems.

For instance, the recently invented Bitcoin cryptocurrency is based on the secure Blockchain system that involves hash functions [1]. This technology, used for information encryption, is pushing forward the demand for more efficient and secure pseudorandom number generators [2] which, in the scope of chaos-based cryptography, were first introduced by Matthews in the 1990s [3]. Contrarily to most algorithms that are used nowadays and based on a limited number of arithmetic or algebraic methods (like elliptic curves), networks of coupled chaotic maps offer quasi-infinite possibilities to generate parallel streams of pseudorandom numbers (PRN) at a rapid pace when they are executed on modern multicore processors. Chaotic maps are able to generate independent and secure pseudorandom sequences (used as information carriers or directly involved in the process of encryption/decryption [4]). However, the majority of well-known chaotic maps are not naturally suitable for encryption [5] and most of them do not exhibit even satisfactory properties for such a purpose.

In this chapter, we explore the novel idea of coupling a symmetric tent map with a logistic map, following several network topologies. We add a specific injection mechanism to capture the escaping orbits. In the goal of extending our results to industrial mathematics, we implement these networks on multicore machines and we test up to 100 trillion iterates of such mappings, in order to make sure that the obtained results are firmly grounded and able to be used in industrial contexts such as e-banking, e-purchasing, or the Internet of Things (IoT).

The chaotic maps, when used in the sterling way, could generate not only chaotic numbers, but also pseudorandom numbers as shown in [6] and as we show in this chapter with more sophisticated numerical experiments.

Various choices of PNR Generators (PRNGs) and crypto-algorithms are currently necessary to implement continuous, reliable security systems. We use a software approach because it is easy to change a cryptosystem to support protection, whereas replacing hardware used for True Random Number Generators would be costly and time-consuming. For instance, after the secure software protocol Wi-Fi Protected Access (WPA) was broken, it was simply updated and no expensive hardware had to be replaced.

It is a very challenging task to design CPRNGs (Chaotic Pseudo Random Number Generators) that are applicable to cryptography: numerous numerical tests must ensure that their properties are satisfactory. We mainly focus on two- to five-dimension maps, although upper dimensions can be very easily explored with modern multicore machines. Nevertheless, in four and five dimensions, the studied CRPNGs are efficient enough for cryptography.

In Sect. 4.2, we briefly recall the dawn and the maturity of researches on chaos. In Sect. 4.3, we explore two-dimensional topologies of networks of coupled chaotic maps. In Sect. 4.4, we study more thoroughly a mapping in higher dimensions (up to 5) far beyond the NIST tests which are limited to a few millions of iterates and which seem not robust enough for industrial applications, although they are routinely used worldwide. In order to check the portability of the computations on multicore architectures, we have implemented all our numerical experiments on several different multicore machines. We conclude this chapter in Sect. 4.5.

4.2 The Dawn and the Maturity of Researches on Chaos

The study of nonlinear dynamics is relatively recent with respect to the long historical development of early mathematics since the Egyptian and the Greek civilizations (and even before). The first alleged artifact of mankind's mathematical thinking goes back to the Upper Paleolithic era. Dating as far back as 22,000 years ago, the Ishango bone is a dark brown bone which happens to be the fibula of a baboon, with a sharp piece of quartz affixed to one end for engraving. It was first thought to be a tally stick, as it has a series of what has been interpreted as tally marks carved in three columns running the length of the tool [7].

Twenty thousand years later, the Rhind Mathematical Papyrus is the best example of Egyptian mathematics. It dates back to around 1650 BC. Its author is the scribe Ahmes who indicated that he copied it from an earlier document dating from the 12th dynasty, around 1800 BC. It is a practical handbook, whose the first part consists of reference tables and a collection of 20 arithmetic and 20 algebraic problems and linear equations. Problem 32 for instance corresponds (in modern notation) to solving $x + \frac{x}{3} + \frac{x}{4} = 2$ for x [8].

Since those early times, mathematics have known great improvements, flourishing in many different fields such as geometry, algebra (both linked, thanks to the invention of Cartesian coordinates by René Descartes [9]), analysis, probability, number and set theory, and so on.

However, nonlinear problems are very difficult to handle, because, as shown by Galois' theory of algebraic equations which provides a connection between field theory and group theory, it is impossible to solve any polynomial equation of degree equal or greater than 5 using only the usual algebraic operations (addition, subtraction, multiplication, division) and the application of radicals (square roots, cube roots, etc.) [10].

The beginning of the study of nonlinear equation systems goes back to the original works of Gaston Julia and Pierre Fatou regarding to one-dimensional maps with a complex variable, nearly a century ago [11, 12]. Compared to thousands of years of mathematical development, a century is a very short period. In France, 30 years later, Igor Gumosky and Christian Mira began their mathematical researches with the help of a computer in 1958 [13]. They developed very elaborate studies of iterations. One of the best-known formulas they published is

$$\begin{cases} x_{n+1} = f(x_n) + by_n \\ y_{n+1} = f(x_{n+1}) - x_n, \end{cases} \quad \text{with } f(x) = ax + 2(1-a)\frac{x^2}{1+x^2} \tag{4.1}$$

which can be considered as a non-autonomous mapping from the plane $\mathbb{R}^2$ onto itself that exhibits esthetic chaos. Surprisingly, slight variations of the parameter value lead to very different shapes of the attractor (Fig. 4.1).

In Ukraine, Alexander Sharkovsky found the intriguing Sharkovsky's order, giving the periods of periodic orbits of such nonlinear maps in 1962, although these results were only published in 1964 [14]. In Japan the Hayashi' School (with disciples like Yoshisuke Ueda and Hiroshi Kawakami), a few years later, was motivated by applications to electric and electronic circuits. Ikeda proposed the Ikeda attractor [15, 16] which is a chaotic attractor for $u \geq 0.6$ (Fig. 4.2).

$$\begin{cases} x_{n+1} = 1 + u(x_n \cos t_n - y_n \sin t_n) \\ y_{n+1} = u(x_n \sin t_n + y_n \cos t_n), \end{cases} \quad \text{with } t_n = 0.4 - \frac{6}{1 + x_n^2 + y_n^2} \tag{4.2}$$

In 1983, Leon O. Chua invented a famous electronic circuit that generates chaos built with only two capacitors, one inductor and one nonlinear negative resistance [17]. Since then, thousands of papers have been published on the general

Fig. 4.1 Gumowski-Mira attractor for parameter values $a = 0.92768$ and $a = 0.93333$

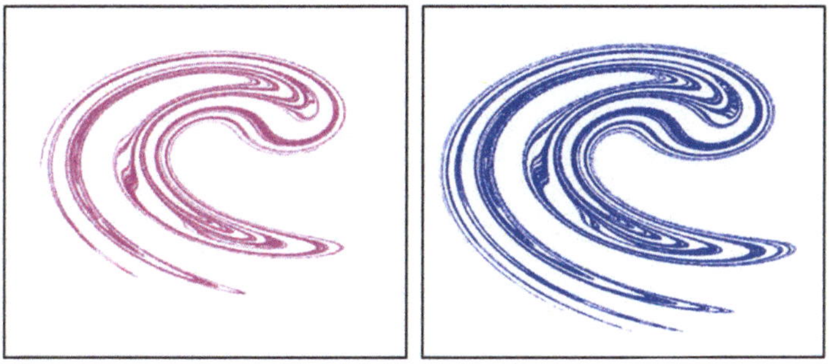

Fig. 4.2 Ikeda attractor for $u = 8.6$ and $u = 8.9$

topic of chaos. However the pace of mathematics is slow, because any progress is based on strictly rigorous proof. Therefore numerous problems still remain unsolved. For example, the long-term dynamics of the Hénon map [18], the first example of a strange attractor for mappings, remains unknown close to the classical parameter values from a strictly mathematical point of view, 40 years after its original publication.

Nevertheless, in spite of this lack of rigorous mathematical results, nowadays, engineers are actively working on applications of chaos for several purposes: global optimization, genetic algorithms, CPRNG, cryptography, and so on. They use nonlinear maps for practical applications without the need of sophisticated theorems. During the last 20 years, several chaotic image encryption methods have been proposed in the literature.

Dynamical systems which present a mixing behavior and that are highly sensitive to initial conditions are called chaotic. Small differences in initial conditions (such as those due to rounding errors in numerical computation) yield widely diverging outcomes for chaotic systems. This effect, popularly known as the butterfly effect, renders long-term predictions impossible in general [19]. This happens even though these systems are deterministic, meaning that their future behavior is fully determined by their initial conditions, with no random elements involved. In other words, the deterministic nature of these systems does not make them predictable. Mastering the global properties of those dynamical systems is a challenging issue nowadays that we try to fix by exploring several network topologies of coupled maps.

In this chapter, after giving some prototypical examples of industrial applications of iterations of nonlinear maps, we focus on the exploration of topologies of coupled nonlinear maps that have a very rich potential of complex behavior. Very long computations on multicore machines are used, generating up to one hundred trillion iterates, in order to assess such topologies. We show the emergence of randomness from chaos and discuss the promising future of chaos theory for cryptographic security.

4.3 Miscellaneous Network Topologies of Coupled Chaotic Maps

4.3.1 Tent-Logistic Entangled Map

In this section we consider only two 1-D maps: the logistic map

$$f_\mu(x) \equiv L_\mu(x) = 1 - \mu x^2 \tag{4.3}$$

and the symmetric tent map

$$f_\mu(x) \equiv T_\mu(x) = 1 - \mu |x| \tag{4.4}$$

both associated to the dynamical system

$$x_{n+1} = f_\mu(x_n), \tag{4.5}$$

where μ is a control parameter which impacts the chaotic degree. Both mappings are sending the one-dimensional interval $[-1,1]$ onto itself.

Since the first study by R. May [20, 21] of the logistic map in the frame of nonlinear dynamical systems, both the logistic (4.3) and the symmetric tent map (4.4) have been fully explored with the aim to easily generate pseudorandom numbers [22].

However, the collapse of iterates of dynamical systems [23] or at least the existence of very short periodic orbits, their non-constant invariant measure, and the easily-recognized shape of the function in the phase space, could lead to avoid the use of such one-dimensional maps (logistic, baker, tent, etc.) or two-dimensional maps (Hénon, Standard, Belykh, etc.) as PRNGs (see [24] for a survey). Yet, the very simple implementation as computer programs of chaotic dynamical systems led some authors to use them as a base for cryptosystems [25, 26]. Even if the logistic and tent maps are topologically conjugates (i.e., they have similar topological properties: distribution, chaoticity, etc.), their numerical behavior differs drastically due to the structure of numbers in computer realization [27].

As said above, both logistic and tent maps are never used in serious cryptography articles because they have weak security properties (collapsing effect) if applied alone. Thus, these maps are often used in modified form to construct CPRNGs [28–30].

Recently, Lozi et al. proposed innovative methods in order to increase randomness properties of the tent and logistic maps over their coupling and sub-sampling [31–33]. Nowadays, hundreds of publications on industrial applications of chaos-based cryptography are available [34–37].

In this chapter, we explore more thoroughly the original idea of combining features of tent (T_μ) and logistic (L_μ) maps to produce a new map with improved properties, through combination in several network topologies. This idea was recently introduced [38, 39] in order to improve previous CPRNGs.

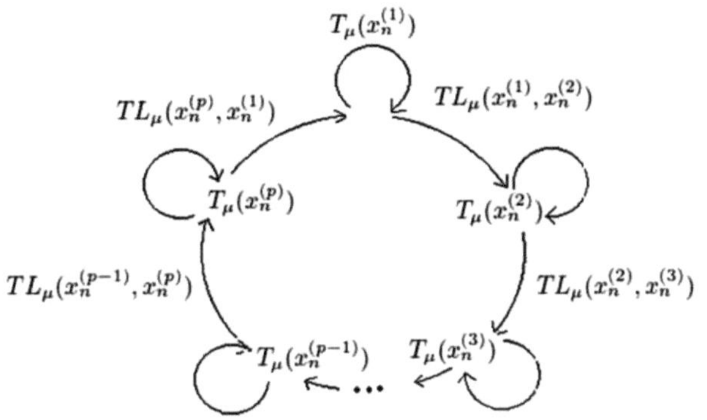

Fig. 4.3 Auto and ring-coupling of the TL_μ and T_μ maps (from [38])

Looking at both Eqs. (4.3) and (4.4), it is possible to reverse the shape of the graph of the tent map T and to entangle it with the graph of the logistic map L. We obtain the combined map

$$f_\mu(x) \equiv TL_\mu(x) = \mu|x| - \mu x^2 = \mu(|x| - x^2) \tag{4.6}$$

When used in more than one dimension, the TL_μ map can be considered as a two-variable map

$$TL_\mu(x^{(i)}, x^{(j)}) = \mu(|x^{(i)}| - (x^{(j)})^2), \quad i \neq j \tag{4.7}$$

Moreover, we can combine again the TL_μ map with T_μ in various ways. If with choose, for instance, a network with a ring shape (Fig. 4.3).

It is possible to define a mapping $M_{\mu,p} : J^p \to J^p$ where $J_p = [-1, 1]^p \subset R^p$:

$$M_{\mu,p} \begin{pmatrix} x_n^{(1)} \\ x_n^{(2)} \\ \cdot \\ \cdot \\ \cdot \\ x_n^{(p)} \end{pmatrix} = \begin{pmatrix} x_{n+1}^{(1)} \\ x_{n+1}^{(2)} \\ \cdot \\ \cdot \\ \cdot \\ x_{n+1}^{(p)} \end{pmatrix} = \begin{pmatrix} T_\mu(x_n^{(1)}) + TL_\mu(x_n^{(1)}, x_n^{(2)}) \\ T_\mu(x_n^{(2)}) + TL_\mu(x_n^{(2)}, x_n^{(3)}) \\ \cdot \\ \cdot \\ \cdot \\ T_\mu(x_n^{(p)}) + TL_\mu(x_n^{(p)}, x_n^{(1)}) \end{pmatrix} \tag{4.8}$$

However, if used in this form, system (4.8) has unstable dynamics and iterated points $x_n^{(1)}, x_n^{(2)}, \ldots, x_n^{(p)}$ quickly spread out. Therefore, to solve the problem of keeping dynamics in the torus $J^p = [-1, 1]^p \subset R^p$, the following injection mechanism has to be used in conjunction with (4.8)

$$\begin{cases} \text{if } (x_{n+1}^{(i)} < -1) \text{ then add } 2 \\ \text{if } (x_{n+1}^{(i)} > 1) \text{ then subtract } 2 \end{cases}, \quad i = 1, 2, \ldots, p. \tag{4.9}$$

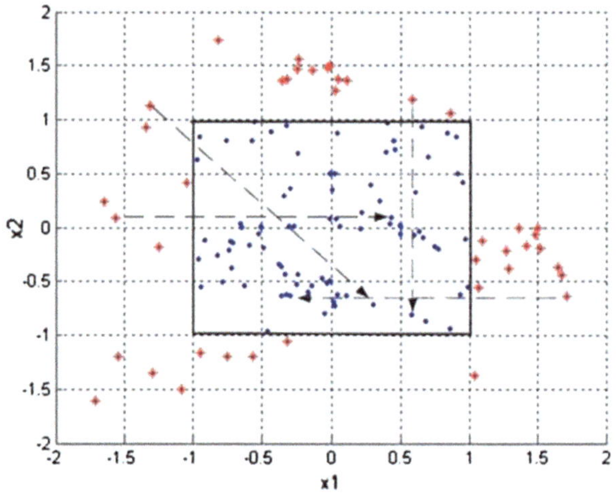

Fig. 4.4 Return mechanism from the $[-2, 2]^p$ torus to $[-1, 1]^p$ (from [38])

Under this injection mechanism, for $1 \leq i \leq p$, points come back from $[-2, 2]^p$ to $[-1, 1]^p$ (Fig. 4.4).

The TL_μ function is a powerful tool to change dynamics. Used in conjunction with T_μ, the map TL_μ makes it possible to establish mutual influence between system components $x_n^{(i)}$ in $M_{\mu,p}$. This multidimensional coupled mapping is interesting because it performs contraction and distance stretching between components, improving chaotic distribution.

The coupling of components has an excellent effect in achieving chaos, because they interact with global system dynamics, being a part of them. Component interaction has a global effect. In order to study this new mapping, we use a graphical approach, however other theoretical assessing functions are also involved.

Note that system (4.8) can be made more generic by introducing constants k^i which generalize considered topologies. Let $\underline{k} = (k^1, k^2, \ldots, k^p)$, we define

$$M_{\mu,p}^{\underline{k}} \begin{pmatrix} x_n^{(1)} \\ x_n^{(2)} \\ \cdot \\ \cdot \\ \cdot \\ x_n^{(p)} \end{pmatrix} = \begin{pmatrix} x_{n+1}^{(1)} \\ x_{n+1}^{(2)} \\ \cdot \\ \cdot \\ \cdot \\ x_{n+1}^{(p)} \end{pmatrix} = \begin{pmatrix} T_\mu(x_n^{(1)}) + k^1 \times TL_\mu(x_n^{(i)}, x_n^{(j)}), & i, j = (1, 2) \text{ or } (2, 1) \\ T_\mu(x_n^{(2)}) + k^2 \times TL_\mu(x_n^{(i)}, x_n^{(j)}) & i, j = (2, 3) \text{ or } (3, 2) \\ \cdot & \\ \cdot & \\ \cdot & \\ T_\mu(x_n^{(p)}) + k^p \times TL_\mu(x_n^{(i)}, x_n^{(j)}) & i, j = (p, 1) \text{ or } (1, p) \end{pmatrix} \tag{4.10}$$

System (4.10) is called alternate if $k^i = (-1)^i$ or $k^i = (-1)^{i+1}$, $1 \leq i \leq p$, or non-alternate if $k^i = +1$, or $k^i = -1$. It can be a mix of alternate and non-alternate if $k^i = +1$ or -1 randomly.

Table 4.1 The sixteen maps defined by Eq. (4.11)

Case	k^1	k^2	i	j	i'	j'
#1	+1	+1	1	2	1	2
#2	+1	−1	1	2	1	2
#3	−1	+1	1	2	1	2
#4	−1	−1	1	2	1	2
#5	+1	+1	2	1	2	1
#6	+1	−1	2	1	2	1
#7	−1	+1	2	1	2	1
#8	−1	−1	2	1	2	1
#9	+1	+1	1	2	2	1
#10	+1	−1	1	2	2	1
#11	−1	+1	1	2	2	1
#12	−1	−1	1	2	2	1
#13	+1	+1	2	1	1	2
#14	+1	−1	2	1	1	2
#15	−1	+1	2	1	1	2
#16	−1	−1	2	1	1	2

4.3.2 Two-Dimensional Network Topologies

We first consider the simplest coupling case, in which only two equations are coupled. The first condition needed to obtain a multidimensional mapping, in the aim of building a new CPRNG, is to obtain excellent uniform distribution of the iterated points. The second condition is that the CPRNG must be assessed positively by the NIST tests [40]. In [38, 39] this two-dimensional case is studied in detail. Using a bifurcation diagram and computation of Lyapunov exponents, it is shown that the best value for the parameter is $\mu = 2$. Therefore, in the rest of this chapter we use this parameter value and we only briefly recall the results found with this value in both of those articles. The general form of $M^{\underline{k}}_{2,2}$ is then

$$M^{\underline{k}}_{2,2}\begin{pmatrix} x_n^{(1)} \\ x_n^{(2)} \end{pmatrix} = \begin{pmatrix} x_{n+1}^{(1)} \\ x_{n+1}^{(2)} \end{pmatrix} = \begin{pmatrix} T_2(x_n^{(1)}) + k^1 \times TL_2(x_n^{(i)}, x_n^{(j)}) \\ T_2(x_n^{(2)}) + k^2 \times TL_2(x_n^{(i')}, x_n^{(j')}) \end{pmatrix} \tag{4.11}$$

with $i,\ j,\ i',\ j' = 1$ or 2, $i \neq j$, and $i' \neq j'$.

Considering this general form, it is possible to define 16 different maps (Table 4.1).

Among this set of maps, we study case #3 and case #13. The map of case #3 is called Single-Coupled alternate due to the shape of the corresponding network and denoted TTL_2^{SC},

$$TTL_2^{SC} = \begin{cases} x_{n+1}^{(1)} = 1 - 2|x_n^{(1)}| - 2(|x_n^{(1)}| - (x_n^{(2)})^2) = T_2(x_n^{(1)}) - TL_2((x_n^{(1)}), (x_n^{(2)})) \\ x_{n+1}^{(2)} = 1 - 2|x_n^{(2)}| + 2(|x_n^{(1)}| - (x_n^{(2)})^2) = T_2(x_n^{(2)}) + TL_2((x_n^{(1)}), (x_n^{(2)})) \end{cases} \quad (4.12)$$

and case #13 is called Ring-Coupled non-alternate and denoted TTL_2^{RC},

$$TTL_2^{RC} = \begin{cases} x_{n+1}^{(1)} = 1 - 2|x_n^{(1)}| + 2(|x_n^{(2)}| - (x_n^{(1)})^2) = T_2(x_n^{(1)}) + TL_2((x_n^{(2)}), (x_n^{(1)})) \\ x_{n+1}^{(2)} = 1 - 2|x_n^{(2)}| + 2(|x_n^{(1)}| - (x_n^{(2)})^2) = T_2(x_n^{(2)}) + TL_2((x_n^{(1)}), (x_n^{(2)})) \end{cases} \quad (4.13)$$

Both systems were selected because they have balanced contraction and stretching processes between components. They allow achieving uniform distribution of the chaotic dynamics. Equations (4.12) and (4.13) are used, of course, in conjunction with injection mechanism (4.9).

The largest torus where points mapped by (4.12) and (4.13) are sent is $[-2, 2]^2$. The confinement from torus $[-2, 2]^2$ to torus $[-1, 1]^2$ of the dynamics obtained by this mechanism is shown in Figs. 4.5 and 4.6: dynamics cross from the negative region (in blue) to the positive one, and conversely to the negative region, if the points stand in the positive regions (in red). Through this operation, the system's dynamics are trapped inside $[-1, 1]^2$. In addition, after this operation is done, the resulting system exhibits more complex dynamics with additional nonlinearity, which is advantageous for chaotic encryption (since it improves security).

A careful distribution analysis of both TTL_2^{SC} and TTL_2^{RC} has been performed using approximated invariant measures.

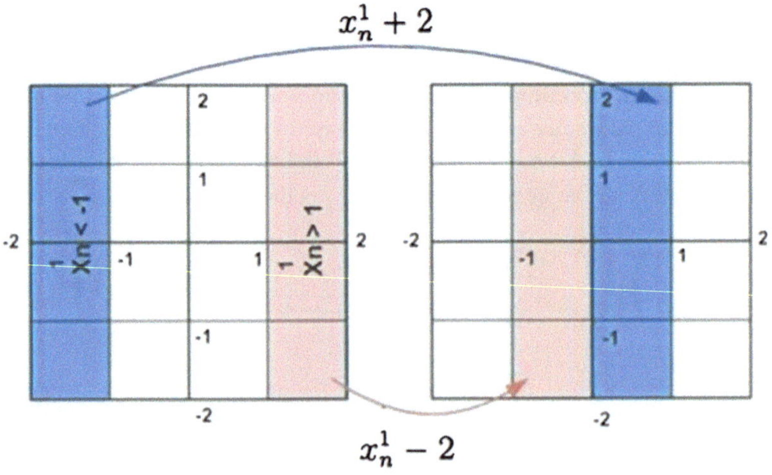

Fig. 4.5 Injection mechanism of the iterates from torus $[-2, 2]^2$ to torus $[-1, 1]^2$. If $x_n^{(1)} > 1$ then $x_n^{(1)} \equiv x_n^{(1)} - 2$; if $x_n^{(1)} < -1$ then $x_n^{(1)} \equiv x_n^{(1)} + 2$ (from [38])

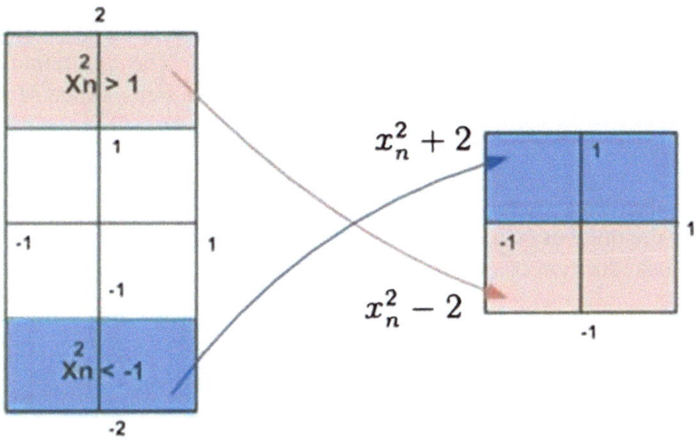

Fig. 4.6 If $x_n^{(2)} > 1$ then $x_n^{(2)} \equiv x_n^{(2)} - 2$; if $x_n^{(2)} < -1$ then $x_n^{(2)} \equiv x_n^{(2)} + 2$ (from [38])

4.3.3 Approximated Invariant Measures

We recall in this section the definition of approximated invariant measures which are important tools for assessing the uniform distribution of iterates. We have previously introduced them for the first studies of the weakly coupled symmetric tent map [22].

We first define an approximation $P_{M,N}(x)$ of the invariant measure, also called the probability distribution function linked to the one-dimensional map f (Eq. (4.5)) when computed with floating numbers (or numbers in double precision). To this goal, we consider a regular partition of M small intervals (boxes) r_i of $J = [-1, 1]$ defined by

$$s_i = -1 + \frac{2i}{M}, \quad i = 0, M, \tag{4.14}$$

$$r_i = [s_i, s_{i+1}[, \quad i = 0, M-2, \tag{4.15}$$

$$r_{M-1} = [s_{M-1}, 1], \tag{4.16}$$

$$J = \bigcup_{0}^{M-1} r_i. \tag{4.17}$$

The length of each box r_i is equal to

$$s_{i+1} - s_i = \frac{2}{M} \tag{4.18}$$

All iterates $f^{(n)}(x)$ belonging to these boxes are collected (after a transient regime of Q iterations decided a priori, i.e., the first Q iterates are discarded). Once the computation of $N + Q$ iterates is completed, the relative number of iterates with respect to N/M in each box r_i represents the value $P_N(s_i)$. The approximated $P_N(x)$ defined in this article is therefore a step function, with M steps. Since M may vary, we define

$$P_{M,N}(s_i) = \frac{1}{2}\frac{M}{N}(\#r_i) \tag{4.19}$$

where $\#r_i$ is the number of iterates belonging to the interval r_i and the constant $1/2$ allows the normalisation of $P_{M,N}(x)$ on the interval J.

$$P_{M,N}(x) = P_{M,N}(s_i), \quad \forall\, x \in r_i \tag{4.20}$$

In the case of p-coupled maps, we are more interested by the distribution of each component $x^{(1)}, x^{(2)}, \ldots, x^{(p)}$ of the vector $X = \begin{pmatrix} x^{(1)} \\ x^{(2)} \\ \cdot \\ \cdot \\ \cdot \\ x^{(p)} \end{pmatrix}$ rather than by the distribution of the variable X itself in J^p. We then consider the approximated probability distribution function $P_{M,N}(x^{(j)})$ associated to one component of X. In this chapter, we use either N_{disc} for M or N_{iter} for N, depending on which is more explicit. The discrepancies E_1 (in norm L_1), E_2 (in norm L_2), and E_∞ (in norm L_∞) between $P_{N_{disc},N_{iter}}(x)$ and the Lebesgue measure, which is the invariant measure associated to the symmetric tent map, are defined by

$$E_{1,N_{disc},Niter}(x) = \| P_{N_{disc},Niter}(x) - 0.5 \|_{L_1} \tag{4.21}$$

$$E_{2,N_{disc},Niter}(x) = \| P_{N_{disc},Niter}(x) - 0.5 \|_{L_2} \tag{4.22}$$

$$E_{\infty,N_{disc},Niter}(x) = \| P_{N_{disc},Niter}(x) - 0.5 \|_{L_\infty} \tag{4.23}$$

In the same way, an approximation of the correlation distribution function $C_{M,N}(x, y)$ is obtained by numerically building a regular partition of M^2 small squares (boxes) of J^2, embedded in the phase subspace (x^l, x^m)

$$s_i = -1 + \frac{2i}{M}, \quad t_j = -1 + \frac{2j}{M}, \quad i, j = 0, M \tag{4.24}$$

$$r_{i,j} = [s_i, s_{i+1}[\times[t_j, t_{j+1}[, \quad i, j = 0, M-2 \tag{4.25}$$

$$r_{M-1,j} = [s_{M-1}, 1] \times [t_j, t_{j+1}], \quad j = 0, M-2 \tag{4.26}$$

$$r_{i,M-1} = [s_i, s_{i+1}[\times[t_{M-1}, 1], \quad j = 0, M-2 \tag{4.27}$$

$$r_{M-1,M-1} = [s_{M-1}, 1] \times [t_{M-1}, 1] \tag{4.28}$$

The measure of the area of each box is

$$(s_{i+1} - s_i).(t_{i+1} - t_i) = \left(\frac{2}{M}\right)^2 \tag{4.29}$$

Once $N + Q$ iterated points (x_n^l, x_n^m) belonging to these boxes are collected, the relative number of iterates with respect to N/M^2 in each box $r_{i,j}$ represents the value $C_N(s_i, t_j)$. The approximated probability distribution function $C_N(x, y)$ defined here is then a two-dimensional step function, with M^2 steps. Since M can take several values in the next sections, we define

$$C_{M,N}(s_i, t_j) = \frac{1}{4}\frac{M^2}{N}(\#r_{i,j}) \tag{4.30}$$

where $\#r_{i,j}$ is the number of iterates belonging to the square $r_{i,j}$ and the constant $1/4$ allows the normalisation of $C_{M,N}(x, y)$ on the square J^2.

$$C_{M,N}(x, y) = C_{M,N}(s_i, t_j) \quad \forall(x, y) \in r_{i,j} \tag{4.31}$$

The discrepancies E_{C_1} (in norm L_1), E_{C_2} (in norm L_2) and E_{C_∞} (in norm L_∞) between $C_{N_{disc},N_{iter}}(x, y)$ and the uniform distribution on the square are defined by

$$E_{C_1,N_{disc},N_{iter}}(x, y) = \|C_{N_{disc},N_{iter}}(x, y) - 0.25\|_{L_1} \tag{4.32}$$

$$E_{C_2,N_{disc},N_{iter}}(x, y) = \|C_{N_{disc},N_{iter}}(x, y) - 0.25\|_{L_2} \tag{4.33}$$

$$E_{C_\infty,N_{disc},N_{iter}}(x, y) = \|C_{N_{disc},N_{iter}}(x, y) - 0.25\|_{L_\infty} \tag{4.34}$$

Finally, let $AC_{N_{disc},N_{iter}}$ be the autocorrelation distribution function which is the correlation function $C_{N_{disc},N_{iter}}$ of (4.31), defined in the delay space $(x_n^{(i)}, x_{n+1}^{(i)})$ instead of the phase (x^l, x^m) space. We define in the same manner than (4.32), (4.33), and (4.34) $E_{C_1,N_{disc},N_{iter}}(x, y)$, $E_{C_2,N_{disc},N_{iter}}(x, y)$, and $E_{C_\infty,N_{disc},N_{iter}}(x, y)$.

4.3.4 Study of Randomness of TTL_2^{SC} and TTL_2^{RC}, and Other Topologies

Using numerical computations, we assess the randomness properties of the two-dimensional maps TTL_2^{SC} and TTL_2^{RC}. If all requirements 1–8 of Fig. 4.7 are verified, the dynamical systems associated to those maps can be considered as pseudorandom and their application to cryptosystems is possible.

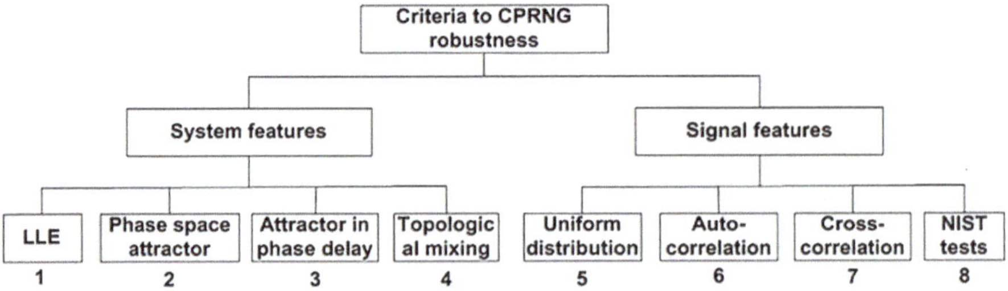

Fig. 4.7 The main criteria for assessing CPRNG (from [34])

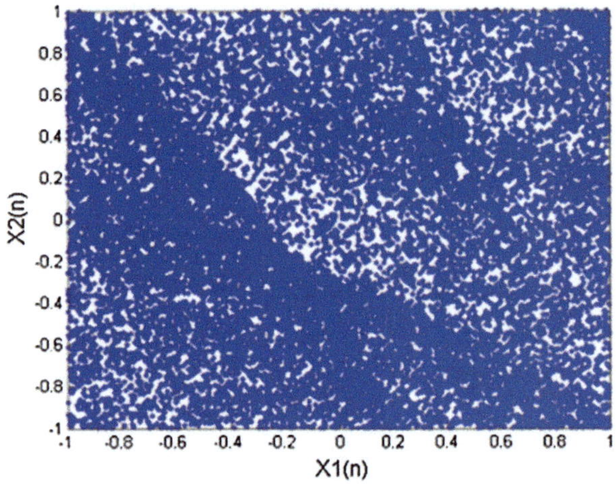

Fig. 4.8 Phase space behavior of TTL_2^{RC} non alternative (4.17), plot of 20, 000 points

Whenever one among the eight criteria is not satisfied for a given map, one cannot consider that the associated dynamical system is a good CPRNG candidate. As said above, when $\mu = 2$, the Lyapunov exponents of both considered maps are positive.

In the phase space, we plot the iterates in the system of coordinates $x_n^{(1)}$ versus $x_n^{(2)}$ in order to analyze the density of the points' distribution. Based on such an analysis, it is possible to assess the complexity of the behavior of dynamics, noticing any weakness or inferring on the nature of randomness. We also use the approximate invariant measures to assess more precisely the distribution of iterates.

The graphs of the attractor in phase space for the TTL_2^{RC} non-alternate (Fig. 4.8) and TTL_2^{SC} alternate (Fig. 4.9) maps are different. The TTL_2^{SC} map has well-scattered points in the whole pattern, but there are some more "concentrated" regions forming curves on the graph. Instead, the map TTL_2^{RC} has good repartition.

Some other numerical results we do not report in this chapter show that even if those maps have good random properties, it is possible to improve mapping randomness by modifying slightly network topologies.

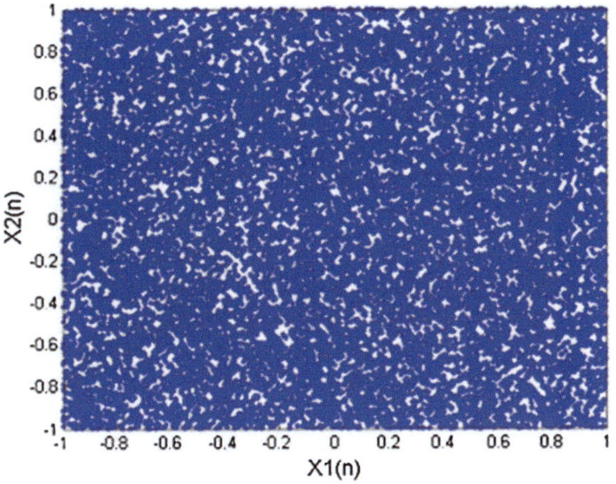

Fig. 4.9 Phase space behavior of TTL_2^{SC} alternative (4.18), plot of 20, 000 points

Equation (4.12) can be rewritten as

$$TTL_2^{SC}(x_n^{(1)}, x_n^{(2)}) = \begin{cases} x_{n+1}^{(1)} &= 1 + 2(x_n^{(2)})^2 - 4|x_n^{(1)}| \\ x_{n+1}^{(2)} &= 1 - 2(x_n^{(2)})^2 + 2(|x_n^{(1)}| - |x_n^{(2)}|) \end{cases} \tag{4.35}$$

In [38], it is shown that if the impact of component $x_n^{(1)}$ is reduced, randomness is improved. Hence, the following $MTTL_2^{SC}$ map is introduced

$$MTTL_2^{SC}(x_n^{(1)}, x_n^{(2)}) = \begin{cases} x_{n+1}^{(1)} &= 1 + 2(x_n^{(2)})^2 - 2|x_n^{(1)}| \\ x_{n+1}^{(2)} &= 1 - 2(x_n^{(2)})^2 + 2(|x_n^{(1)}| - |x_n^{(2)}|) \end{cases} \tag{4.36}$$

and the injection mechanism (4.9) is used as well, but it is restricted to three phases:

$$\begin{cases} \text{if } (x_{n+1}^{(1)} > 1) \text{ then subtract } 2 \\ \text{if } (x_{n+1}^{(2)} < -1) \text{ then add } 2 \\ \text{if } (x_{n+1}^{(2)} > 1) \text{ then subtract } 2 \end{cases} \tag{4.37}$$

This injection mechanism allows the regions containing iterates to match excellently (Fig. 4.10).

The change of topology leading to $MTTL_2^{SC}$ greatly improves the density of iterates in the phase space (Fig. 4.11) where 10^9 points are plotted. The point distribution of iterates in phase delay for the variable $x^{(2)}$ is quite good as well (Fig. 4.12). On both pictures, a grid of 200×200 boxes is generated to use the box counting method defined in Sect. 4.3.3. Moreover, the largest Lyapunov exponent is equal to 0.5905, indicating a strong chaotic behavior.

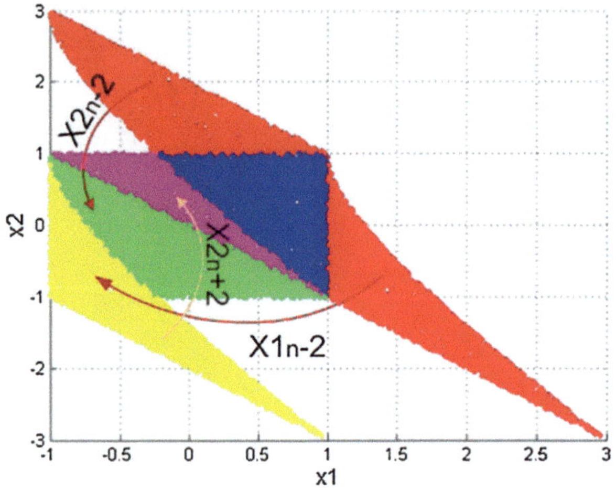

Fig. 4.10 Injection mechanism (4.21) of the $MTTL_2^{SC}$ alternative map (From [38])

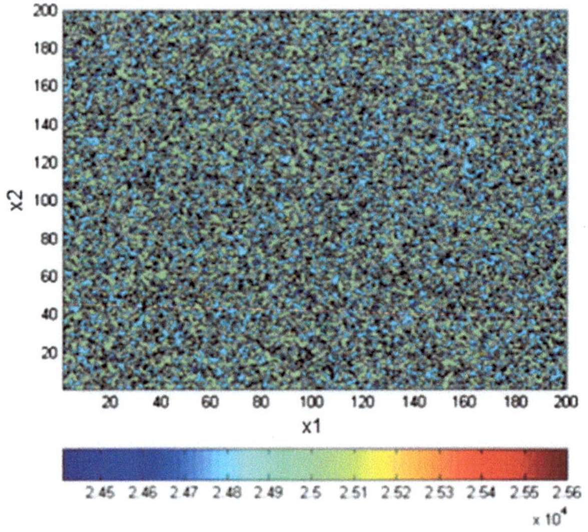

Fig. 4.11 Approximate density function of the $MTTL_2^{SC}$ alternative map, on the $(x^{(1)}, x^{(2)})$ plane (from [38])

However, regarding the phase delay for the variable $x^{(1)}$, results are not satisfactory. We have plotted in Fig. 4.13 10^9 iterates of $MTTL_2^{SC}$ in the delay plane, and in Fig. 4.14 the same iterates using the counting box method.

When such a great number of iterates is computed, one has to be cautious with raw graphical methods because irregularities of the density repartition are masked due to the huge number of plotted points. Therefore, these figures highlight the necessity of using the tools we have defined in Sect. 4.3.3.

Nevertheless, NIST tests were used to check randomness properties of $MTTL_2^{SC}$. Since they only require binary sequences, we generated 4×10^6 iterates whose 5×10^5 first ones were cut off. The rest of the sequence was converted to binary form according to the IEEE-754 standard (32-bit single-precision floating point).

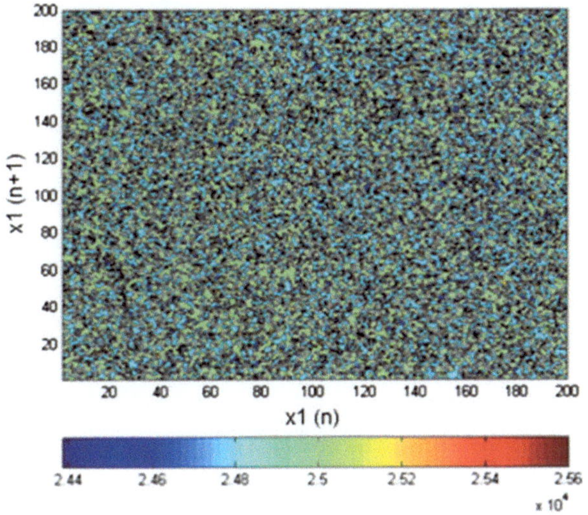

Fig. 4.12 Approximate density function of the $MTTL_2^{SC}$ alternative map, on the $(x_n^{(1)}, x_{n+1}^{(1)})$ plane (from [38])

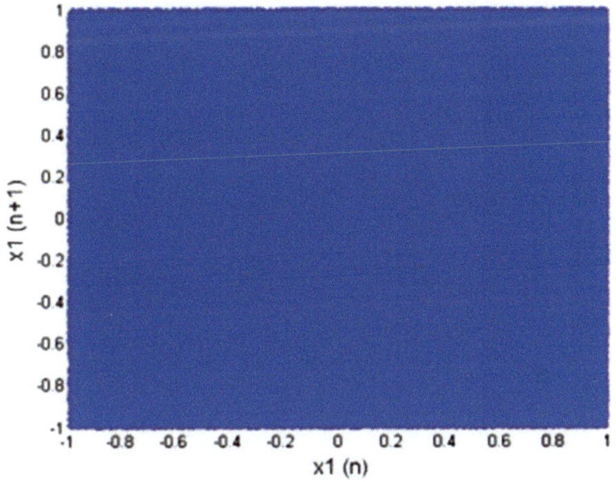

Fig. 4.13 Plot of one billion iterates of $MTTL_2^{SC}$ in the delay plane

Both variables of the generator successfully passed NIST tests, demonstrating strong randomness and robustness against numerous statistical attacks with respect to these tests (Figs. 4.15 and 4.16).

As said in the introduction, networks of coupled chaotic maps offer quasi-infinite possibilities to generate parallel streams of pseudorandom numbers. For example, in [39], the following modification of $MTTL_2^{SC}$ is also studied and shows good randomness properties

$$NTTL_2^{SC}(x_n^{(1)}, x_n^{(2)}) = \begin{cases} x_{n+1}^{(1)} &= 1 - 2|x_n^{(2)}| = T_2(x_n^{(2)}) \\ x_{n+1}^{(2)} &= 1 - (2x_n^{(2)})^2 - 2(|x_n^{(2)}| - |x_n^{(1)}|) \\ &= L_2(x_n^{(2)}) + T_2(x_n^{(2)}) - T_2(x_n^{(1)}) \end{cases} \tag{4.38}$$

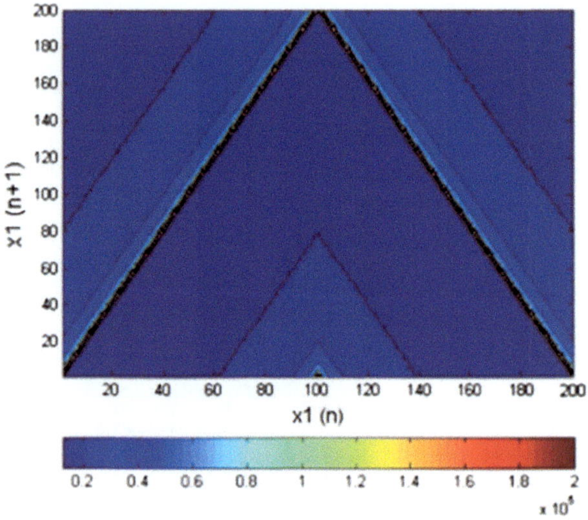

Fig. 4.14 Plot of one billion iterates of $MTTL_2^{SC}$ using the counting box method

RESULTS FOR THE UNIFORMITY OF P-VALUES AND THE PROPORTION OF PASSING SEQUENCES

generator is <data/Modified TL_{\mu}^{SC} alternative map_x1.txt>

C1	C2	C3	C4	C5	C6	C7	C8	C9	C10	P-VALUE	PROPORTION	STATISTICAL TEST
8	8	11	9	10	8	11	15	11	9	0.897763	100/100	Frequency
13	13	12	7	11	10	12	9	5	8	0.678686	99/100	BlockFrequency
6	7	5	12	16	12	12	9	14	7	0.191687	100/100	CumulativeSums
8	10	12	6	14	12	9	6	12	11	0.678686	100/100	Runs
14	11	12	10	15	5	6	13	8	6	0.236810	99/100	LongestRun
9	6	13	10	7	10	11	11	12	11	0.897763	97/100	Rank
11	12	6	19	4	11	11	13	8	5	0.037566	97/100	FFT
7	9	13	14	12	9	9	11	7	9	0.816537	100/100	NonOverlappingTemplate
10	11	15	10	11	9	12	6	11	5	0.595549	98/100	OverlappingTemplate
11	10	5	7	5	13	16	5	13	15	0.058984	100/100	Universal
14	6	11	10	7	9	13	12	8	10	0.739918	98/100	ApproximateEntropy
2	9	7	8	5	7	5	5	8	7	0.689019	63/63	RandomExcursions
5	8	4	4	6	4	4	11	6	11	0.222869	63/63	RandomExcursionsVariant
12	10	12	13	7	8	7	7	6	18	0.171867	99/100	Serial
9	13	11	12	7	9	7	16	7	9	0.534146	99/100	LinearComplexity

Fig. 4.15 Successful results of $NIST$ tests for the $MTTL_2^{SC}$ alternate map for the variable $x^{(1)}$ (from [38])

4.4 Numerical Study of a Particular Realisation of the $M^{k}_{\overline{\mu},p}$ Map in Higher Dimension

4.4.1 Mapping in Higher Dimension

Higher dimensional systems make it possible to achieve better randomness and uniform point distribution, because more perturbations and nonlinear mixing are involved. In this section, we focus on a particular realization of the $M^{k}_{\overline{\mu},p}$ map (4.10) from dimension two to dimension five.

RESULTS FOR THE UNIFORMITY OF P-VALUES AND THE PROPORTION OF PASSING SEQUENCES

generator is <data/Modified TL_{\mu}^{SC} alternative map_x2.txt>

C1	C2	C3	C4	C5	C6	C7	C8	C9	C10	P-VALUE	PROPORTION	STATISTICAL TEST
18	6	8	12	9	6	7	10	11	13	0.191687	98/100	Frequency
12	7	12	7	3	11	13	10	13	12	0.366918	98/100	BlockFrequency
15	14	8	6	8	13	7	10	9	10	0.494392	98/100	CumulativeSums
12	15	11	8	7	12	9	5	8	13	0.474986	98/100	Runs
9	12	13	13	9	14	9	6	8	7	0.637119	100/100	LongestRun
8	12	8	10	13	15	10	6	7	11	0.616305	98/100	Rank
8	12	9	15	9	8	17	9	9	4	0.181557	99/100	FFT
7	12	7	12	6	9	15	12	7	13	0.437274	100/100	NonOverlappingTemplate
9	12	11	3	16	8	10	13	10	8	0.289667	99/100	OverlappingTemplate
9	13	10	6	8	8	11	10	11	14	0.816537	99/100	Universal
7	24	9	7	7	8	8	17	7	6	0.000347	98/100	ApproximateEntropy
2	4	2	5	5	7	2	13	4	8	0.011791	52/52	RandomExcursions
5	4	8	5	2	1	8	6	4	9	0.191687	52/52	RandomExcursionsVariant
6	10	8	7	15	15	15	8	8	8	0.236810	100/100	Serial
7	9	11	11	6	15	7	11	8	15	0.419021	99/100	LinearComplexity

Fig. 4.16 *NIST* tests for the variable $x^{(2)}$ (from [38])

Usually, three or four dimensions are complex enough to create robust random sequences as we show here. Thus, it is advantageous if the system can increase its dimension. Since the $MTTL_2^{SC}$ alternative map cannot be nested in higher dimensions, we describe how to improve randomness and to obtain the best distribution of points, and how to produce more complex dynamics than the $TTL_2^{SC}(x^{(2)}, x^{(1)})$ alternative map in dimension greater than 2. Let

$$TTL_2^{RC,pD} = \begin{cases} x_{n+1}^{(1)} = 1 - 2|x_n^{(1)}| + 2(|x_n^{(2)}| - (x_n^{(1)})^2) \\ x_{n+1}^{(2)} = 1 - 2|x_n^{(2)}| + 2(|x_n^{(3)}| - (x_n^{(2)})^2) \\ \vdots \\ x_{n+1}^{(p)} = 1 - 2|x_n^{(p)}| + 2(|x_n^{(1)}| - (x_n^{(p)})^2) \end{cases} \tag{4.39}$$

be this realization.

We show in Figs. 4.17 and 4.18 successful NIST tests for $TTL_2^{RC,pD}$ in 3-D and 4-D, for the variable $x^{(1)}$.

C1	C2	C3	C4	C5	C6	C7	C8	C9	C10	P-VALUE	PROPORTION	STATISTICAL TEST
8	14	8	9	10	9	11	12	6	13	0.779188	100/100	Frequency
11	9	9	8	6	15	7	13	9	13	0.574903	100/100	BlockFrequency
14	6	13	7	11	5	10	11	9	14	0.401199	100/100	CumulativeSums
12	10	7	7	16	8	13	7	13	7	0.366918	99/100	CumulativeSums
16	9	7	11	14	12	6	13	7	5	0.181557	100/100	Runs
13	9	14	11	11	8	9	12	5	8	0.678686	100/100	LongestRun
14	9	7	8	9	16	9	12	6	10	0.455937	100/100	Rank
13	4	9	11	7	4	10	12	19	11	0.037566	100/100	FFT
14	8	8	9	8	15	11	11	8	8	0.699313	100/100	NonOverlappingTemplate
14	15	12	10	6	9	13	7	3	11	0.162606	99/100	OverlappingTemplate
8	7	11	16	9	12	10	9	7	11	0.678686	100/100	Universal
13	11	10	12	6	12	12	14	6	4	0.304126	97/100	ApproximateEntropy
5	5	6	9	2	7	5	8	9	6	0.637119	62/62	RandomExcursions
6	2	4	9	6	11	6	5	6	7	0.407091	62/62	RandomExcursionsVariant
13	8	15	8	12	9	7	15	8	5	0.275709	99/100	Serial
13	6	15	12	11	6	15	8	8	6	0.213309	99/100	Serial
9	6	8	13	8	11	10	11	12	12	0.883171	99/100	LinearComplexity

Fig. 4.17 *NIST* test for $TTL_2^{RC,3D}$ for $x^{(1)}$ (from [38])

C1	C2	C3	C4	C5	C6	C7	C8	C9	C10	P-VALUE	PROPORTION	STATISTICAL TEST
7	5	12	14	10	9	12	16	8	7	0.289667	99/100	Frequency
7	7	9	10	6	10	14	8	10	19	0.137282	99/100	BlockFrequency
8	2	9	16	13	9	13	9	7	14	0.090936	99/100	CumulativeSums
5	8	14	11	11	11	14	5	10	11	0.437274	99/100	CumulativeSums
6	16	13	11	9	10	8	7	11	9	0.554420	100/100	Runs
9	13	6	9	14	10	8	11	12	8	0.779188	99/100	LongestRun
9	8	14	6	12	12	8	10	8	13	0.719747	100/100	Rank
10	10	17	5	9	13	14	10	6	6	0.153763	99/100	FFT
9	7	9	13	9	10	10	14	6	13	0.719747	100/100	NonOverlappingTemplate
5	9	12	7	7	12	12	13	12	11	0.637119	99/100	OverlappingTemplate
12	16	8	7	9	10	7	12	8	11	0.616305	99/100	Universal
8	16	6	12	11	13	5	7	13	9	0.249284	99/100	ApproximateEntropy
4	8	4	6	8	5	7	8	9	7	0.804337	66/66	RandomExcursions
4	7	7	8	2	8	6	8	7	9	0.602458	66/66	RandomExcursionsVariant
11	10	10	18	6	5	11	12	10	7	0.213309	100/100	Serial
8	11	10	10	12	11	10	9	9	10	0.998821	98/100	Serial
10	7	13	11	8	7	11	14	11	8	0.798139	99/100	LinearComplexity

Fig. 4.18 $NIST$ test for $TTL_2^{RC,4D}$ for $x^{(1)}$ (from [38])

4.4.2 Numerical Experiments

All NIST tests for dimensions three to five for every variable are successful, showing that these realizations in 3-D up to 5-D are good CPRNGs. In addition to those tests, we study the mapping more thoroughly, far beyond the NIST tests which are limited to a few million iterates and which seem not robust enough for industrial mathematics, although they are routinely used worldwide.

In order to check the portability of the computations on multicore architectures, we have implemented all our numerical experiments on several different multicore machines.

4.4.2.1 Checking the Uniform Repartition of Iterated Points

We first compute the discrepancies E_1 (in norm L_1), E_2 (in norm L_2)n and E_∞ (in norm E_∞) between $P_{N_{disc},N_{iter}}(x)$ and the Lebesgue measure which is the uniform measure on the interval $J = [-1, 1]$. We set $M = N_{iter} = 200$, and vary the number N_{iter} of iterated points in the range 10^4 to 10^{14}. From our knowledge, this article is the first one that checks such a huge number of iterates (in conjunction with [39]). We compare $E_{1,200,N_{iter}}(x^{(1)})$ for $TTL_2^{RC,pD}$ with $p = 2$ to 5 (Table 4.2, Fig. 4.19).

As shown in Fig. 4.19, $E_{1,200,N_{iter}(x^{(1)})}$ decreases steadily when N_{iter} increases. However, the decreasing process is promptly (with respect to N_{iter}) bounded below for $p = 2$. This is also the case for other values of p, however, the boundary decreases with p, therefore showing better randomness properties for higher dimensional mappings.

Table 4.3 compares $x^{(1)}, x^{(2)}, \ldots, x^{(p)}$ for $TTL_2^{RC,5D}$, for different values of N_{iter}. It is obvious that the same quality of randomness is obtained for each one of them, contrarily to the results obtained for $MTTL_2^{SC}$.

Table 4.2 $E_{1,200,N_{iter}}(x^{(1)})$ for $TTL_2^{RC,pD}$ with $p = 2$ to 5

N_{iter}	$p = 2$	$p = 3$	$p = 4$	$p = 5$
10^4	1.5631	1.5553	1.5587	1.5574
10^5	0.55475	0.5166	0.51315	0.5154
10^6	0.269016	0.159306	0.158548	0.158436
10^7	0.224189	0.050509	0.0501934	0.0505558
10^8	0.219427	0.0164173	0.0159175	0.0160018
10^9	0.218957	0.00640196	0.00505021	0.00509754
10^{10}	0.218912	0.00420266	0.00160505	0.00160396
10^{11}	0.218913	0.00392507	0.000513833	0.000505591
10^{12}	0.218913	0.00389001	0.000189371	0.000160547
10^{13}	0.218914	0.00388778	0.000112764	5.04473e-05
10^{14}	0.218914	0.003887	0.000101139	1.59929e-05

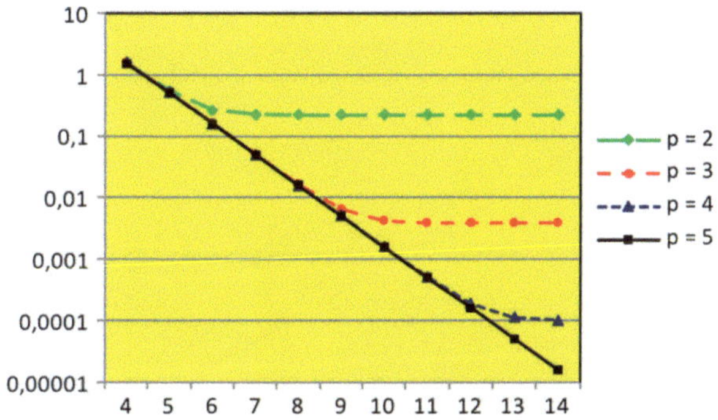

Fig. 4.19 Graph of $E_{1,200,N_{iter}}(x^{(1)})$ for $TTL_2^{RC,pD}$ with $p = 2$ to 5, with respect to N_{iter} (horizontal axis, logarithmic value)

Table 4.3 $E_{1,200,N_{iter}}(x^{(i)})$ for $TTL_2^{RC,5D}$ for $i = 1$ to 5

N_{iter}	$x^{(1)}$	$x^{(2)}$	$x^{(3)}$	$x^{(4)}$	$x^{(5)}$
10^4	1.5574	1.55725	1.556	1.5585	1.55925
10^5	0.5154	0.51061	0.5098	0.51494	0.51293
10^6	0.158436	0.159162	0.159564	0.159864	0.159926
10^7	0.0505558	0.0504866	0.0503746	0.0505688	0.0505268
10^8	0.0160018	0.0158328	0.0158498	0.0160336	0.01591
10^9	0.00509754	0.0050514	0.00505756	0.00501442	0.00503467
10^{10}	0.00160396	0.00159738	0.00160099	0.00159454	0.00159916
10^{11}	0.000505591	0.000506327	0.000507006	0.000504258	0.000507526
10^{12}	0.000160547	0.000159192	0.000160014	0.000159213	0.000159159
10^{13}	5.04473e-05	5.03574e-05	5.05868e-05	5.04694e-05	5.01681e-05
10^{14}	1.59929e-05	1.60291e-05	1.59282e-05	1.59832e-05	1.60775e-05

Table 4.4 Comparison between $E_{1,200,N_{iter}}(x^{(1)})$, $E_{2,200,N_{iter}}(x^{(1)})$, and $E_{\infty,200,N_{iter}}(x^{(1)})$ for $TTL_2^{RC,5D}$

N_{iter}	Norm L_1	Norm L_2	Norm L_∞
10^4	1.5574	2.0038	19
10^5	0.5154	0.635522	3.4
10^6	0.158436	0.199731	0.96
10^7	0.0505558	0.0633486	0.256
10^8	0.0160018	0.02007	0.0896
10^9	0.00509754	0.00638219	0.02688
10^{10}	0.00160396	0.00200966	0.008672
10^{11}	0.000505591	0.000631963	0.0027444
10^{12}	0.000160547	0.000201102	0.0008602
10^{13}	5.04473e-05	6.32233e-05	0.00026894
10^{14}	1.59929e-05	2.00533e-05	9.89792e-05

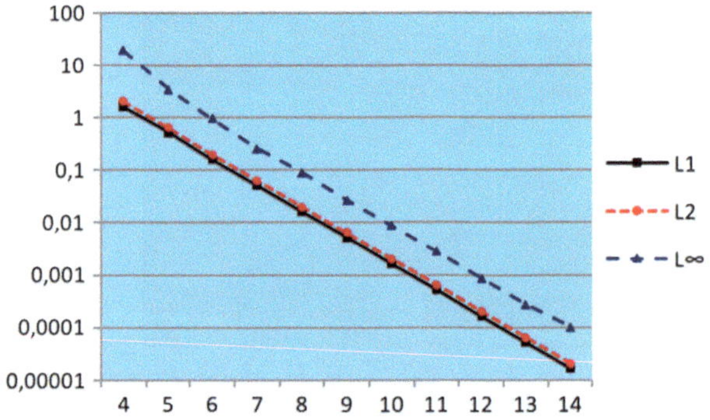

Fig. 4.20 Comparison between $E_{1,200,N_{iter}}(x^{(1)})$, $E_{2,200,N_{iter}}(x^{(1)})$, and $E_{\infty,200,N_{iter}}(x^{(1)})$ (vertical axis) for $TTL_2^{RC,5D}$ with respect to N_{iter} (horizontal axis, logarithmic value)

The comparisons between $E_{1,200,N_{iter}}(x^{(1)})$, $E_{2,200,N_{iter}}(x^{(1)})$, and $E_{\infty,N_{iter}}(x^{(1)})$ for $TTL_2^{RC,5D}$ in Table 4.4 and Fig. 4.20 show that

$$E_{1,200,N_{iter}}(x^{(1)}) < E_{2,200,N_{iter}}(x^{(1)}) < E_{\infty,N_{iter}}(x^{(1)}) \quad (4.40)$$

for every value of N_{iter}.

4.4.2.2 Autocorrelation Study in the Delay Space

In this section, we assess autocorrelation errors $E_{AC_1,N_{disc},N_{iter}}(x, y)$, $E_{AC_2,N_{disc},N_{iter}}(x, y)$, and $E_{AC_\infty,disc,N_{iter}}(x, y)$, defined by Equations (4.32), (4.33), and (4.34), in the delay space. As in Sect. 4.4.2.1, we have performed the experi-

Table 4.5 Comparison between $E_{AC_1,200,N_{iter}}(x_n^{(1)}, x_{n+1}^{(1)})$, $E_{AC_1,200,N_{iter}}(x_n^{(1)}, x_{n+2}^{(1)})$, and $E_{AC_1,200,N_{iter}}(x_n^{(1)}, x_{n+3}^{(1)})$ for $TTL_2^{RC,2D}$

N_{iter}	$(x_n^{(1)}, x_{n+1}^{(1)})$	$(x_n^{(1)}, x_{n+2}^{(1)})$	$(x_n^{(1)}, x_{n+3}^{(1)})$
10^4	1.55955	1.57265	1.5515
10^5	0.55199	0.699355	0.547539
10^6	0.269654	0.519675	0.250936
10^7	0.224104	0.49941	0.198634
10^8	0.21938	0.497011	0.193007
10^9	0.218949	0.496766	0.192309
10^{10}	0.218914	0.496808	0.192253
10^{11}	0.218915	0.496793	0.192247
10^{12}	0.218913	0.496797	0.192245
10^{13}	0.218914		
10^{14}	0.218914		

Table 4.6 Comparison between $E_{AC_1,200,N_{iter}}(x_n^{(1)}, x_{n+1}^{(1)})$, $E_{AC_1,200,N_{iter}}(x_n^{(1)}, x_{n+2}^{(1)})$, and $E_{AC_1,200,N_{iter}}(x_n^{(1)}, x_{n+3}^{(1)})$ for $TTL_2^{RC,3D}$

N_{iter}	$(x_n^{(1)}, x_{n+1}^{(1)})$	$(x_n^{(1)}, x_{n+2}^{(1)})$	$(x_n^{(1)}, x_{n+3}^{(1)})$
10^4	1.55575	1.5528	1.5489
10^5	0.51516	0.512514	0.514889
10^6	0.160148	0.158843	0.159728
10^7	0.0505148	0.0515855	0.0550998
10^8	0.0164343	0.0190644	0.0269715
10^9	0.00640451	0.0113919	0.0221408
10^{10}	0.00420824	0.0103092	0.0216388
10^{11}	0.003926197	0.0102078	0.0215621
10^{12}	0.00388937	0.0101965	0.0215576
10^{13}	0.00388768		
10^{14}	0.003887		

ments for $M = 20$ to 20, 000, however, in this chapter, we only present the results for $M = 200$. We first compare $E_{AC_1,200,N_{iter}}(x_n^{(1)}, x_{n+1}^{(1)})$ with $E_{AC_1,200,N_{iter}}(x_n^{(1)}, x_{n+2}^{(1)})$ and $E_{AC_1,200,N_{iter}}(x_n^{(1)}, x_{n+3}^{(1)})$ for $TTL_2^{RC,pD}$ when the dimension of the system is within the range $p = 2$ to 5 (Tables 4.5, 4.6, 4.7 and 4.8). It is possible to see that better randomness properties are obtained for higher dimensional mappings.

The comparison between $E_{AC_1,200,N_{iter}}(x_n^{(1)}, x_{n+1}^{(1)})$, $E_{AC_2,200,N_{iter}}(x_n^{(1)}, x_{n+1}^{(1)})$, and $E_{AC_\infty,200,N_{iter}}(x_n^{(1)}, x_{n+1}^{(1)})$ for $TTL_2^{RC,5D}$ in Table 4.9 shows that numerically

Table 4.7 Comparison between $E_{AC_1,200,N_{iter}}(x_n^{(1)}, x_{n+1}^{(1)})$, $E_{AC_1,200,N_{iter}}(x_n^{(1)}, x_{n+2}^{(1)})$, and $E_{AC_1,200,N_{iter}}(x_n^{(1)}, x_{n+3}^{(1)})$ for $TTL_2^{RC,4D}$

N_{iter}	$(x_n^{(1)}, x_{n+1}^{(1)})$	$(x_n^{(1)}, x_{n+2}^{(1)})$	$(x_n^{(1)}, x_{n+3}^{(1)})$
10^4	1.5571	1.5518	1.54985
10^5	0.51115	0.510784	0.511188
10^6	0.158472	0.159263	0.159292
10^7	0.0503522	0.0506053	0.0506126
10^8	0.0159245	0.0159484	0.015918
10^9	0.00502109	0.00502642	0.00502197
10^{10}	0.00159193	0.00161135	0.00162232
10^{11}	0.00051438	0.000532052	0.0005489
10^{12}	0.000189418	0.000217634	0.000276982
10^{13}	0.000112771		
10^{14}	0.000101139		

Table 4.8 Comparison between $E_{AC_1,200,N_{iter}}(x_n^{(1)}, x_{n+1}^{(1)})$, $E_{AC_1,200,N_{iter}}(x_n^{(1)}, x_{n+2}^{(1)})$, and $E_{AC_1,200,N_{iter}}(x_n^{(1)}, x_{n+3}^{(1)})$ for $TTL_2^{RC,5D}$

N_{iter}	$(x_n^{(1)}, x_{n+1}^{(1)})$	$(x_n^{(1)}, x_{n+2}^{(1)})$	$(x_n^{(1)}, x_{n+3}^{(1)})$
10^4	1.5577	1.5531	1.54975
10^5	0.51372	0.511144	0.513918
10^6	0.15872	0.158775	0.158022
10^7	0.0503658	0.0504011	0.0501632
10^8	0.0159765	0.0159229	0.0159837
10^9	0.00509015	0.00502869	0.00503495
10^{10}	0.00159581	0.00159398	0.00158143
10^{11}	0.000505068	0.000506309	0.000502137
10^{12}	0.000160547	0.000159144	0.000159246
10^{13}	5.0394e-05		
10^{14}	1.59929e-05		

$$E_{AC_1,200,N_{iter}}(x_n^{(1)}, x_{n+1}^{(1)}) < E_{AC_2,200,N_{iter}}(x_n^{(1)}, x_{n+1}^{(1)}) < E_{AC_\infty,200,N_{iter}}(x_n^{(1)}, x_{n+1}^{(1)}) \tag{4.41}$$

Equation (4.41) is not only valid for $M = 200$, but also for other values of M and every component of X.

In order to illustrate the numerical results displayed in these tables, we plot in Fig. 4.21 the repartition of iterates of $TTL_2^{RC,5D}$ in the delay plane $(x_n^{(1)}, x_{n+1}^{(1)})$, using the box counting method. On a grid of 200×200 boxes ($N_{iter} = M = 200$),

Table 4.9 Comparison between $E_{AC_1,200,N_{iter}}(x_n^{(1)}, x_{n+1}^{(1)})$, $E_{AC_2,200,N_{iter}}(x_n^{(1)}, x_{n+1}^{(1)})$, and $E_{AC_\infty,200,N_{iter}}(x_n^{(1)}, x_{n+1}^{(1)})$ for $TTL_2^{RC,5D}$

N_{iter}	Norm L_1	Norm L_2	Norm L_∞
10^4	1.5577	2.0012	19
10^5	0.51372	0.633959	3.8
10^6	0.15872	0.199793	0.88
10^7	0.0503658	0.0631425	0.26
10^8	0.0159765	0.0200503	0.084
10^9	0.00509015	0.00636626	0.02528
10^{10}	0.00159581	0.00199936	0.008604
10^{11}	0.000505068	0.000633088	0.0025432
10^{12}	0.000160547	0.000201102	0.0008602
10^{13}	5.0394e-05	6.31756e-05	0.000280168
10^{14}	1.59929e-05	2.00533e-05	9.89792e-05

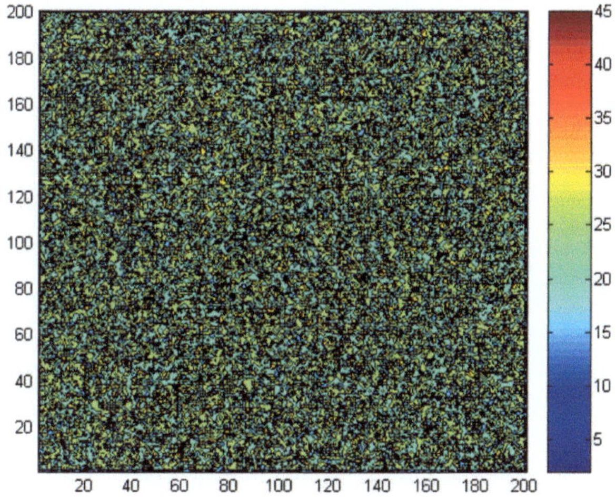

Fig. 4.21 Repartition of iterates in the delay plane $(x_n^{(1)}, x_{n+1}^{(1)})$ of $TTL_2^{RC,5D}$ with the box counting method; 10^6 points are generated on a grid of 200×200 boxes, the horizontal axis is $x_n^{(1)}$, and the vertical axis is $x_{n+1}^{(1)}$

we have generated 10^6 points. The horizontal axis is $x_n^{(1)}$, and the vertical axis is $x_{n+1}^{(1)}$. In order to check very carefully the repartition of the iterates of $TTL_2^{RC,5D}$, we have also plotted the repartition in the delay planes $(x_n^{(1)}, x_{n+2}^{(1)})$, $(x_n^{(1)}, x_{n+3}^{(1)})$, and $(x_n^{(1)}, x_{n+4}^{(1)})$ (Figs. 4.22, 4.23, and 4.24). This repartition is uniform everywhere as shown also in Table 4.8.

We find the same regularity for every component $x^{(2)}$, $x^{(3)}$, $x^{(4)}$, and $x^{(5)}$, as shown in Figs. 4.25, 4.26, 4.27, 4.28, and in Table 4.10.

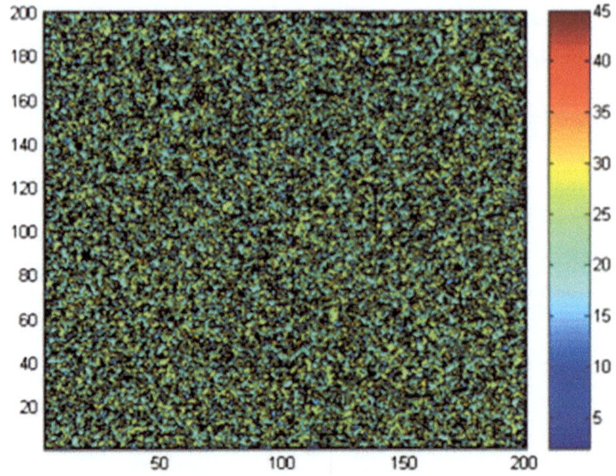

Fig. 4.22 Repartition of iterates in the delay plane $(x_n^{(1)}, x_{n+2}^{(1)})$ of $TTL_2^{RC,5D}$, as in Fig. 4.21

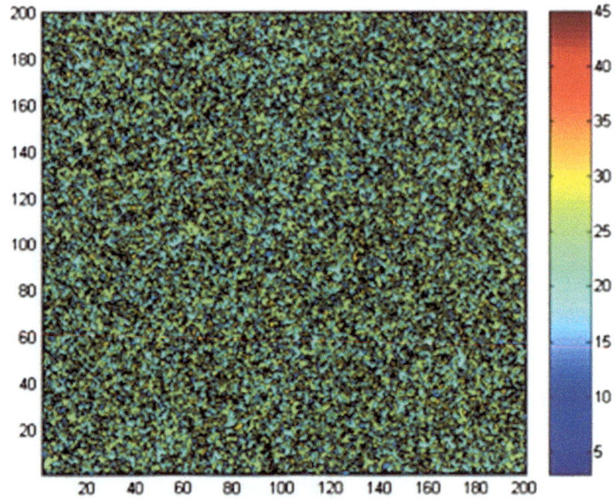

Fig. 4.23 Repartition of iterates in the delay plane $(x_n^{(1)}, x_{n+3}^{(1)})$ of $TTL_2^{RC,5D}$, as in Fig. 4.21

4.4.2.3 Autocorrelation Study in the Phase Space

Finally, in this section, we assess the autocorrelation errors $E_{C_1,N_{disc},N_{iter}}(x, y)$, $E_{C_2,N_{disc},N_{iter}}(x, y)$, and $E_{C_\infty,N_{disc},N_{iter}}(x, y)$, defined by Eqs. (4.32), (4.33), and (4.34), in the phase space. We checked all combinations of the components. Due to space limitations, we only provide part of the numerical computations we have performed to carefully check the randomness of $TTL_2^{RC,pD}$ for $p = 2, 5$ and $i = 1, p$. Like in the previous section, we only provide the results for $M = 200$. We first compare $E_{C_1,200,N_{iter}}(x_n^{(1)}, x_n^{(2)})$, $E_{C_2,200,N_{iter}}(x_n^{(1)}, x_n^{(2)})$, and $E_{C_\infty,200,N_{iter}}(x_n^{(1)}, x_n^{(2)})$ (Table 4.11), and our other results verified that

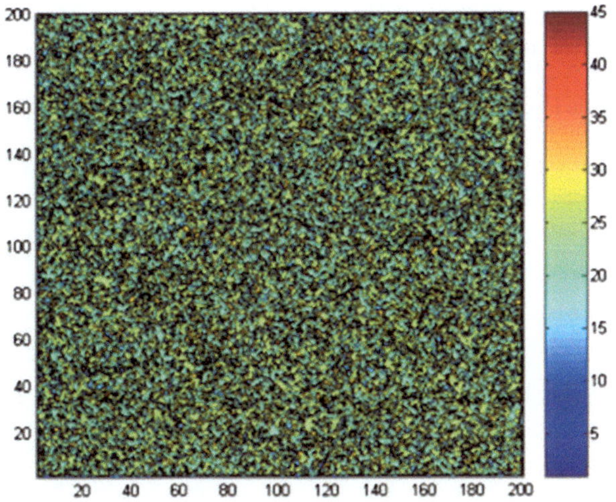

Fig. 4.24 Repartition of iterates in the delay plane $(x_n^{(1)}, x_{n+4}^{(1)})$ of $TTL_2^{RC,5D}$, as in Fig. 4.21

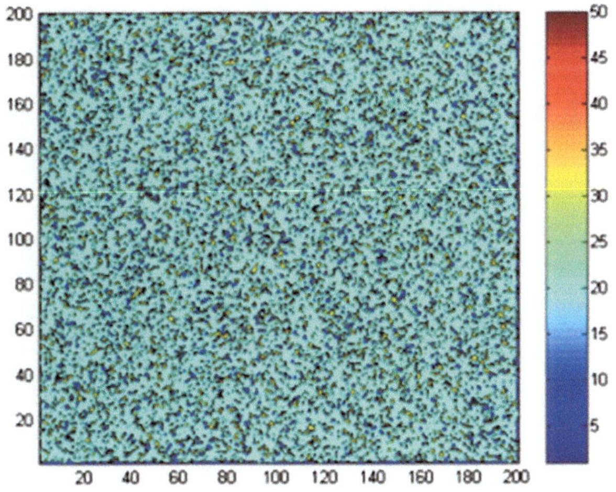

Fig. 4.25 Repartition of iterates in the delay plane $(x_n^{(2)}, x_{n+1}^{(2)})$ of $TTL_2^{RC,5D}$; box counting method, 10^6 points are generated on a grid of 200 × 200 boxes, the horizontal axis is $x_n^{(2)}$, and the vertical axis is $x_{n+1}^{(2)}$

$$E_{C_1,N_{disc},N_{iter}}(x_n^{(1)}, x_n^{(2)}) < E_{C_2,N_{disc},N_{iter}}(x_n^{(1)}, x_n^{(2)}) < E_{C_\infty,N_{disc},N_{iter}}(x_n^{(1)}, x_n^{(2)}) \tag{4.42}$$

We have also assessed the autocorrelation errors $E_{C_1,N_{disc},N_{iter}}(x_n^{(i)}, x_n^{(j)})$ for $i, j = 1, 5, i \neq j$, and various values of the number of iterates for $TTL_2^{RC,5D}$ (Table 4.12). We have performed the same experiments for $E_{C_1,N_{disc},N_{iter}}(x_n^{(1)}, x_n^{(2)})$ for $p = 1, 5$ (Table 4.13).

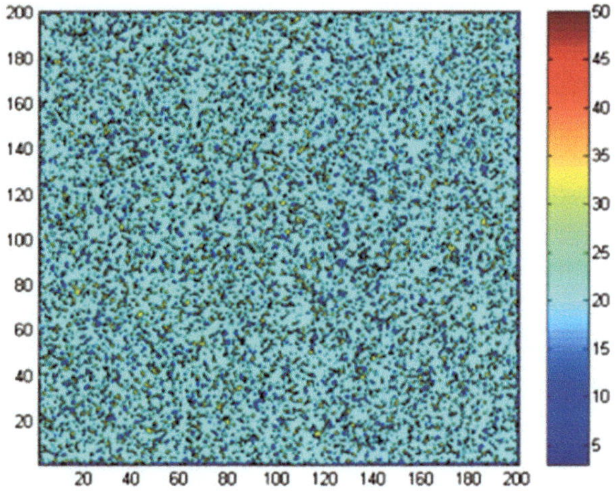

Fig. 4.26 Repartition of iterates in the delay plane $(x_n^{(3)}, x_{n+1}^{(3)})$ of $TTL_2^{RC,5D}$, as in Fig. 4.25

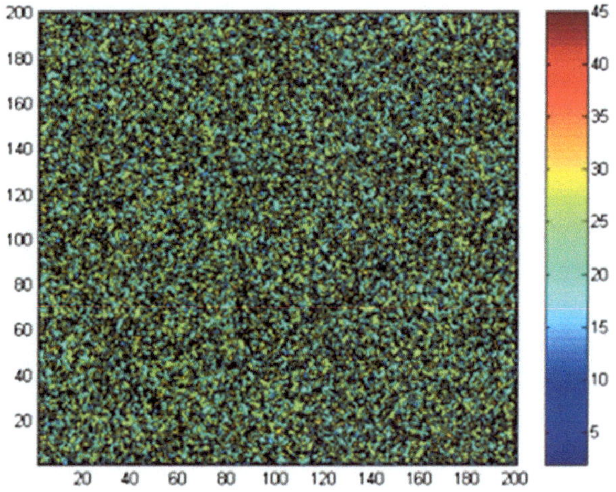

Fig. 4.27 Repartition of iterates in the delay plane $(x_n^{(4)}, x_{n+1}^{(4)})$ of $TTL_2^{RC,5D}$, as in Fig. 4.25

Our numerical experiments all show a similar trend: $TTL_2^{RC,pD}$ is a good candidate for a CPRNG, and the randomness performance of such mappings increases in higher dimensions.

4.4.2.4 Checking the Influence of Discretization in Computation of Approximated Invariant Measures

In order to verify that the computations we have performed using the discretization $M = N_{disc} = 200$ of the phase space and the delay space in the numerical experi-

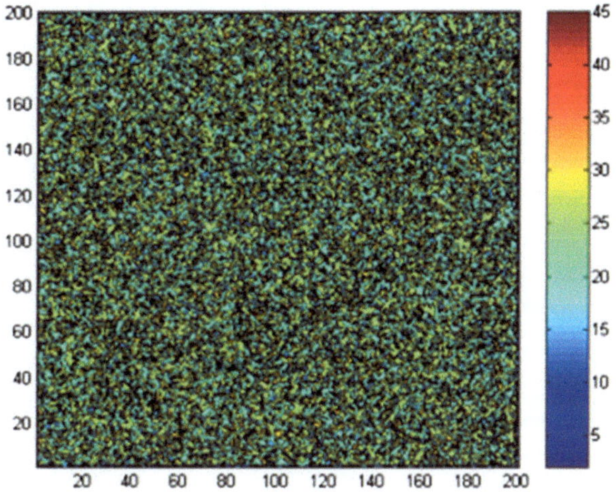

Fig. 4.28 Repartition of iterates in the delay plane $(x_n^{(5)}, x_{n+1}^{(5)})$ of $TTL_2^{RC,5D}$, as in Fig. 4.25

Table 4.10 Comparison between $E_{AC_1,200,N_{iter}}(x_n^{(i)}, x_{n+1}^{(i)})$, $E_{AC_1,200,N_{iter}}(x_n^{(i)}, x_{n+2}^{(i)})$, and $E_{AC_1,200,N_{iter}}(x_n^{(i)}, x_{n+3}^{(i)})$ for $TTL_2^{RC,5D}$ for $i = 1$ to 5

N_{iter}	i	$(x_n^{(i)}, x_{n+1}^{(i)})$	$(x_n^{(i)}, x_{n+2}^{(i)})$	$(x_n^{(i)}, x_{n+3}^{(i)})$
10^4	1	1.5577	1.5531	1.54975
	2	1.5577	1.5526	1.5508
	3	1.5577	1.5542	1.54445
	4	1.5577	1.5533	1.5468
	5	1.5577	1.5541	1.5504
10^8	1	0.0159765	0.0159229	0.0159837
	2	0.0159765	0.0159999	0.0158293
	3	0.0159765	0.0159047	0.0159605
	4	0.0159765	0.0159269	0.0159282
	5	0.0159765	0.0160591	0.0159274
10^{12}	1	0.000160547	0.000159144	0.000159246
	2	0.000159192	0.000159635	0.000159064
	3	0.000160014	0.00015892	0.000160555
	4	0.000159213	0.000159696	0.000159215
	5	0.000159159	0.000158831	0.000160007

ments do not introduce artifacts, we have performed the same computations varying also the value of $M = N_{disc} = 20, 200, 2000, 20000$, for $TTL_2^{RC,4D}$ (Table 4.14 and Fig. 4.29). The results show a normal regularity following the increasing value of N_{disc}.

Table 4.11 Comparison between $E_{AC_1,200,N_{iter}}(x_n^{(1)}, x_n^{(2)})$, $E_{AC_2,200,N_{iter}}(x_n^{(1)}, x_n^{(2)})$, and $E_{AC_\infty,200,N_{iter}}(x_n^{(1)}, x_n^{(2)})$ for $TTL_2^{RC,5D}$

N_{iter}	Norm L_1	Norm L_2	Norm L_∞
10^4	1.55915	2.00818	15
10^5	0.514	0.633448	3.4
10^6	0.158058	0.198943	0.96
10^7	0.0505508	0.0634574	0.308
10^8	0.0160114	0.0200538	0.0852
10^9	0. 00507915	0.0063595	0.02716
10^{10}	0.0015927	0.00199644	0.008128
10^{11}	0.000506086	0.000633916	0.0025712
10^{12}	0.000158795	0.000199203	0.00089288
10^{13}	5.03666e-05	6.30356e-05	0.000270156
10^{14}	1.60489e-05	2.00692e-05	8.53124e-05

Table 4.12 Comparison between $E_{C_1,200,N_{iter}}(x_n^{(i)}, x_n^{(j)})$, for $i, j = 1$ to 5, $i \neq j$, and for various values of number of iterates for $TTL_2^{RC,5D}$

N_{iter}	10^6	10^8	10^{10}	10^{12}	10^{14}
$x(1), x(2)$	0.158058	0.0160114	0.0015927	0.000158795	1.60489e-05
$x(1), x(3)$	0.158956	0.0159261	0.00159456	0.000159326	1.73852e-05
$x(1), x(4)$	0.15943	0.0160321	0.00160091	0.000160038	1.74599e-05
$x(1), x(5)$	0.159074	0.0158962	0.00160204	0.000159048	1.59133e-05
$x(2), x(3)$	0.15825	0.0159754	0.00159442	0.000160659	1.60419e-05
$x(2), x(4)$	0.159248	0.0159668	0.00159961	0.000160313	1.73507e-05
$x(2), x(5)$	0.15889	0.0160116	0.0015934	0.000160462	1.73496e-05
$x(3), x(4)$	0.159136	0.0158826	0.00158123	0.000158758	1.59451e-05
$x(3), x(5)$	0.159216	0.0159341	0.00161268	0.000159079	1.75013e-05
$x(4), x(5)$	0.158918	0.0160516	0.0016008	0.000159907	1.59445e-05

4.4.2.5 Computation Time of PRNs

The numerical experiments performed in this section have involved several multicore machines. We show in Table 4.15 different computation times (in seconds) for the generation of N_{iter} PRNs for $TTL_2^{RC,pD}$ with $p = 2$ to 5, and various values of the number of iterates (N_{iter}). The machine used is a laptop computer with a Core i7 4980HQ processor with eight logical cores.

Table 4.16 shows the computation time of only one PRN in the same experiment. Time is expressed in 10^{-10} s.

Table 4.13 Comparison between $E_{C_1,200,N_{iter}}(x_n^{(i)}, x_n^{(j)})$, for $TTL_2^{RC,pD}$ for $p = 2, \ldots, 5$, and various values of the number of iterates

N_{iter}	$p = 2$	$p = 3$	$p = 4$	$p = 5$
10^4	1.5624	1.5568	1.55725	1.55915
10^5	0.57955	0.5163	0.51083	0.514
10^6	0.330084	0.160282	0.158256	0.158058
10^7	0.294918	0.0509584	0.0504002	0.0505508
10^8	0.291428	0.0176344	0.0157924	0.0160114
10^9	0.291012	0.00911485	0.00506758	0.00507915
10^{10}	0.291025	0.00783204	0.00159046	0.0015927
10^{11}	0.291033	0.00771201	0.000521561	0.000506086
10^{12}	0.291036	0.00769998	0.000209109	0.000158795
10^{13}		0.00769867	0.000150031	5.03666e-05
10^{14}		0.00769874	0.000144162	1.60489e-05

Table 4.14 Comparison between $E_{C_1,N_{disc},N_{iter}}(x_n^{(1)}, x_n^{(2)})$,for $TTL_2^{RC,4D}$ $M = N_{disc} =$ 20, 200, 2000, 20000, and various values of the number of iterates

N_{iter}	$N_{disc} = 20$	$N_{disc} = 200$	$N_{disc} = 2000$	$N_{disc} = 20000$
10^4	0.1508	1.55725	1.99501	1.99995
10^5	0.04894	0.51083	1.95066	1.9995
10^6	0.015544	0.158256	1.55759	1.99501
10^7	0.005487	0.0504002	0.512542	1.95062
10^8	0.00159524	0.0157924	0.158971	1.55763
10^9	0.000517392	0.00506758	0.0504555	0.513028
10^{10}	0.000205706	0.00159046	0.0159528	0.159054
10^{11}	0.000147202	0.000521561	0.0050481	0.0504422
10^{12}		0.000209109		
10^{13}		0.000150031		
10^{14}		0.000144162		

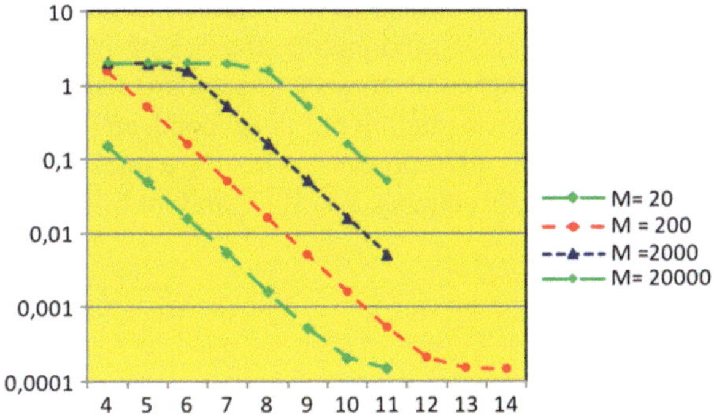

Fig. 4.29 Comparison between $E_{C_1,N_{disc},N_{iter}}$ $(x_n^{(1)}, y_n^{(2)})$, for $TTL_2^{RC,4D}$, $M = N_{disc} =$ 20, 200, 2000, 20, 000, and various values of the number of iterates

Table 4.15 Comparison of computation times (in second) for the generation of N_{iter} PRNs for $TTL_2^{RC,pD}$ with $p = 2$ to 5, and various values of N_{iter} iterates

N_{iter}	$p = 2$	$p = 3$	$p = 4$	$p = 5$
10^4	0.000146	0.000216	0.000161	0.000142
10^5	0.000216	0.000277	0.000262	0.000339
10^6	0.001176	0.002403	0.001681	0.002467
10^7	0.011006	0.016195	0.018968	0.022351
10^8	0.113093	0.161776	0.166701	0.227638
10^9	1.09998	1.58949	1.60441	2.29003
10^{10}	11.4901	18.0142	18.537	26.1946
10^{11}	123.765	183.563	185.449	257.244

Table 4.16 Comparison of computation times (in 10^{-10} s) for the generation of only one PRN for $TTL_2^{RC,pD}$ with $p = 2$ to 5, and various values of the number of iterates

N_{iter}	$p = 2$	$p = 3$	$p = 4$	$p = 5$
10^4	73.0	72.0	40.25	28.4
10^5	10.8	9.233	6.55	6.78
10^6	5.88	8.01	4.2025	4.934
10^7	5.503	5.39833	4.742	4.702
10^8	5.65465	4.0444	4.16753	4.55276
10^9	5.4999	5.2983	4.01103	4.58006
10^{10}	5.74505	4.50335	4.63425	5.23892
10^{11}	6.18825	6.11877	4.63622	5.14488

These results show that the pace of computation is very high. When $TTL_2^{RC,5D}$ is the mapping tested, and the machine used is a laptop computer with a Core i7 4980HQ processor with 8 logical cores, computing 10^{11} iterates with five parallel streams of PRNs leads to around 2 billion PRNs being produced per second. Since these PRNs are computed in the standard double precision format, it is possible to extract from each 50 random bits (the size of the mantissa being 52 bits for a double precision floating-point number in standard IEEE-754). Therefore, $TTL_2^{RC,5D}$ can produce 100 billion random bits per second, an incredible pace! With a machine with 4 Intel Xeon E7–4870 processors having a total of 80 logical cores, the computation is twice as fast, producing 2×10^{11} random bits per second.

4.5 Conclusion

In this chapter, we thoroughly explored the novel idea of combining features of a tent map (T_μ) and a logistic map (L_μ) to produce a new map with improved properties, through combination in several network topologies. This idea was recently introduced [38, 39] in order to improve previous CPRNGs. We have summarized the previously explored topologies in dimension two. We have presented new results of numerical experiments in higher dimensions (up to five) for the mapping $TTL_2^{RC,pD}$ on multicore machines and shown that $TTL_2^{RC,5D}$ is a very good CPRNG which is fit for industrial applications. The pace of generation of random bits can be incredibly high (up to 200 billion random bits per second).

References

1. Delahaye, J.-P.: Cryptocurrencies and blockchains. Inferences **2**, 4 (2016)
2. Menezes, A.J., Van Oorschot, P.C.: Handbook of Applied Cryptography. CRC Press, Boca Raton (1996)
3. Matthews, R.: On the derivation of chaotic encryption algorithm. Cryptologia **13**(1), 29–42 (1989)
4. Lozi, R., Cherrier, E.: Noise-resisting ciphering based on a chaotic multi-stream pseudorandom number generator. In: Proceedings of the 2011 International Conference for Internet Technology and Secured Transactions (ICITST), Abu Dhabi, pp. 91–96 (2011)
5. Li, C.-Y., Chen, Y.-H., Chang, T.-Y., Deng, L.-Y., Kiwing, T.: Period extension and randomness enhancement using high-throughput reseeding-mixing PRNG. IEEE Trans. Very Large Scale Integr. VLSI Syst. **20**(2), 385–389 (2012)
6. Noura, H., Assad, S.E., Vladeanu, C.: Design of a fast and robust chaos-based cryptosystem for image encryption. In: 8th International Conference on Communications (COMM 2010), pp. 423–426 (2010)
7. Bogoshi, J., Naidoo, K., Webb, J.: The oldest mathematical artifact. Math. Gazette **71**(458), 294 (1987)
8. Smith, D.E.: History of Mathematics, vol. I, pp. 47–49. Dover Publication Inc., New-York (1923)
9. Descartes, R.: Discours de la méthode. La géométrie (1637)
10. Galois, E.: Mémoire sur les conditions de résolubilité des équations par radicaux (mémoire manuscrit de 1830). J. Math Pure et Appl. **10**, 471–433 (1845)
11. Julia, G.: Mémoire sur l'itération des fonctions rationnelles. Journal de mathématiques pures et appliquées **8**(1), 47–246 (1918)
12. Fatou, P.: Sur l'itération des fonctions transcendantes entières. Acta Math. **47**, 337–370 (1926)
13. Gumowski, I., Mira, C.: Recurrence and Discrete Dynamics systems. Lecture Notes in Mathematics. Springer, Berlin (1980)
14. Sharkovskiĭ, A.N.: Coexistence of cycles of a continuous map of the line into itself. Intern. J. Bifurc. Chaos, **5**(5), 1263–1273 (1995). Ukrainian Math. J. **16**, 61–71 (1964). [in Russian]
15. Ikeda, K.: Multiple-valued stationary state and its instability of the transmitted light by a ring cavity system. Opt. Commun. **30**, 257–261 (1979)
16. Ikeda, K., Daido, H., Akimoto, O.: Optical turbulence: chaotic behavior of transmitted light from a ring cavity. Phys. Rev. Lett. **45**, 709–712 (1980)
17. Chua, L.O., Kumoro, M., Matsumoto, T.: The double scroll family. IEEE Trans. Circuit Syst. **32**(11), 1055–1058 (1984)

18. Hénon, M.A.: Two-dimensional mapping with a strange attractor. Commun. Math. Phys. **50**, 69–77 (1976)
19. Lorenz, E.N.: Deterministic nonperiodic flow. J. Atmos. Sci. **20**, 130–141 (1963)
20. May, R.M.: Stability and Complexity of Models Ecosystems. Princeton University Press, Princeton (1973)
21. May, R.: Biological populations with nonoverlapping generations: stable points, stable cycles, and chaos. Sci. New Ser. **186**(4164), 645–647 (1974)
22. Lozi, R.: Giga-periodic orbits for weakly coupled tent and logistic discretized maps. In: Siddiqi, A.H., Duff, I.S., Christensen, O. (eds.) Modern Mathematical Models Methods and Algorithms for Real-World Systems, pp. 80–124. Anamaya Publishers, New Delhi, India (2006)
23. Yuan, G., Yorke, J.A.: Collapsing of chaos in one dimensional maps. Phys. D **136**, 18–30 (2000)
24. Lozi, R.: Can we trust in numerical computations of chaotic solutions of dynamical systems? Topol. Dyn. Chaos, **A**(84), 63–98 (2013). Letellier, C., Gilmore, R. (eds.) World Scientific Series on Nonlinear Sciences
25. Baptista, M.S.: Cryptography with chaos. Phys. Lett. A **240**, 50–54 (1998)
26. Ariffin, M.R.K., Noorani, M.S.M.: Modified baptista type chaotic cryptosystem via matrix secret key. Phys. Lett. A **372**, 5427–5430 (2008)
27. Lanford III, O.E.: Informal remarks on the orbit structure of discrete approximations to chaotic maps. Exp. Math. **7**, 317–324 (1998)
28. Wong, W.K., Lee, L.P., Wong, K.W.: A modified chaotic cryptographic method. In: Communications and Multimedia Security Issues of the New Century, pp. 123–126 (2001)
29. Nejati, H., Beirami, A., Massoud, Y.: A realizable modified tent map for true random number generation. Circuits Syst. MWSCAS **10**, 621–624 (2008)
30. Lozi, R.: Mathematical chaotic circuits: an efficient tool for shaping numerous architectures of mixed chaotic/pseudo random number generator. In: Matoušek, M. (ed.) Proceedings of the Mendel 2014, pp. 163–176 (2014)
31. Lozi, R.: Emergence of randomness from chaos. Int. J. Bifurc. Chaos, **22**(2), 1250021–1/1250021–15 (2012)
32. Rojas, A., Taralova, I., Lozi, R.: New alternate ring-coupled map for multirandom number generation. J. Nonlinear Syst. Appl. **4**(1), 64–69 (2013)
33. Garasym, O., Lozi, R., Taralova, I.: Robust PRNG based on homogeneously distributed chaotic dynamics. J. Phys: Conf. Ser. **692**, 012011 (2016)
34. Jallaouli, O., Assad, S.E., Chetto, M., Lozi, R.: Design and analyses of two stream ciphers based on chaotic coupling and multiplexing techniques. Multimedia tools and applications, 27 pp, 29 June 2017. published online
35. Garasym, O., Taralova, I., Lozi, R.: Application of observer-based chaotic synchronization and identifiability to the original CSK model for secure information transmission. Indian J. Ind. Appl. Math. **6**(1), 1–26 (2015)
36. Farajallah, M., Assad, S.E., Deforges, O.: Fast and secure chaos-based cryptosystem for images. Int. J. Bifurc. Chaos **26**(2), 1650021 (2016)
37. Taralova, I., Lozi, R., Assad, S.E.: Chaotic generator synthesis: dynamical and statistical analysis. In: International IEEE Conference for Internet Technology And Secured Transactions, pp. 56–59 (2012)
38. Garasym, O., Taralova, I., Lozi, R.: New nonlinear CPRNG based on tent and logistic map. In: Jinhu Lü, G.C., Yu, X.Y.W. (eds.) Complex Systems and Networks, Dynamics, Controls and Application, pp. 131–162. Springer, Berlin (2016). Springer: Understanding Complex Systems
39. Garasym, O., Lozi, J.-P., Lozi, R.: How useful randomness for cryptography can emerge from multicore-implemented complex networks of chaotic maps? J. Differ. Equ. Appl., 1–39, February 2017. published online
40. Rukhin, A., Soto, J., Nechvatal, J., Barker, E., Leigh, S., Levenson, M., Banks, D., Heckert, A., Dray, J., Vo, S.: Statistical test suite for random and pseudorandom number generators for cryptographic applications. NIST special publication (2010)

Chapter 5
Challenges in Optimal Control Problems for Gas and Fluid Flow in Networks of Pipes and Canals: From Modeling to Industrial Applications

Falk M. Hante, Günter Leugering, Alexander Martin, Lars Schewe and Martin Schmidt

Abstract We consider optimal control problems for the flow of gas or fresh water in pipe networks as well as drainage or sewer systems in open canals. The equations of motion are taken to be represented by the nonlinear isothermal Euler gas equations, the water hammer equations, or the St. Venant equations for flow. We formulate model hierarchies and derive an abstract model for such network flow problems including pipes, junctions, and controllable elements such as valves, weirs, pumps, as well as compressors. We use the abstract model to give an overview of the known results and challenges concerning equilibria, well-posedness, controllability, and optimal control. A major challenge concerning the optimization is to deal with switching on–off states that are inherent to controllable devices in such applications combined with continuous simulation and optimization of the gas flow. We formulate the corresponding mixed-integer nonlinear optimal control problems and outline a decomposition approach as a solution technique.

Keywords Networks · Pipes · Canals · Euler and St. Venant equations
Hierarchy of models · Domain decomposition · Controllability · Optimal control

---

F.M. Hante (✉) · G. Leugering
Lehrstuhl Angewandte Mathematik II, Friedrich-Alexander-Universität Erlangen-Nürnberg (FAU), Cauerstr. 11, 91058 Erlangen, Germany
e-mail: falk.hante@fau.de

G. Leugering
e-mail: guenter.leugering@fau.de

A. Martin · L. Schewe · M. Schmidt
Discrete Optimization, Friedrich-Alexander-Universität Erlangen-Nürnberg (FAU), Cauerstr. 11, 91058 Erlangen, Germany
e-mail: alexander.martin@fau.de

L. Schewe
e-mail: lars.schewe@fau.de

M. Schmidt
Energie Campus Nürnberg, Fürther Str. 250, 90429 Nürnberg, Germany
e-mail: mar.schmidt@fau.de

P. Manchanda et al. (eds.), *Industrial Mathematics and Complex Systems*,
Industrial and Applied Mathematics, DOI 10.1007/978-981-10-3758-0_5

5.1 Introduction

The optimization and control of networked transport systems are becoming an increasingly important branch of industrial applied mathematics. In particular, gas flow in pipe networks including providers, customers, valves, compressor stations, and the like provides a grand challenge with respect to customer satisfaction, low-cost operation of the network, legal restrictions, pressure and flow restrictions, sensitivities with respect to temperature, and market conditions. Given the fact that pipe systems involve easily thousands of pipes, valves, and a number of compressor stations, which, in turn, are whole factories all by themselves, turns the overall problem into a multiscale problem in time and space.

While the physical quantities are typically viewed as continuous entities, decisions are not. The decisions of switching a compressor on or closing a valve are 0–1 processes. On the other hand, having switched on a compressor based on some decision-enhancing argument, the compressor as physical entity is controlled by a continuous profile ranging from the idle state to the desired state. Similarly, the operation of valves, release elements, or tanks for fresh water or sewage water systems is again a combination of discrete or integer controls and continuous controls. Pressurized flow problems appear also in hot steam pipes in power plants, where in addition to the transportation problem nonlinear fluid-structure interactions and a variety of design problems are important.

What has been said so far exactly applies to other transportation systems in civil engineering, such as in fresh water pressure-flow pipe networks as well as sewer systems with the free surface flow in open or closed canals that, in turn, may switch to pressurized flow under severe weather conditions. Again, opening a weir or a sluice gate in the possibly polluted waste water networks or river regulatory systems as well as operating valves, tanks, purification plants, or pumps in fresh water systems involves discrete and continuous optimization variables and cost or merit functions to be optimized. In conclusion, one ends up with a vastly complex, discrete-continuous multilevel, and multicriteria optimization problem involving systems of time-dependent partial differential equations, ordinary differential equations, as well as algebraic equality and inequality constraints for the governing state variables as well as control constraints. On top of that, the problem formulations are typically inexact, as parameters (e.g., wall roughness and other material properties) are unknown or uncertain. Knowledge about initial and equilibrium conditions are lacking as well. This indicates that data plays a predominant role in the applicability of the mathematical methods. Finally, all what is done in controlling, operating, and planning of such a complex system should be done in real time or for a large number of instances, respectively. An example for the different aspects to tackle such a problem is given in [51], where these aspects are discussed for gas networks.

It is obvious that a mathematical program cannot cope with all these difficulties and challenges. Nevertheless, it is also obvious that the mathematics community should be aware of these challenges and particular of those leading to new and interesting mathematics. The particular instant that the Indian Society of Industrial Applied

Fig. 5.1 A gas compressor

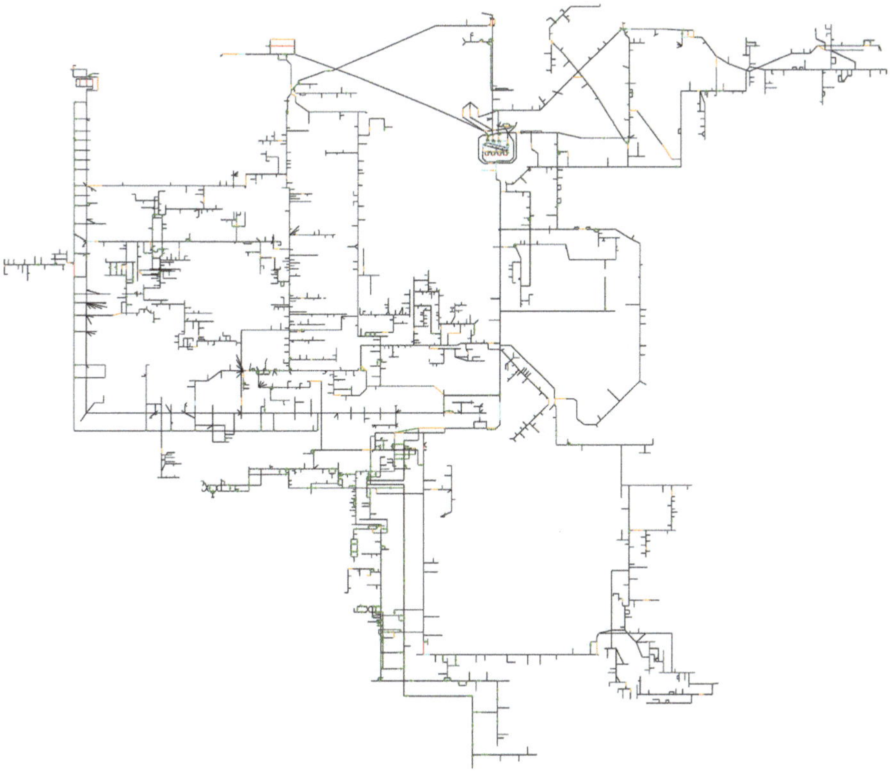

Fig. 5.2 A real-world gas network of Germany's largest gas transport company Open Grid Europe GmbH. Lines correspond to pipes or active elements like compressors as given in Fig. 5.1. Connection points of these lines correspond to simple junctions or entry as well as exit customers

Mathematics (ISIAM) held an international workshop at Sharda University in January 2016 and is now committed to publishing a thematic volume regarding industrial applied mathematics is an opportunity to provide a survey article on problems that are grand challenges both for the Indian society and the Indian mathematical community. The authors sincerely hope that this article provides some hints and stipulations where to concentrate future research resources.

The article is organized as follows: In Sect. 5.2, we first embark on the modeling of gas flow. We start with a rather general system of equations and then derive a hierarchy of simpler models until we arrive at algebraic relations for which even explicit formulae are known. We then provide a network modeling for the corresponding systems of equations, where we introduce boundary conditions at so-called simple nodes (inflow and outflow nodes) and transmission conditions at interior nodes, where either pipes meet or valves, compressors, and the like are coupled to pipes. The node conditions involve discrete and continuous control variables. The same program is then pursued for fresh and waste water systems. It becomes obvious that all the systems can be put into a common abstract framework, namely systems of switching nonlinear hyperbolic balance laws on metric graphs. Clearly, such hybrid formulations are non-standard from the point of view of dynamical systems (PDEs, ODEs, Integro-PDEs, etc.). We then discuss some system-theoretical results in Sect. 5.3 that are needed for optimal control by discussing the existence of equilibria, linearizations around such an equilibrium, Riemann invariants, and discretization techniques. The topic of the final Sect. 5.4 is then how to apply these results and techniques to optimal control problems. Here, we also show computational results on problems from real-world applications. We provide, in a sense, a road-map from modeling to optimal control, where in addition to the dynamical system, side constraints for the states and the controls have to be satisfied throughout the operation. At each step, we pose open questions and refer to known results.

5.2 Modeling of Flow in Pipes and Open Canals

In this section, we introduce three example problems and their common generalization. For every problem, we first state the model for a single pipe or canal and then introduce a network model that also contains active, i.e., controllable, elements. Apart from this common structure, we emphasize different aspects of the models in our examples. For instance, the gas network example contains a discussion of a fine-grained model hierarchy, whereas the sewage example contains a derivation of the model equations.

Before we start with the different examples, we fix some notation common to all models. We consider networked systems that we commonly model by a metric graph $G = (N, E)$ with nodes $N = \{n_1, n_2, \ldots, n_{|N|}\}$ and edges $E = \{e_1, e_2, \ldots, e_{|E|}\}$. Each edge e_i represents a pipe or canal as a one-dimensional object of normalized length 1, and we therefore associate to each edge an interval $[0, 1]$. Moreover, we associate with each edge a direction pointing from $x = 0$ to $x = 1$. For what follows, we introduce the edge-node-incidence matrix $D \in \mathbb{Z}^{|E|\times|N|}$ with entries

$$d_{ij} = \begin{cases} -1, & \text{if node } n_j \text{ is the left node of the edge } e_i, \\ +1, & \text{if node } n_j \text{ is the right node of the edge } e_i, \\ 0, & \text{else.} \end{cases}$$

The set of edges that are connected to a node j is denoted by $\mathcal{I}_j := \{i = 1, \dots, |E| : d_{ij} \neq 0\}$ and the set of in- and outgoing edges are given by $\mathcal{I}_j^+ := \{i \in \mathcal{I}_j : d_{ij} = 1\}$ and $\mathcal{I}_j^- := \{i \in \mathcal{I}_j : d_{ij} = -1\}$. Finally, for each node we introduce the edge degree $d_j := |\mathcal{I}_j|$.

We subdivide the set of nodes further, depending on their role in the network. To this end, we introduce three sets of node indices:

- the set $\mathcal{J}_\alpha$ corresponds to nodes that are active, i.e., controllable, e.g., valves, compressors, and pumps;
- the set $\mathcal{J}_\beta$ corresponds to boundary nodes at which gas or water enters or exits the system; and
- the set $\mathcal{J}_\pi$ corresponds to nodes that are passive in the sense that they do not belong to one of the sets above. We call such nodes also junctions.

The set $\mathcal{J}_\alpha$ will typically be subdivided further depending on the discussed model. For nodes n_j with $j \in \mathcal{J}_\alpha$, we assume that $d_j = 2$ with one incoming edge with index $i \in \mathcal{I}_j^+$ and one outgoing edge with index $k \in \mathcal{I}_j^-$. For all other node types, we make no assumptions on their edge degree. We set $\mathcal{J} = \mathcal{J}_\alpha \cup \mathcal{J}_\beta \cup \mathcal{J}_\pi$.

5.2.1 Gas Flow

In this section, we describe the modeling of gas flow. We start by presenting a hierarchy of models for a single pipe in Sect. 5.2.1.1 and afterward discuss a model for an entire network with valves and compressors in Sect. 5.2.1.2.

5.2.1.1 A Single Pipe

The Euler equations for the flow of gas are given by a system of nonlinear hyperbolic partial differential equations (PDEs), which represent the motion of a compressible non-viscous fluid or a gas. They consist of the continuity equation, the balance of moments, and the energy equation. The full set of equations is given by (see, e.g., [10, 57, 58, 70])

$$\begin{aligned} \partial_t \rho + \partial_x(\rho v) &= 0, \\ \partial_t(\rho v) + \partial_x(p + \rho v^2) &= -\frac{\lambda}{2D}\rho v\,|v| - g\rho h', \\ \partial_t\left(\rho(\frac{1}{2}v^2 + e)\right) + \partial_x\left(\rho v(\frac{1}{2}v^2 + e) + pv\right) &= -\frac{k_w}{D}\,(T - T_w)\,. \end{aligned} \tag{5.1}$$

Here, ρ denotes the density, v the velocity of the gas, T its temperature, and p the pressure. We further denote with g the gravitational constant, with $h' = h'(x)$ the slope of the pipe, with λ the friction coefficient of the pipe, with D the diameter, with k_w the heat coefficient, with $T_w = T_w(x)$ the temperature of the wall, and the variable $e = c_v T + gh$ denotes the internal energy, where c_v is the specific heat. The conserved, respectively balanced, quantities of the system are the flux $q = a\rho v$ (where a is the cross-sectional area of the pipe), the density ρ, and the total energy $E = \rho(1/2v^2 + e)$. In addition to the Eq. (5.1) we use the constitutive law for a real gas

$$p = R_s \rho T z(p, T),$$

where $z = z(p, T)$ is the real gas, or compressibility, factor and R_s is the specific gas constant. Note that $z = 1$ holds for an ideal gas. The Eq. (5.1) allow for three characteristics corresponding to the eigenvalues of the Jacobi matrix of the flux function that are given by

$$\lambda_1 = v - c, \quad \lambda_2 = v, \quad \lambda_3 = v + c,$$

where c is the speed of sound, i.e., $c^2 = \partial_\rho p$ (for constant entropy). For a natural gas, this is approximately 340 ms^{-1}. While the first and third characteristics are genuinely nonlinear, the second is linear degenerate. For the linear degenerate contact discontinuities evolve. We consider pipes of finite length ℓ and by a reparameterization $x \mapsto x\ell$ we may assume having (5.1) for $x \in (0, 1)$. The characteristics determine the direction and velocity of acoustic waves inducing the gas flow in the pipe and, hence, the number of boundary conditions that have to be imposed at the ends of the pipe. In particular, in the subsonic case ($|v| < c$) that we consider in the sequel and with positive flow direction of the gas, the first two characteristics are oriented such that the first is right and the second is left going. In this case, two boundary conditions have to be imposed on the left and one at the right end of the pipe.

We consider here the isothermal case only but note, however, that the temperature may have a significant effect: Long pipes may develop large temperature gradients depending on the weather conditions. In the isothermal case ($T \equiv$ const), the energy equation becomes obsolete. Thus, we obtain

$$\begin{aligned} \partial_t \rho + \partial_x(\rho v) &= 0, \\ \partial_t(\rho v) + \partial_x(p + \rho v^2) &= -\frac{\lambda}{2D}\rho v\,|v| - g\rho h'. \end{aligned} \tag{5.2}$$

In this case, there are two characteristics $\lambda_1 = v - c$ and $\lambda_2 = v + c$ such that in the common subsonic case we have one in- and one outgoing characteristic, and, hence, one boundary condition at each boundary point. In the particular case $z(p) \equiv$ const, we obtain a constant speed of sound $c = \sqrt{p/\rho}$.

It is often more convenient to express the state variables in a different way. In particular, often the flux q and the pressure p in a pipe are used. Here we have $q = a\rho v$ and $p = c^2\rho$. With this, we can rewrite System (5.2) as follows:

$$
\begin{aligned}
\partial_t p + \frac{c^2}{a}\partial_x q &= 0, \\
\partial_t q + \partial_x \left(ap + \frac{c^2}{a}\frac{q^2}{p} \right) &= -\frac{\lambda c^2}{2Da}\frac{q\,|q|}{p} - \frac{ga}{c^2}h' p.
\end{aligned} \tag{5.3}
$$

We now write this system in terms of vectors. To this end, we define

$$
y := \begin{pmatrix} p \\ q \end{pmatrix}, \quad F(y) := \begin{pmatrix} \frac{c^2}{a} q \\ ap + \frac{c^2}{a}\frac{q^2}{p} \end{pmatrix}, \quad S(y; x) := \begin{pmatrix} 0 \\ -\frac{\lambda c^2}{2Da}\frac{q|q|}{p} - \frac{ga}{c^2}h' p \end{pmatrix}. \tag{5.4}
$$

Then, System (5.3) can be rewritten as a first-order system of nonlinear hyperbolic balance equations

$$
\partial_t y + \partial_x F(y) = S(y; x).
$$

For small velocities $|v| \ll c$, we arrive at the semilinear model

$$
\begin{aligned}
\partial_t p + \frac{c^2}{a}\partial_x q &= 0, \\
\partial_t q + a\partial_x p &= -\frac{\lambda c^2}{2Da}\frac{q\,|q|}{p} - \frac{ga}{c^2}h' p.
\end{aligned} \tag{5.5}
$$

This model exhibits the simple characteristics $\lambda_1 = -c$ and $\lambda_2 = c$. If in addition $\partial_t q$ is small, one obtains the quasi-stationary (friction dominated) model, see [10],

$$
\begin{aligned}
\partial_t p + \frac{c^2}{a}\partial_x q &= 0, \\
a\partial_x p &= -\frac{\lambda c^2}{2Da}\frac{q\,|q|}{p} - \frac{ga}{c^2}h' p.
\end{aligned} \tag{5.6}
$$

Finally, when considering the stationary case, all derivatives with respect to time vanish and we obtain

$$
\begin{aligned}
\frac{c^2}{a}\partial_x q &= 0, \\
a\partial_x p &= -\frac{\lambda c^2}{2Da}\frac{q\,|q|}{p} - \frac{ga}{c^2}h' p.
\end{aligned} \tag{5.7}
$$

With constant compressibility factor $z \equiv$ const and by further neglecting the gravity term, we get that flux q is constant (hence, determined by the boundary data) and the remaining momentum equation turns into the algebraic model

$$
p_{\text{out}} = \sqrt{p_{\text{in}}^2 - \frac{\lambda c^2 \ell}{Da^2} q\,|q|}, \tag{5.8}
$$

where p_{out} and p_{in} are the pressure at the end and the inlet of the pipe, respectively. The algebraic model (5.8) is discussed, e.g., in [67] and in chapter [26] of the recent book [51].

Remark 5.1 In view of the vectorial notation (5.4), we may embed the hierarchy of models (5.3), (5.5), (5.6), and (5.7) into one format. For this, it is only necessary to introduce

$$\begin{aligned} M^1 &:= \begin{pmatrix} 1 & 0 \\ 0 & 1 \end{pmatrix}, \quad F^1(y) := \begin{pmatrix} \frac{c^2}{a} q \\ ap + \frac{c^2}{a} \frac{q^2}{p} \end{pmatrix}, \\ M^2 &:= \begin{pmatrix} 1 & 0 \\ 0 & 1 \end{pmatrix}, \quad F^2(y) := \begin{pmatrix} \frac{c^2}{a} q \\ ap \end{pmatrix}, \\ M^3 &:= \begin{pmatrix} 1 & 0 \\ 0 & 0 \end{pmatrix}, \quad F^3(y) := \begin{pmatrix} \frac{c^2}{a} q \\ ap \end{pmatrix}, \\ M^4 &:= \begin{pmatrix} 0 & 0 \\ 0 & 0 \end{pmatrix}, \quad F^4(y) := \begin{pmatrix} \frac{c^2}{a} q \\ ap \end{pmatrix}. \end{aligned} \tag{5.9}$$

Then, we can write

$$M^j \partial_t y + \partial_x F^j(y) = S(y; x), \quad j = 1, 2, 3, 4.$$

The above hierarchy and even further intermediate models can also be obtained from asymptotic analysis; see [10].

5.2.1.2 Networks with Pipes, Valves, and Compressors

In order to formulate a complete model for an entire network on a finite time horizon, we have to specify some continuity conditions. First, the pressure variables $p_i(n_j)$ coincide for all incident edges $i \in \mathscr{I}_j$. We express these transmission conditions at all passive nodes by imposing

$$p_i(n_j, t) = p_k(n_j, t), \quad j \in \mathscr{J}_\pi, \ i, k \in \mathscr{I}_j, \ t \in (0, T).$$

The nodal balance equation for the fluxes can be written as the classical Kirchhoff-type condition at non-boundary nodes:

$$\sum_{i \in \mathscr{I}_j} d_{ij} q_i(n_j, t) = 0, \quad j \in \mathscr{J} \setminus \mathscr{J}_\beta.$$

We now turn to the active, i.e., controllable, nodes $j \in \mathscr{J}_\alpha$. These model compressors ($\mathscr{J}_c$) and valves ($\mathscr{J}_v$). The main problem in gas flow is the inherent pressure drop due to friction at the interior pipe surface. This significant pressure drop necessitates compressor stations within the network. Clearly, such compressor stations are costly and expensive to operate. Therefore, typically only few such stations appear in the given network. For example, the German gas network contains about 70 such stations with a power of approximately 2400 MW. The pressure at the outlet of such a station can be up to over 100 bar. The description of compressors is typically established via characteristic diagrams based on measured specific changes in adiabatic enthalpy H_{ad} of the compression process. This quantity depends on the pressure and the temperature and is given by

$$H_{\text{ad}} = z(p_L, T_L)T_L R_{\text{s}} \frac{\kappa}{\kappa - 1}\left(\left(\frac{p_R}{p_L}\right)^{\frac{\kappa-1}{\kappa}} - 1\right),$$

where the isentropic exponent κ is itself pressure and temperature dependent, but is often taken to be a compressor specific constant, e.g., $\kappa = 1.29$. Here, T_L denotes the temperature at the inlet of the compressor. Accordingly, p_L and p_R denote the pressures at the inlet and outlet of the compressor. After introducing a switching variable $s_j^c(t) \in \{0, 1\}$ and the shorthand notation $\bar{\kappa}(q_k) = \text{sign}(q_k(n_j, t))(\kappa - 1)/\kappa$, we obtain a model for a compressor node with index $j \in \mathscr{J}_{\text{c}}$ for all $t \in (0, T)$:

$$0 = s_j^c(t)\left[u_j - C|q_k(n_j, t)|\left(\left(\frac{p_k(n_j, t)}{p_i(n_j, t)}\right)^{\bar{\kappa}(q_k)} - 1\right)\right]$$
$$+ (1 - s_j^c(t))\left[p_i(n_j, t) - p_k(n_j, t)\right].$$

For valves, the model is considerably simpler. With the switching variable $s_j^c(t) \in \{0, 1\}$, the model for a valve node with index $j \in \mathscr{J}_{\text{v}}$ for all $t \in (0, T)$ reads

$$s_j^v(t)\left(p_i(n_j, t) - p_k(n_j, t)\right) + (1 - s_j^v(t))q_i(n_j, t) = 0.$$

In total, we arrive at the following system given in Model 1.

5.2.2 *Fresh Water Systems*

In this section, we describe the modeling of fresh water flow. We again derive a hierarchy of models for a single pipe in Sect. 5.2.2.1 and afterward discuss a model for an entire network with valves and pumps in Sect. 5.2.2.2.

5.2.2.1 A Single Pipe

In order to obtain a model hierarchy for pressurized pipe flow of water similar to the one we have seen for gas flow we consider the fundamental equations of conservation of mass and conservation of momentum for incompressible flow

Model 1: Gas network model; $x \in (0,1)$ and $t \in (0,T)$

$$
\begin{aligned}
\partial_t p_i(x,t) + \frac{c^2}{a_i}\partial_x q_i(x,t) &= 0, \quad i \in \mathscr{I},\\
\partial_t q_i(x,t) + \partial_x\left(a p_i(x,t) + \frac{c^2}{a_i}\frac{q_i(x,t)^2}{p_i(x,t)}\right)&\\
= -\frac{\lambda c^2}{2D_i a_i}\frac{q_i(x,t)|q_i(x,t)|}{p_i(x,t)} - \frac{ga}{c^2}h_i' p_i(x,t), &\quad i \in \mathscr{I},\\
p_i(n_j,t) = p_k(n_j,t), &\quad j \in \mathscr{J}_\pi,\ i,k \in \mathscr{I}_j,\\
\sum_{i \in \mathscr{I}_j} d_{ij} q_i(n_j,t) = 0, &\quad j \in \mathscr{J} \setminus \mathscr{J}_\beta,\\
s_j^v(t)\,(p_i(n_j,t) - p_k(n_j,t)) + (1 - s_j^v(t)) q_i(n_j,t) = 0, &\quad j \in \mathscr{J}_v,\ i,k \in \mathscr{I}_j,\\
s_j^c(t)\left[u_j - C|q_k(n_j,t)|\left(\left(\frac{p_k(n_j,t)}{p_i(n_j,t)}\right)^{\tilde{\kappa}(q_k)} - 1\right)\right]&\\
+ (1 - s_j^c(t))\,[p_i(n_j,t) - p_k(n_j,t)] = 0, &\quad j \in \mathscr{J}_c,\ i,k \in \mathscr{I}_j,\\
g_j(p_i(n_j,t), q_i(n_j,t)) = u_j(t), &\quad j \in \mathscr{J}_\beta,\ i \in \mathscr{I}_j,\\
p_i(x,0) = p_{i,0}(x),\ q_i(x,0) = q_{i,0}(x), &\quad i \in \mathscr{I},
\end{aligned}
$$

$$
\begin{aligned}
\partial_t(\rho a) + \partial_x(\rho u a) &= 0,\\
\partial_t(\rho u a) + \partial_x(\rho a u^2) + a\partial_x p &= -ga\rho\left(\frac{d}{dx}z + \frac{\lambda}{2gD}u|u|\right),
\end{aligned}
$$

where ρ is the density, u is the fluid velocity, and p is the pressure. Here, a is the cross-sectional area of the pipe, D its diameter, and z its elevation above a reference level. One introduces the piezometric height $h(t,x) = z(x) + p(t,x)/(g\rho_0)$, where ρ_0 is the density of water in free surface flow at reference level, the flux $q = ua$ and one assumes $c^2 = \partial_\rho p$, where c is the speed of sound in fresh water at normalized conditions. With these variables, we can verify for $\rho = \rho_0$ that

$$
\begin{aligned}
\partial_t h &= -\frac{c^2}{g\rho_0 a}\partial_x q,\\
a\partial_x p &= ga\rho_0\partial_x(h - z).
\end{aligned}
$$

Thus, we arrive at

$$
\begin{aligned}
\partial_t h + \frac{c^2}{ga}\partial_x q &= 0,\\
\partial_t q + \frac{1}{a}\partial_x q^2 + ga\partial_x h &= -\frac{\lambda}{2aD}|q|q.
\end{aligned}
\tag{5.10}
$$

For a pipe of finite length ℓ we may again employ a reparameterization $x \mapsto x\ell$, having (5.10) for $x \in (0,1)$. Moreover, we may again introduce a vectorial notation

$$y := \begin{pmatrix} h \\ q \end{pmatrix}, \quad F(y) := \begin{pmatrix} \frac{c^2}{ga} q \\ \frac{1}{a} q^2 + gah \end{pmatrix}, \quad S(y; x) := \begin{pmatrix} 0 \\ -\frac{\lambda}{2aD} q\,|q| \end{pmatrix}.$$

Then (5.10) can be rewritten as a first-order system of nonlinear hyperbolic balance equations

$$\partial_t y + \partial_x F(y) = S(y; x).$$

As in the case of gas flow, one may deduce a number of simplifications and obtain a hierarchy of models. First, we may neglect the nonlinear term $\frac{1}{a} \partial_x q^2$ in the momentum equation in order to arrive at a semilinear model called water hammer equations [1], i.e.,

$$\begin{aligned} \partial_t h + \frac{c^2}{ga} \partial_x q &= 0, \\ \partial_t q + ga \partial_x h &= -\frac{\lambda}{2aD} |q| q. \end{aligned} \tag{5.11}$$

We may also neglect the temporal dynamics in the second equation to end up with the quasi-stationary model

$$\begin{aligned} \partial_t h + \frac{c^2}{ga} \partial_x q &= 0, \\ ga \partial_x h &= -\frac{\lambda}{2aD} |q| q. \end{aligned} \tag{5.12}$$

In the stationary case, we have

$$\begin{aligned} \frac{c^2}{ga} \partial_x q &= 0, \\ ga \partial_x h &= -\frac{\lambda}{2aD} |q| q. \end{aligned} \tag{5.13}$$

As this implies $q = q_0 = \text{const}$, we have the formula

$$h_{\text{in}} - h_{\text{out}} = \frac{\lambda L}{2ga^2 D} q_0 |q_0|.$$

Remark 5.2 As we did for the gas case, we also embed the hierarchy of models (5.10)–(5.13) into one format. With $M^1, \ldots, M^4$ as in (5.9) and

$$F^1(y) := \begin{pmatrix} \frac{c^2}{ga} q \\ \frac{1}{a} q^2 + gah \end{pmatrix}, \quad F^2(y) := F^3(y) := F^4(y) := \begin{pmatrix} \frac{c^2}{ga} q \\ gah \end{pmatrix}$$

we can write

$$M^j \partial_t y + \partial_x F^j(y) = S(y; x), \quad j = 1, 2, 3, 4.$$

5.2.2.2 Pipe Networks

On a finite time horizon $(0, T)$, let us consider a fresh water pipeline system including valves and pumps. The pressure increase of a pump expressed in terms of the piezometric height $\Delta h = h_R - h_L$ for given flow q and piezometric heights h_L and h_R corresponding to the pressure at the inlet and outlet can be described by

$$\Delta h = u^2 \left(\alpha - \beta \left(\frac{q}{u}\right)^{\gamma}\right),$$

where pump-dependent $\alpha > 0$ is the maximal pressure increase, γ and β are efficiency parameters, and u is the relative speed subject to our control [63]. Valves are modeled in a straightforward sense similarly to the gas case. Thus, letting J_v and J_p denote the set of node indices corresponding to valves and compressors, respectively, we obtain the network model given in Model 2.

Model 2: Fresh water network model; $x \in (0,1)$ and $t \in (0,T)$

$$\begin{aligned}
\partial_t h_i(x,t) + \frac{c_i^2}{g a_i}\partial_x q_i(x,t) &= 0, \quad i \in \mathcal{I},\\
\partial_t q_i(x,t) + \frac{1}{a_i}\partial_x q_i^2(x,t) + g a_i \partial_x h_i(x,t) &= -\frac{\lambda_i}{2 a_i D_i}|q_i(x,t)|q_i(x,t) \quad i \in \mathcal{I},\\
h_i(n_j,t) &= h_k(n_j,t), \quad j \in \mathcal{J}_\pi,\ i,k \in \mathcal{I}_j,\\
\sum_{i \in \mathcal{I}_j} d_{ij} q_i(n_j,t) &= 0, \quad j \in \mathcal{J} \setminus \mathcal{J}_\beta,\\
s_j^v(t)\left(h_i(n_j,t) - h_k(n_j,t)\right) + (1 - s_j^v(t)) q_i(n_j,t) &= 0, \quad j \in \mathcal{J}_v,\ i,k \in \mathcal{I}_j,\\
s_j^p(t)\left[h_k(n_j,t) - h_i(n_j,t) - u_j^2\left(\alpha_j - \beta_j\left(\frac{q_k(n_j,t)}{u_j}\right)^{\gamma_j}\right)\right]&\\
+ (1 - s_j^p(t))\left[h_i(n_j,t) - h_k(n_j,t)\right] &= 0, \quad j \in \mathcal{J}_p,\ i,k \in \mathcal{I}_j,\\
g_j(h_i(n_j,t), q_i(n_j,t)) &= u_j(t), \quad j \in \mathcal{J}_\beta,\ i \in \mathcal{I}_j\\
h_i(x,0) = p_{i,0}(x),\ q_i(x,0) &= q_{ij}(x), \quad i \in \mathcal{I},
\end{aligned}$$

5.2.3 *Modeling Sewage Flow*

The third type of models concerns the flow of water in open canals and, in particular, in networks of such canals. The latter are often considered as sewer systems. More precisely, sewage flow is modeled as a wave of shallow water running through a long, slender, and prismatic canal. While the shape of the canal profile is often of minor theoretical interest, we have to deal with nontrivial canal shapes in practical applications and, therefore, we describe a canal and its properties in a more general setting.

5.2.3.1 A Single Canal

To model a single canal we may again choose a one-dimensional model because a canal is long and relatively thin (small aspect ratio) and the flow changes significantly only along the flow direction of the canal. The floor of the canal is elevated by a (assumed smooth) floor function z_0 and the shape profile of the canal is characterized by the canal width function $w(h)$, describing the width of the canal in dependence of the filling height, and is assumed to fulfill the following well-shapedness property: Namely, the canal width function $w(h)$ is called well-shaped if there exists $\varepsilon^w_{\min}$ and $\varepsilon^w_{\max}$ both in $\mathbb{R}^+$ with $\varepsilon^w_{\max} > \varepsilon^w_{\min} > 0$ and $w(h) \in \mathscr{C}^1(\mathbb{R}^+; [\varepsilon^w_{\min}, \varepsilon^w_{\max}])$. We now focus on sewage flowing through a well-shaped canal $X \subset \mathbb{R}$. The motion of the liquid is observed over a time interval $\Theta \subset \mathbb{R}^+$ and can be described by physical quantities, which we call primary variables: These variables consist of the water height h and the velocity along the canal V. In the case of pollution, the primary variables are completed by the vector $\boldsymbol{\rho} \in \mathbb{R}^r$ representing concentrations of chemical solutes transported by the sewage. We have to remark that $\boldsymbol{\rho}(t,x) \in (\mathbb{R}^+_0)^r$ for all $(t,x) \in \Theta \times X$ would be a reasonable restriction, as negative concentrations have no physical meaning. Nevertheless, this restriction is not required for the correctness of the mathematical derivations and is therefore neglected. Based on these primary variables and the canal width function $w(h)$, we introduce some additional, so-called secondary variables, consisting of the wetted cross-sectional area of the sewage $A(t,x)$, the flow rate of the sewage $Q(t,x)$, and, in case of pollution, the vector of r amounts of substances $\mathbf{R}(t,x)$ is used to describe the mass of pollution in a cross-sectional area. These are defined as

$$
\begin{aligned}
A(t,x) &:= \int_0^{h(t,x)} w(z)\,\mathrm{d}z,\\
Q(t,x) &:= V(t,x)\int_0^{h(t,x)} w(z)\,\mathrm{d}z,\\
\mathbf{R}(t,x) &:= \boldsymbol{\rho}(t,x)\int_0^{h(t,x)} w(z)\,\mathrm{d}z.
\end{aligned}
$$

We use the vector notation in order to distinguish explicitly from the scalar case. In order to derive the physical balance laws describing the dynamics of the flow variables, we introduce a small but arbitrary part of the time-space domain, which is called control volume and is defined as $\Theta_c \times X_c := (t^0, t^1) \times (x^0, x^1) \subset \Theta \times X$. We can now state the system in terms of the variables $(A, Q, \mathbf{R})$ instead of $(h, V, \boldsymbol{\rho})$. Indeed, by $A = \int_0^h w(z)\,\mathrm{d}z$ we can interpret $A = A(h)$ and $\partial_h A(h) = w(h)$. Our assumption that the canal is well-shaped then implies that $A(h)$ is bijective. We have

$$
h'(A) = \frac{1}{w(h(A))}, \quad h(A) = \int_0^A \frac{1}{w(h(a))}\,\mathrm{d}a.
$$

The inversion of the other variables provides

$$V(A,Q)=\frac{Q}{A},\quad \rho(A,\mathbf{R})=\frac{1}{A}\mathbf{R}.$$

We use this to define the hydrostatic pressure function η as a function of A,

$$\eta(A):=g\int_0^{h(A)}(h(A)-z)w(z)\,\mathrm{d}z,$$

and its derivative is given by

$$\eta'(A)=gAh'(A)=\frac{gA}{w(h(A))}>0,\quad A\in\mathbb{R}^+,$$

where g is, as before, the acceleration due to gravity. Let us now assume that the quantities A, Q, and $\mathbf{R}$ are continuously differentiable functions with respect to time and space. We arrive at the mass balance equation in integral form

$$\int_{\Theta_c}\int_{X_c}\partial_t A(t,x)+\partial_x Q(t,x)\,\mathrm{d}x\,\mathrm{d}t=\int_{\Theta_c}\int_{X_c}s_M(t,x)\,\mathrm{d}x\,\mathrm{d}t, \tag{5.14}$$

where $s_M(t,x)$ is a lateral in- or outflow along the canal. Similarly, the momentum balance is equivalent to

$$\begin{aligned}&\int_{\Theta_c}\int_{X_c}\partial_t Q(t,x)+\partial_x\Big(\frac{Q^2(t,x)}{A(t,x)}+\eta(A(t,x))\Big)\,\mathrm{d}x\,\mathrm{d}t\\ &=\int_{\Theta_c}\int_{X_c}s_P(A(t,x),Q(t,x),x)\,\mathrm{d}x\,\mathrm{d}t,\end{aligned}$$

where $s_P(A,Q,x)$ is the friction term. Moreover, in case of pollution, the corresponding balance reads as

$$\int_{\Theta_c}\int_{X_c}\partial_t\mathbf{R}(t,x)+\partial_x\frac{Q(t,x)}{A(t,x)}\mathbf{R}(t,x)\,\mathrm{d}x\,\mathrm{d}t=\int_{\Theta_c}\int_{X_c}\mathbf{s}_S(\mathbf{R}(t,x),t,x)\,\mathrm{d}x\,\mathrm{d}t. \tag{5.15}$$

As $\Theta_c\times X_c$ is chosen arbitrarily, we can conclude that Eqs. (5.14) and (5.15) must hold in a pointwise sense in $\Theta\times X$. For a canal of length ℓ, using a reparameterization $x\mapsto x\ell$, this leads to a system of hyperbolic equations on $(0,1)$, which we call augmented shallow water equations in conservation form:

$$\begin{aligned}
\partial_t A + \partial_x Q &= s_M(t,x),\\
\partial_t Q + \partial_x \left(\frac{Q^2}{A} + \eta(A)\right) &= -g\left(A z_0' + \frac{\lambda Q(x,t)|Q(x,t)|}{2DA}\right) =: s_P(A,Q,x),\\
\partial_t \mathbf{R} + \partial_x \left(\frac{Q}{A}\mathbf{R}\right) &= \mathbf{s}_S(\mathbf{R},t,x),
\end{aligned} \tag{5.16}$$

where $s_S(\mathbf{R},t,x)$ is a lateral in- or outflow term for the pollutant. We can put this in a vector format as follows

$$\partial_t \begin{pmatrix} A \\ Q \\ \mathbf{R} \end{pmatrix} + \partial_x \begin{pmatrix} Q \\ \frac{Q^2}{A} + \eta(A) \\ \frac{Q}{A}\mathbf{R} \end{pmatrix} = \begin{pmatrix} s_M(t,x) \\ s_P(A,Q,x) \\ \mathbf{s}_S(\mathbf{R},t,x) \end{pmatrix}.$$

For convenience, we set

$$y(t,x) := \begin{pmatrix} A(t,x) \\ Q(t,x) \\ \mathbf{R}(t,x) \end{pmatrix}, \quad F(y) := \begin{pmatrix} Q \\ \frac{Q^2}{A} + \eta(A) \\ \frac{Q}{A}\mathbf{R} \end{pmatrix}, \quad S(y,t,x) := \begin{pmatrix} s_M(t,x) \\ s_P(A,Q,x) \\ \mathbf{s}_S(\mathbf{R},t,x) \end{pmatrix}$$

and arrive at the system of hyperbolic balance laws:

$$\partial_t y(t,x) + \partial_x F(y(t,x)) = S(y(t,x),t,x). \tag{5.17}$$

Remark 5.3 We add that the system variables may be switched to V, A. Then we have,

$$\partial_t \begin{pmatrix} A \\ V \\ \mathbf{R} \end{pmatrix} + \partial_x \begin{pmatrix} AV \\ \frac{V^2}{2} + gh(A) \\ V\mathbf{R} \end{pmatrix} = \begin{pmatrix} s_M(t,x) \\ s_{P,1}(A,V,x) \\ \mathbf{s}_S(\mathbf{R},t,x) \end{pmatrix},$$

where $s_{P,1}(A,V,x)$ is a suitably modified friction term. If we set

$$y(t,x) := \begin{pmatrix} A(t,x) \\ V(t,x) \\ \mathbf{R}(t,x) \end{pmatrix}, \quad F(y) := \begin{pmatrix} AV \\ \frac{V^2}{2} + gh(A) \\ V\mathbf{R} \end{pmatrix}, \quad S(y,t,x) := \begin{pmatrix} s_M(t,x) \\ s_{P,1}(A,V,x) \\ \mathbf{s}_S(\mathbf{R},t,x) \end{pmatrix},$$

we arrive again at a format as in (5.17). The quasilinear format then reads as

$$\partial_t \begin{pmatrix} A \\ V \\ \mathbf{R} \end{pmatrix} + \begin{pmatrix} V & A & 0 \\ \frac{g}{w(h(A))} & V & 0 \\ 0 & \mathbf{R} & V \end{pmatrix} \partial_x \begin{pmatrix} A \\ V \\ \mathbf{R} \end{pmatrix} = \begin{pmatrix} cs_M(t,x) \\ s_{P,1}(A,V,x) \\ \mathbf{s}_S(\mathbf{R},t,x) \end{pmatrix}.$$

In this system, the first two equations resemble the classical shallow water equations, which are completely independent from the substance amounts $\mathbf{R}$. The last r equations regarding the transport of the substance amounts are also called transport equations of passive scalars.

Remark 5.4 As in the preceding examples, we can also derive a stationary variant of the Eq. (5.16) and write these two models in a common format. With

$$M^1 := \begin{pmatrix} 1 & 0 & 0 \\ 0 & 1 & 0 \\ 0 & 0 & I_r \end{pmatrix},$$

where I_r is the $r \times r$ identity matrix, M^4 the $(2+r) \times (2+r)$ zero matrix and

$$F^1(y) := F^4(y) := \begin{pmatrix} Q \\ \frac{Q^2}{A} + \eta(A) \\ \frac{Q}{A}\mathbf{R} \end{pmatrix}$$

we can write

$$M^j \partial_t y + \partial_x F^j(y) = S(y; t, x), \quad j = 1, 4.$$

5.2.3.2 Shallow Water Equations on Networks

On a finite time horizon $(0, T)$, we now consider an urban drainage network consisting of a set of nodes representing canal junctions possibly involving active elements such as slice gates or pumps and a set of edges representing prismatic sewer canals. As the pipe model, we use the shallow water equations discussed in the preceding section. To connect the pipes, we need adequate coupling conditions which occur as boundary conditions for each canal. The boundary conditions at the canal boundaries, $y_i(n_j, t)$, $j \in \mathcal{J}_\beta$, are given for all $t \in (0, T)$. At the other nodes n_j, i.e., nodes n_j with $j \in \mathcal{J} \setminus \mathcal{J}_\beta$, the states have to satisfy transmission conditions. The most important of these conditions is again Kirchhoff's junction rule, which guarantees that no mass is lost as the liquid flows across the vertices n_j:

$$\sum_{i \in \mathcal{I}_j} d_{ij} Q_i(n_j, t) = 0, \quad t \in (0, T).$$

For passive nodes, Kirchhoff's junction rule is completed with another coupling condition stating continuity of free surface height

$$h_i(n_j, t) = h_k(n_j, t), \quad j \in \mathcal{J}_\pi, \; i, k \in \mathcal{I}_j, \; t \in (0, T),$$

or continuity of particle velocity

$$\frac{Q_i(n_j,t)}{A_i(n_j,t)} = \frac{Q_k(n_j,t)}{A_k(n_j,t)}, \quad j \in \mathscr{J}_\pi,\ i,k \in \mathscr{I}_j,\ t \in (0,T). \tag{5.18}$$

Active nodes can be subdivided into two types: gates ($\mathscr{J}_{\mathrm{g}}$) and pumps ($\mathscr{J}_{\mathrm{p}}$). At a sluice gate, we have an upstream water level h_1 and a downstream level $h_2 \le h_1$. The actual height of the gate is h_0. Considering a simple geometry of the gate area, we have a width b and hydraulic constant κ that we do not want to elaborate upon further. With this, the flow through the gate is given by

$$Q = \kappa b h_0 \sqrt{h_1 - h_2}.$$

In our context, we identify the gate as a boundary condition between two consecutive canals. We control the height h_0 and put the coefficients into the definition of the control that we then call $u_j(t)$, where j is the index for the active node n_j with $j \in \mathscr{J}_{\mathrm{g}}$. Thus, for $i,k \in \mathscr{I}_j$ and $t \in (0,T)$ we have

$$Q_i(n_j,t) = u_j(t)\,\mathrm{sign}(h_i(n_j,t) - h_k(n_j,t))\sqrt{|h_i(n_j,t) - h_k(n_j,t)|}.$$

We again introduce a decision variable $s_j^g(t) \in \{0,1\}$ such that if the gate is turned off (not active) $s_j^g(t) = 0$ and otherwise $s_j^g(t) = 1$ holds. Thus, for $i,k \in \mathscr{I}_j$ and $t \in [0,T]$ we have

$$\begin{aligned} 0 = s_j^g(t)&\left(Q_i(n_j,t) - u_j(t)\,\mathrm{sign}(h_i(n_j,t) - h_k(n_j,t))\sqrt{|h_i(n_j,t) - h_k(n_j,t)|}\right)\\ &+ (1 - s_j^g(t))\left(h_i(n_j,t) - h_k(n_j,t)\right). \end{aligned}$$

Pumps can be included in the modeling in a similar way. There are a number of models with increasing accuracy when compared to real data. See [61] for an account of models that are represented as transmission conditions between two adjacent canals. Clearly, the simplest such model is when the flow rate is set equal to the pump rate and there appears a transmission condition

$$s_j^p(t)(Q_i(n_j,t) - \hat{Q}_j) + (1 - s_j^p(t))(h_i(n_j,t) - h_k(n_j,t)) = 0, \quad j \in \mathscr{J}_{\mathrm{p}},\ t \in (0,T).$$

Combining these parts then leads to the network model given in Model 3, where, for concreteness, we choose (5.18) as the coupling condition.

Model 3: Sewage network model; $x \in (0,1)$ and $t \in (0,T)$

$$
\begin{aligned}
\partial_t A_i(x,t) + \partial_x Q_i(x,t) &= 0 \quad i \in \mathscr{I},\\
\partial_t Q_i(x,t) + \partial_x \left(\frac{Q_i^2(x,t)}{A_i(x,t)} + \eta_i(A_i(x,t)) \right) &= s_{P,i}(A_i(x,t), Q_i(x,t), x) \quad i \in \mathscr{I},\\
\partial_t \mathbf{R}_i(x,t) + \partial_x \left(\frac{Q_i(x,t)}{A_i(x,t)} \mathbf{R}_i(x,t) \right) &= \mathbf{s}_{S,i}(\mathbf{R}_i(x,t), t, x) \quad i \in \mathscr{I},\\
\frac{Q_i(n_j,t)}{A_i(n_j,t)} &= \frac{Q_k(n_j,t)}{A_k(n_j,t)} \quad j \in \mathscr{J}_\pi,\ i,k \in \mathscr{I}_j,\\
\sum_{i \in \mathscr{I}_j} d_{ij} Q_i(y_i)(n_j) &= 0, \quad j \in \mathscr{J} \setminus \mathscr{J}_\beta,\\
(1 - s_j^g(t))\,(h_i(n_j,t) - h_k(n_j,t)) &\\
+ s_j^g(t) \Bigg(Q_i(n_j,t) - u_j(t)\,\mathrm{sign}(h_i(n_j,t) &\\
- h_k(n_j,t)) \sqrt{|h_i(n_j,t) - h_k(n_j,t)|} \Bigg) &= 0, \quad j \in \mathscr{J}_{\mathrm{g}},\ i,k \in \mathscr{I}_j,\\
s_j^p(t)(Q_i(n_j,t) - \hat{Q}_j)] + (1 - s_j^p(t))(h_i(n_j,t) - h_k(n_j,t)) &= 0, \quad j \in \mathscr{J}_{\mathrm{p}},\ i,k \in \mathscr{I}_j,\\
g(Q_i(n_j,t), A_i(n_j,t)) &= u_j, \quad j \in \mathscr{J}_\beta,\ i \in \mathscr{I}_j,\\
Q_i(x,0) = Q_{i,0}(x),\ A_i(x,0) &= A_{i,0}(x), \quad i \in \mathscr{I},
\end{aligned}
$$

5.2.4 Abstract Model

The modeling in this section has revealed that in all cases of interest, say on the level of a quasilinear formulation, we can write all models in a common abstract setting as

$$
\begin{aligned}
\partial_t y_i + A_i(y_i)\partial_x y_i &= S_i(y_i), \quad i \in \mathscr{I},\ (x,t) \in (0,1) \times (0,T),\\
E_i(y_i)(n_j) &= E_k(y_k)(n_j), \quad j \in \mathscr{J}_\pi,\ i,k \in \mathscr{I}_j,\ t \in (0,T),\\
\sum_{i \in \mathscr{I}_j} d_{ij} Q_i(y_i)(n_j) &= 0, \quad j \in \mathscr{J} \setminus \mathscr{J}_\beta,\ t \in (0,T),\\
C_j(y_i(n_j), y_k(n_j), s_j, u_j) &= 0, \quad j \in \mathscr{J}_\alpha,\ i,k \in \mathscr{I}_j,\ t \in (0,T),\\
B_i(y_i)(n_j) &= u_j, \quad j \in \mathscr{J}_\beta,\ i \in \mathscr{I}_j,\ t \in (0,T),\\
y_i(\cdot,0) &= y_{i0}, \quad i \in \mathscr{I}.
\end{aligned}
\tag{5.19}
$$

The following three examples give a detailed overview how the preceding models fit into this abstract framework.

Example 5.1 We begin with gas networks, where we have

$$y_i = \begin{pmatrix} p_i \\ q_i \end{pmatrix}, \quad F_i(y_i) = \begin{pmatrix} \frac{c_i^2}{a_i} q_i \\ ap + \frac{c_i^2}{a_i} \frac{q_i^2}{p_i} \end{pmatrix}, \quad A_i(y_i) = DF_i(y_i) = \begin{pmatrix} 0 & \frac{c_i^2}{a_i} \\ a - \frac{c_i^2}{a_i} \frac{q_i^2}{p_i^2} & 2\frac{c_i^2}{a_i} \frac{q_i}{p_i} \end{pmatrix}$$

and

$$E_i(y_i) = p_i, \quad Q_i(y_i) = q_i.$$

At active nodes $j \in \mathscr{J}_\alpha = \mathscr{J}_\text{v} \cup \mathscr{J}_\text{c}$ we impose valve or compressor conditions. Thus, for $j \in \mathscr{J}_\text{v}$ we have

$$C_j(y_i(n_j), y_k(n_j), s_j, u_j) = s_j^\text{v}(t)(p_i(n_j, t) - p_k(n_j, t)) + (1 - s_j^\text{v}(t))q_i(n_j, t)$$

and for $j \in \mathscr{J}_\text{c}$ we have

$$\begin{aligned} & C_j(y_i(n_j), y_k(n_j), s_j, u_j) \\ =& s_j^c(t) \left[u_j - C|q_k(n_j, t)| \left(\left(\frac{p_k(n_j, t)}{p_i(n_j, t)} \right)^{\bar{\kappa}(q_k)} - 1 \right) \right] \\ & + (1 - s_j^c(t)) \left[p_i(n_j, t) - p_k(n_j, t) \right]. \end{aligned}$$

Example 5.2 For fresh water systems, we have

$$y_i = \begin{pmatrix} h_i \\ q_i \end{pmatrix}, \quad F_i(y_i) = \begin{pmatrix} \frac{c_i^2}{g a_i} q_i \\ \frac{1}{a_i} q_i^2 + g a_i h_i \end{pmatrix}, \quad A_i(y_i) = DF_i(y_i) = \begin{pmatrix} 0 & \frac{c_i^2}{g a_i} \\ g a_i & \frac{2}{a_i} q_i \end{pmatrix}$$

and

$$E_i(y_i) = h_i, \quad Q_i(y_i) = q_i.$$

At active nodes $j \in \mathscr{J}_\alpha = \mathscr{J}_\text{p} \cup \mathscr{J}_\text{v}$, we impose pump or valve conditions. Thus, for $j \in \mathscr{J}_\text{p}$ we have

$$\begin{aligned} & C_j(y_i(n_j), y_k(n_j), s_j, u_j) \\ =& s_j^p(t) \left[h_i(n_j, t) - h_k(n_j, t) - u_j^2 \left(\alpha_j - \beta_j \left(\frac{q_k(n_j, t)}{u_j} \right)^{\gamma_j} \right) \right] \\ & + (1 - s_j^p(t)) \left[h_i(n_j, t) - h_k(n_j, t) \right] \end{aligned}$$

and for $j \in \mathscr{J}_\text{v}$ we have

$$C_j(y_i(n_j), y_k(n_j), s_j, u_j) = s_j^\text{v}(t) \left(h_i(n_j, t) - h_k(n_j, t) \right) + (1 - s_j^\text{v}(t))q_i(n_j, t).$$

Example 5.3 Finally, we consider sewer systems. There, the pipe model can be brought into the desired form via

$$y_i := \begin{pmatrix} A_i(t,x) \\ Q_i(t,x) \end{pmatrix}, \quad F_i(y_i) := \begin{pmatrix} Q_i \\ \frac{Q_i^2}{A_i} + \eta(A_i) \end{pmatrix},$$

as well as

$$A_i(y_i) := DF_i(y_i) = \begin{pmatrix} 0 & 1 \\ -\frac{Q_i^2}{A_i^2} + \frac{gA_i}{w(h(A_i))} & 2\frac{Q_i}{A_i} \end{pmatrix},$$

and for the coupling conditions, we set

$$E_i(y_i) := \frac{Q_i}{A_i}, \quad Q_i(y_i) := Q_i.$$

At active nodes $j \in \mathcal{J}_\alpha = \mathcal{J}_\mathrm{p} \cup \mathcal{J}_\mathrm{g}$ we impose pump or gate conditions. Thus, for $j \in \mathcal{J}_\mathrm{p}$ we have

$$C_j(y_i(n_j), y_k(n_j), s_j, u_j) = s_j^p(t)(Q_i(n_j,t) - \hat{Q}_j) + (1 - s_j^p(t))(h_i(n_j,t) - h_k(n_j,t)),$$

and for $j \in \mathcal{J}_\mathrm{g}$, we have

$$\begin{aligned} &C_j(y_i(n_j), y_k(n_j), s_j, u_j) \\ &= (1 - s_j^g(t)) \left(h_i(n_j,t) - h_k(n_j,t)\right) \\ &\quad + s_j^g(t)\left(Q_i(n_j,t) - u_j(t)\,\mathrm{sign}(h_i(n_j,t) - h_k(n_j,t))\sqrt{|h_i(n_j,t) - h_k(n_j,t)|}\right). \end{aligned}$$

Our framework can also be extended to a setting where we switch between models. From the point of view of efficiency in the context of large-scale applications like, e.g., real-world gas or water networks, we would like to take into account model adaptivity. That is to say, in a network region with very little dynamics we would like to invoke a stationary model, in regions where moderate dynamics govern the process, a semilinear time-dependent model may be appropriate, whereas in regions with significant dynamics, the fully nonlinear system needs to be taken into account. Thus, we have a set of mass matrices

$$M_i^{s_i^m(t)}, \quad s_i^m(t) \in \{0, 1, 2, \ldots, m_i\}$$

and a set of system matrices

$$A_i^{s_i^m(t)}(y_i), \quad s_i^m(t) \in \{0, 1, 2, \ldots, m_i\}.$$

In all models, we keep the source terms as they are essential in the applications discussed here, yielding

$$
\begin{aligned}
M_i^{s_i^m} \partial_t y_i + A_i^{s_i^m}(y_i)\partial_x y_i &= S_i(y_i), && i \in \mathscr{I},\\
E_i(y_i)(n_j) &= E_k(y_k)(n_j), && j \in \mathscr{J}_\pi,\ i,k \in \mathscr{I}_j,\\
\sum_{i\in\mathscr{I}_j} d_{ij} Q_i(y_i)(n_j) &= 0, && j \in \mathscr{J} \setminus \mathscr{J}_\beta,\\
C_j(y_i(n_j), y_k(n_j), s_j, u_j) &= 0, && j \in \mathscr{J}_\alpha,\ i,k \in \mathscr{I}_j,\\
B_i(y_i)(n_j) &= u_j, && j \in \mathscr{J}_\beta,\ i \in \mathscr{I}_j,\\
y_i(\cdot,0) &= y_{i0}, && i \in \mathscr{I},
\end{aligned}
\tag{5.20}
$$

where $x \in (0,1)$ and $t \in (0,T)$. Such model adaptivity with s_i^m taken as adjoint-based error estimators and hence state depending switching rules can numerically be exploited to speed up simulations [21]. However, in one way or another, systems of Type (5.20) appear also naturally in the context of gas and water network simulation and control or, in a more general notion, in energy networks. We also want to add that one also may have to consider model switching in the transmission conditions to ensure well-posedness.

Remark 5.5 To the best knowledge of the authors, there is no mathematical analysis available for Model 1, 2, and 3 covering all nonlinearities and mixed regularities due to the switching functions $s_j(t) \in \{0,1\}$ for $j \in \mathscr{J}_\alpha$. This holds even for the simplest possible network, namely a two-link system with one controllable device (e.g., a valve or a compressor) at the connection point and of course extends to the abstract system (5.19). Even for smooth relaxations of $s_j(\cdot)$, no published result seems to be available, though we belief that the theory of Li Tatsien can be applied in this case—at least for tree-like graphs. As a matter of fact, once the corresponding problem is understood for a star-like graph, the tree network can typically be handled using a so-called peeling technique; see [53, 59].

What has been said of course also applies to problem (5.20) including model switching. Note that, for state depending switching rules, one can no longer guarantee a classical notion of continuous dependency of the solution on parameters. Rather, one has to work with set-valued solutions and discuss upper semicontinuity of the solution set. How this can be realized for semilinear equations on networks and implications thereof are discussed in [45, 46].

5.3 System-Theoretical Results

In this section, we collect some relevant system-theoretical facts that apply to our abstract model (5.19) and point out open problems. For fixed integer variables, we show how to derive equilibria for such a system using the example of gas networks and discuss how linearization can be used to investigate solutions in a neighborhood of such an equilibrium. We then study Riemann invariants for the system that are the basis for well-posedness, controllability, and reachability results. We close the section by discussing discretization and piecewise linearization to obtain simplified models that can cope with integer variables.

5.3.1 *Equilibria and Linearization*

It has become amply clear that in all applications discussed in Sect. 5.2 we arrive at the common abstract model (5.19). An elementary question is the existence and characterization of equilibria Y, i.e., a solution of

$$\begin{aligned} A_i(Y_i)\partial_x Y_i &= S_i(Y_i), \quad i \in \mathscr{I}, \\ E_i(Y_i)(n_j) &= E_k(Y_k)(n_j), \quad j \in \mathscr{J}_\pi,\ i,k \in \mathscr{I}_j, \\ \sum_{i\in\mathscr{I}_j} d_{ij} Q_i(Y_i)(n_j) &= 0, \quad j \in \mathscr{J} \setminus \mathscr{J}_\beta, \\ C_j(Y_i(n_j), Y_k(n_j), s_j, u_j) &= 0, \quad j \in \mathscr{J}_\alpha,\ i,k \in \mathscr{I}_j, \\ B_i(Y_i)(n_j) &= u_j, \quad j \in \mathscr{J}_\beta,\ i \in \mathscr{I}_j, \end{aligned} \tag{5.21}$$

for $x \in (0, 1)$, a constant $s = (s_j)_{j\in\mathscr{J}_\alpha}$, and constant $u = (u_j)_{j\in\mathscr{J}_\alpha\cup\mathscr{J}_\beta}$. In order to provide some evidence that the analytical description of such equilibria is possible but can be quite involved, we exemplarily study here the stationary solutions of the isothermal Euler equations in a single horizontal pipe. The case of non-horizontal pipes and results concerning tree-like networks can be found in [37]. An analysis concerning more general networks including cycles is available in [41, 42].

Example 5.4 Consider the isothermal Euler equation (5.2). For every stationary state, the flow rate q is constant. Hence, the density ρ satisfies the ordinary differential equation

$$(a^2c^2\rho^2 - q^2)\rho_x = -\frac{1}{2}\theta q|q|\rho - a^2\rho^3 gh',$$

where $\theta = \lambda/D$. Separation of variables yields

$$\int \frac{a^2c^2\rho^2 - q^2}{\frac{1}{2}\theta q|q|\rho + a^2\rho^3 gh'}\rho_x \,\mathrm{d}x = -x + \hat{C}.$$

For horizontal pipes (i.e., for $h' = 0$), we get a constant solution ρ if $q = 0$ and for $q \neq 0$ we have

$$\int \left(\frac{2a^2c^2}{\theta q|q|}\rho - \frac{2\,\mathrm{sign}(q)}{\theta\rho}\right)\rho_x \,\mathrm{d}x = -x + \hat{C}.$$

This yields the implicit solution

$$\frac{a^2c^2}{\theta q|q|}\rho^2 - \frac{\text{sign}(q)}{\theta}\ln(\rho^2) = -x + \hat{C}. \tag{5.22}$$

By multiplication with θq we obtain

$$\frac{a^2c^2}{|q|}\rho^2 - |q|\ln(\rho^2) = \theta\, q(-x + \hat{C})$$

and, hence, we have the equation

$$\frac{1}{|q|}a^2c^2\rho^2 - |q|\ln(a^2c^2\rho^2) + |q|\ln(a^2c^2) = \theta\, q\,(-x + \hat{C}).$$

Therefore,

$$\left(ac\frac{\rho}{q}\right)^2 - \ln\left(\left(ac\frac{\rho}{q}\right)^2\right) = \theta\,\text{sign}(q)\,(-x + \hat{C}) - \ln(a^2c^2) + \ln(q^2).$$

With the auxiliary variable $\xi = (ac\rho/q)^2$, for which in the subcritical case $\xi \in (1, \infty)$, we obtain

$$-\xi + \ln(\xi) = \theta\,\text{sign}(q)\,(x - \hat{C}) + \ln(a^2c^2) - \ln(q^2).$$

The application of the exponential function on both sides of the equation yields

$$\exp(-\xi + \ln(\xi)) = \exp\left(\theta\,\text{sign}(q)\,(x - \hat{C}) + \ln(a^2c^2) - \ln(q^2)\right). \tag{5.23}$$

Let $W_{-1}(x)$ denote a special branch of the Lambert W function defined as the inverse function of $x \mapsto x\,\exp(x)$ for $x \in (-\infty, -1)$. Thus $W_{-1}(x) \le -1$ is defined for $x \in (-1/e, 0)$. For $x \in [-1/e, 0)$ we get the equation

$$W_{-1}(x) = \ln(-x) - \ln(-W_{-1}(x)).$$

Then we obtain from (5.23)

$$-\xi = W_{-1}\left(-\frac{a^2c^2}{q^2}\exp\left(\theta\,\text{sign}(q)\,(x - \hat{C})\right)\right).$$

Hence, resubstituting ξ and solving for ρ we get

$$\rho = |q|\frac{1}{ac}\sqrt{-\,W_{-1}\left(-\frac{a^2c^2}{q^2}\exp\left(\theta\,\text{sign}(q)(x - \hat{C})\right)\right)}. \tag{5.24}$$

Note that the value of $\hat{C}$ can be computed from the boundary values. For example, with $\rho_0 = \rho(0)$, Eq. (5.22) implies

$$\hat{C} = \operatorname{sign}(q)\,\frac{1}{\theta}\left(\left(ac\frac{\rho_0}{q}\right)^2 - \ln(\rho_0^2)\right). \tag{5.25}$$

The Lambert W function $W_{-1}(x)$ can be computed to arbitrary precision or approximated by

$$W_{-1}(x) \approx \ln(-x) - \ln(-(\ln(-x) - \ln(-\ln(-x) - \cdots))),$$

see [12].

An example of a pressure-flow relation for stationary solutions obtained by such an approximation compared to the stationary solution obtained from the lowest level in the model hierarchy is plotted in Fig. 5.3. It shows the typical behavior of the considered dynamics that is largely determined by the source term.

It becomes apparent from the discussion in the above example that equilibria may become singular, i.e., there is a critical length. This has severe practical implications, as gas pipes need to be calibrated in order to avoid such singular behavior. This becomes a critical issue for very long under-water pipes.

Stationary states are of great interest in the industrial context, as one is interested in small variations around such equilibria if it is not possible to stay there. In that respect, the variation y of the stationary state Y is of interest. Now, assume the sum $\hat{y} := Y + y$ satisfies (5.19). By (5.21), we have

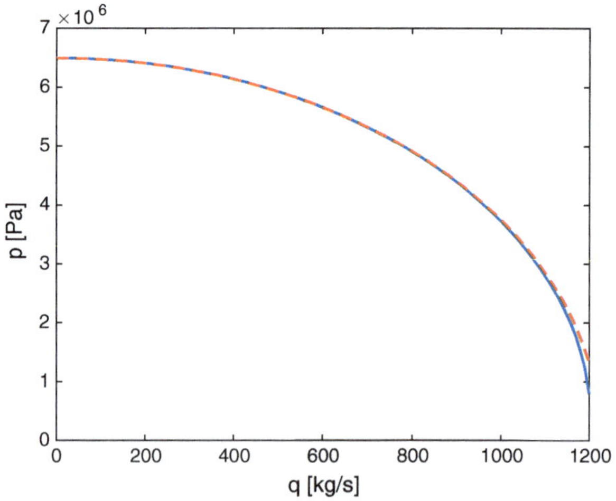

Fig. 5.3 The stationary pressure $p = c^2\rho$ for $h' = 0$ with $p(0) = 6500\,\text{kPa}$, $c = 340\,\text{m}\,\text{s}^{-1}$, $D = 1\,\text{m}$, and $\lambda = 0.005$ on a pipe of length $\ell = 30\,\text{km}$ in dependency of $q \in [0, 1200]\,\text{kg}\,\text{s}^{-1}$, obtained by numerically solving Eqs. (5.24), (5.25) (*solid line*) and the approximation resulting from (5.8) (*dashed line*)

$$
\begin{aligned}
\partial_t y_i + A_i(Y_i + y_i)\partial_x y_i &= -A_i(Y_i + y_i)\partial_x Y_i + S_i(Y_i + y_i), \quad i \in \mathscr{I},\\
E_i(Y_i + y_i)(n_j) &= E_k(Y_k + y_k)(n_j), \quad j \in \mathscr{J}_\pi,\ i, k \in \mathscr{I}_j,\\
\sum_{i \in \mathscr{I}_j} d_{ij} Q_i(Y_i + y_i)(n_j) &= 0, \quad j \in \mathscr{J} \setminus \mathscr{J}_\beta,\\
C_j((Y_i + y_i)(n_j), (Y_k + y_k)(n_j), s_j, u_j) &= 0, \quad j \in \mathscr{J}_\alpha,\ i, k \in \mathscr{I}_j,\\
B_i(Y_i + y_i)(n_j) &= u_j, \quad j \in \mathscr{J}_\beta,\ i \in \mathscr{I}_j,\\
y_i(\cdot, 0) &= y_{i0} - Y_{i0}, \quad i \in \mathscr{I},
\end{aligned}
$$

where, as usual $x \in (0, 1),\ t \in (0, T)$. We define

$$
\begin{aligned}
-A_i(Y_i + y_i)\partial_x Y_i + S_i(Y_i + y_i) &=: \hat{S}_i(y_i),\\
E_i(Y_i + y_i)(n_j) &=: \hat{E}_i(y_i)(n_j),\\
Q_i(Y_i + y_i)(n_j) &=: \hat{Q}_i(y_i)(n_j),\\
C_j((Y_i + y_i)(n_j), (Y_k + y_k)(n_j), s_j, u_j) &=: \hat{C}_j(y_i)(n_j), (y_k)(n_j), s_j, u_j).
\end{aligned}
\tag{5.26}
$$

We clearly see that, $\hat{y}$ satisfies a modified version of (5.21), where we replace each operator with its counterpart from (5.26). Moreover, we get

$$
\hat{S}_i(0) = 0.
$$

This shows that the perturbation y of the equilibrium, which is not assumed small, satisfies the original system with a source term that vanishes for the zero perturbation.

If the perturbations of an equilibrium are considered small, then one arrives at a linear model. To this end, we fix the switching structure s and the controls u at the equilibrium Y. As changing the switching structure s cannot be considered as a small variation, we concentrate on variations $v = (v_j)_{j \in \mathscr{J}_\beta}$ of continuous boundary controls. A Taylor approximation in Y for all terms in (5.19) then yields

$$
\begin{aligned}
\partial_t y_i + A_i(Y_i)\partial_x y_i &= DS_i(Y_i)y_i, \quad i \in \mathscr{I},\\
DE_i(Y_i)y_i(n_j) &= DE_k(Y_k)y_k(n_j), \quad j \in \mathscr{J}_\pi,\ i, k \in \mathscr{I}_j,\\
\sum_{i \in \mathscr{I}_j} d_{ij} DQ_i(Y_i)y_i(n_j) &= 0, \quad j \in \mathscr{J} \setminus \mathscr{J}_\beta,\\
DC_j(Y_i(n_j), Y_k(n_j), s_j, u_j)(y_i, y_k)(n_j) &= 0, \quad j \in \mathscr{J}_\alpha,\ i, k \in \mathscr{I}_j,\\
DB_i(Y_i)y_i(n_j) &= v_j, \quad j \in \mathscr{J}_\beta,\ i \in \mathscr{I}_j,\\
y_i(\cdot, 0) &= y_{i0} - Y_{i0}, \quad i \in \mathscr{I}.
\end{aligned}
$$

Questions regarding well-posedness, controllability, stabilizability, and optimal control for these linear systems on general graphs have been considered in the literature to a certain degree of maturity; see, e.g., [16, 52, 53, 64] and the discussion in Sect. 5.4.

Remark 5.6 We pose some open questions:

- For the general abstract situation, the existence of an equilibrium Y to (5.19) through (5.21) appears to be an open question.
- In general, it appears interesting to obtain full information of the set of equilibria, e.g., connectedness or convexity, also in the case of compressors or pumps.
- How does an equilibrium for a given switching structure behave once the switching structure changes?
- What is the sensitivity of equilibria with respect to parameter changes in general?

5.3.2 Riemann Invariants

Solutions of (5.19) can be analyzed in small neighborhoods of a given equilibrium Y. The method of choice is the concept of semi-global classical solutions in the sense of Li Tatsien [59]. In order to apply the theory given in [59], one needs to transform System (5.19) into a new coordinate system which reveals a diagonal hyperbolic differential expression. To this end, Riemann invariants are very useful. Fortunately, in the applications, the edgewise 2-by-2 hyperbolic balance laws admit such Riemann invariants. We consider the equations in quasilinear form:

$$\partial_t y_i + A_i(y_i)\partial_x y_i = S_i(y_i), \quad i \in \mathscr{I}, \tag{5.27}$$

and we assume that

$$A_i(y_i) \text{ has two eigenvalues } \lambda_i^- < 0 < \lambda_i^+. \tag{5.28}$$

This condition is typically fulfilled in our examples: In the case of gas and fresh water networks, it corresponds to the assumption that the flow is subsonic, in the case of sewage networks it corresponds to the assumption that the flow is subcritical. We denote the corresponding left eigenvectors by $\ell_i^\pm(y_i)$ while the right eigenvectors are denoted by $r_i^\pm(y_i)$. We impose

$$\ell_i^\pm r_i^\pm = 0, \quad \|r_i^\pm\| = \|\ell_i^\pm\| = 1.$$

By definition, the Riemann invariants $\xi_i^\pm(y_i)$ satisfy the equation

$$\nabla \xi_i^\pm = \ell_i^\pm.$$

We apply $\ell_i^\pm$ from the left of (5.27) and obtain

$$\ell_i^\pm \partial_t y_i + \lambda_i^\pm \ell_i^\pm \partial_x y_i = \ell_i^\pm S_i(y_i).$$

Clearly, using the Riemann invariants $\xi_i^{\pm}$, we obtain

$$\partial_t \xi_i^{\pm} = \ell_i^{\pm} \partial_t y_i, \quad \partial_x \xi_i^{\pm} = \ell_i^{\pm} \partial_x y_i$$

and, therefore, we arrive at the system

$$\partial_t \xi_i^{\pm} + \lambda_i^{\pm}(y_i) \partial_x \xi_i^{\pm} = \ell_i^{\pm} S_i(y_i).$$

Thus, the main part (the one including spatial derivatives) is diagonalized with respect to $\xi_i^{\pm}$. Clearly, the coupling still is present via the state variables y_i and via the source terms. In case of a perturbed equilibrium $Y + y$, we have eigenvalues $\lambda_i^{\pm}(Y_i + y_i)$ of $A_i(Y_i + y_i)$ and left and right eigenvectors $\ell_i^{\pm}(Y_i + y_i)$ and $r_i^{\pm}(Y_i + y_i)$, respectively. Accordingly, $\xi_i^{\pm}(Y_i + y_i)$ satisfy

$$\partial_t \xi_i^{\pm} + \lambda_i^{\pm}(Y_i + y_i) \partial_x \xi_i^{\pm} = \ell_i^{\pm}(Y_i + y_i) \tilde{S}_i(Y_i + y_i) =: S_i^{\pm}(y_i). \tag{5.29}$$

We assume that we have a diffeomorphism H_i such that

$$y_i = (y_i^1, y_i^2)^\top = H(\xi_i^+, \xi_i^-) = (h_{1i}(\xi_i^+, \xi_i^-), h_{2i}(\xi_i^+, \xi_i^-))^\top,$$

together with

$$H^{-1}(y_i) = (\xi_i^+, \xi_i^-)^\top = (h_{1i}^{-1}(y_i^1, y_i^2), h_{2i}^{-1}(y_i^1, y_i^2))^\top.$$

We now partition the system into Riemann invariants with labels "$-$" and "$+$": $\xi^- := (\xi_1^-, \dots, \xi_n^-)^\top$ and $\xi^+ := (\xi_1^+, \dots, \xi_n^+)^\top$. We further introduce the diagonal matrix

$$\begin{aligned} \Lambda(\xi^+, \xi^-) := \operatorname{diag}(&\lambda_1^-(H_1(\xi_1^+, \xi_1^-)), \dots, \lambda_n^-(H_n(\xi_n^+, \xi_n^-)), \\ &\lambda_1^+(H_1(\xi_1^+, \xi_1^-)), \dots, \lambda_n^+(H_n(\xi_n^+, \xi_n^-))) \end{aligned}$$

and split Λ into $\Lambda = (\Lambda^-, \Lambda^+)^\top$ with

$$\Lambda^- = \operatorname{diag}(\lambda_1^-(H_1(\xi_1^+, \xi_1^-)), \dots, \lambda_n^-(H_n(\xi_n^+, \xi_n^-)))$$

and

$$\Lambda^+ = \operatorname{diag}(\lambda_1^+(H_1(\xi_1^+, \xi_1^-)), \dots, \lambda_n^+(H_n(\xi_n^+, \xi_n^-))).$$

Moreover, we introduce the system source vector

$$S(\xi^+, \xi^-) := (S_1^-(\xi_1^+, \xi_1^-), \dots, S_n^-(\xi_n^+, \xi_n^-), S_1^+(\xi_1^+, \xi_1^-), \dots, S_n^+(\xi_n^+, \xi_n^-))^\top.$$

Then, (5.29) can be written as

$$\partial_t \xi^{\pm} + \Lambda^{\pm}(\xi^{+}, \xi^{-})\partial_x \xi^{-} = S^{\pm}(\xi^{+}, \xi^{-}).$$

We would like to express the boundary and nodal conditions in terms of the new variables $\xi^{\pm}$. In fact, we would like to have a two-point boundary value problem. Clearly, one can impose boundary conditions at $x = 0$ for ξ^{+}, while at $x = 1$ we may impose boundary conditions for ξ^{-}. Thus, we are aiming at a reformulation of the boundary and nodal conditions in the following way,

$$\begin{aligned} \xi^{+}(0,t) &= G^{1}(\xi^{-}(0,t); s, u) + R^{1}(t; s, u), \\ \xi^{-}(1,t) &= G^{2}(\xi^{+}(1,t); s, u) + R^{2}(t; s, u), \end{aligned}$$

such that, finally, the entire system (5.19) can be put into the standard format

$$\begin{aligned} \partial_t \xi^{\pm} + \Lambda^{\pm}(\xi^{+}, \xi^{-})\partial_x \xi^{-} &= S^{\pm}(\xi^{+}, \xi^{-}), \\ \xi^{+}(0,t) &= G^{1}(\xi^{-}(0,t); s, u) + R^{1}(t; s, u), \\ \xi^{-}(1,t) &= G^{2}(\xi^{+}(1,t); s, u) + R^{2}(t; s, u), \\ \xi^{+}(\cdot, 0) &= \xi_0^{+}, \\ \xi^{-}(\cdot, 0) &= \xi_0^{-}. \end{aligned} \tag{5.30}$$

It is not obvious, however, how the nodal conditions included in (5.19) can be transformed into the format of (5.30). We will use the particular structure, namely the continuity conditions and the Kirchhoff-type balance condition as well as the boundary conditions between two consecutive edges including a valve and a compressor or pump, respectively,

$$\begin{aligned} E_i(Y_i + y_i)(n_j) &= E_k(Y_k + y_k)(n_j), \quad j \in \mathcal{J}_\pi,\ i,k \in \mathcal{I}_j,\ t \in (0,T), \\ \sum_{i \in \mathcal{I}_j} d_{ij} Q_i(Y_i + y_i)(n_j) &= 0, \quad j \in \mathcal{J} \setminus \mathcal{J}_\beta,\ t \in (0,T), \\ C_j((Y_i + y_i)(n_j), (Y_k + y_k)(n_j), s_j, u_j) &= 0, \quad j \in \mathcal{J}_\alpha,\ i,k \in \mathcal{I}_j,\ t \in (0,T), \\ B_i(Y_i + y_i)(n_j) &= u_j, \quad j \in \mathcal{J}_\beta,\ i \in \mathcal{I}_j,\ t \in (0,T). \end{aligned}$$

We need to express the equilibrium Y_i by the Riemann invariants $\tilde{\xi}_i^{\pm}$ and $y_i = (y_i^1, y_i^2)^\top$ by the Riemann invariants $\xi_i^{\pm}$ using the mappings H and H^{-1}. In order to proceed, we first consider a node n_j, $j \in \mathcal{J}_\pi$, with $d_j = m$. At such a node we have the junction condition $P_j = P_j(\xi^{+}, \xi^{-}) = 0$ with $P_j(\xi^{+}, \xi^{-})$ given by

$$\begin{pmatrix} E_1(\tilde{\xi}_1^{+} + \xi_1^{+}, \tilde{\xi}_1^{-} + \xi_1^{-})(n_j) - E_m(\tilde{\xi}_m^{+} + \xi_m^{+}, \tilde{\xi}_m^{-} + \xi_m^{-})(n_j) \\ E_2(\tilde{\xi}_2^{+} + \xi_2^{+}, \tilde{\xi}_2^{-} + \xi_2^{-})(n_j) - E_m(\tilde{\xi}_m^{+} + \xi_m^{+}, \tilde{\xi}_m^{-} + \xi_m^{-})(n_j) \\ \vdots \\ E_{m-1}(\tilde{\xi}_{m-1}^{+} + \xi_{m-1}^{+}, \tilde{\xi}_{m-1}^{-} + \xi_{m-1}^{-})(n_j) - E_m(\tilde{\xi}_m^{+} + \xi_m^{+}, \tilde{\xi}_m^{-} + \xi_m^{-})(n_j) \\ \sum_{i \in \mathcal{I}_i} d_{ij} Q_i(\tilde{\xi}_i^{+} + \xi_i^{+}, \tilde{\xi}_i^{-} + \xi_i^{-})(n_j) \end{pmatrix}.$$

We consider the Jacobian of $P_j(\xi^+, \xi^-)$ with respect to ξ^+ evaluated at $(0, 0)$ and abbreviate

$$\partial_{\xi_i^+} E_i(\tilde{\xi}_i^+ + \xi_i^+, \tilde{\xi}_i^- + \xi_i^-)(n_j)|_{(\xi_i^+,\xi_i^-)=(0,0)} =: \partial_{\xi_i^+} \tilde{E}_i,$$
$$\partial_{\xi_i^+} d_{ij} Q_i(\tilde{\xi}_i^+ + \xi_i^+, \tilde{\xi}_i^- + \xi_i^-)(n_j)|_{(\xi_i^+,\xi_i^-)=(0,0)} =: \partial_{\xi_i^+} \tilde{Q}_i.$$

This yields the Jacobian

$$D_{\xi^+} P_j(0, 0) = \begin{pmatrix} \partial_{\xi_1^+} \tilde{E}_1 & & & & -\partial_{\xi_m^+} \tilde{E}_m \\ & \partial_{\xi_2^+} \tilde{E}_2 & & & -\partial_{\xi_m^+} \tilde{E}_m \\ & & \ddots & & \vdots \\ & & & \partial_{\xi_{m-1}^+} \tilde{E}_{m-1} & -\partial_{\xi_m^+} \tilde{E}_m \\ \partial_{\xi_1^+} \tilde{Q}_1 & \partial_{\xi_2^+} \tilde{Q}_2 & \cdots & \partial_{\xi_{m-1}^+} \tilde{Q}_{m-1} & \partial_{\xi_m^+} \tilde{Q}_m \end{pmatrix}. \tag{5.31}$$

Assuming that $D_{\xi^+} P_j(0, 0)$ is invertible, by the implicit function theorem, there exists a function G^j such that

$$\xi^+(n_j) = G^j(\xi^-(n_j)).$$

Remark 5.7 By the same arguments, one can consider the Jacobian $D_{\xi^-} P_j(0, 0)$ of P_j with respect to ξ^- at the point $(0, 0)$. By the construction of the quantities E_i and Q_i it is clear that, once $D_{\xi^+} P_j(0, 0)$ is invertible, the same applies to $D_{\xi^-} P_j(0, 0)$. Thus,

$$\nabla_{\xi^-} G^j(0) = (\det D_{\xi^+} P_j(0, 0))^{-1} \det D_{\xi^-} P_j(0, 0).$$

We now look at a serial node n_j, $j \in \mathcal{J}_\alpha$, containing active elements such as valves and compressors or pumps, respectively. We have the equation

$$C_j((Y_i + y_i)(n_j), (Y_k + y_k)(n_j), s_j, u_j) = 0, \quad j \in \mathcal{J}_\alpha,\ i, k \in \mathcal{I}_j,\ t \in (0, T).$$

Upon using the Riemann invariants, this turns into

$$\begin{aligned} &C_j((\tilde{\xi}_i^+ + \xi_i^+, \tilde{\xi}_i^- + \xi_i^-)(n_j), ((\tilde{\xi}_k^+ + \xi_k^+), \tilde{\xi}_k^- + \xi_k^-))(n_j), s_j, u_j) \\ =:\ &\tilde{C}_j(\xi_i^+, \xi_i^-, \xi_k^+, \xi_i^-; s, u) = 0. \end{aligned}$$

In addition, at such nodes, we have the equation

$$\begin{aligned} &d_{ij} Q_i(\tilde{\xi}_i^+ + \xi_i^+, \tilde{\xi}_i^- + \xi_i^-)(n_j)) + d_{kj} Q_k(\tilde{\xi}_k^+ + \xi_k^+, \tilde{\xi}_k^- + \xi_k^-))(n_j)) \\ =:\ &\tilde{Q}_i(\xi_i^+, \xi_i^-) + \tilde{Q}_k(\xi_k^+, \xi_k^-) = 0. \end{aligned}$$

Therefore, the full nodal condition for nodes containing active elements reads

$$W_j(\xi_i^+, \xi_i^-, \xi_k^+, \xi_k^-; s, u) := \begin{pmatrix} \tilde{C}_j(\xi_i^+, \xi_i^-, \xi_i^+, \xi_i^-; s, u) \\ \tilde{Q}_i(\xi_i^+, \xi_i^-) + \tilde{Q}_k(\xi_k^+, \xi_k^-) \end{pmatrix} = 0.$$

Thus,

$$D_{\xi^+} W_j(0, 0, 0, 0; s, u) = \begin{pmatrix} \partial_{\xi_i^+} \tilde{C}_j(s, u) & \partial_{\xi_i^+} \tilde{C}_j(s, u) \\ \partial_{\xi_i^+} \tilde{Q}_i & \partial_{\xi_k^+} \tilde{Q}_k \end{pmatrix}. \tag{5.32}$$

We assume again that $D_{\xi^+} W_j(0, 0, 0, 0; s, u)$ is invertible for all choices of s, u. In this case, there is also a function G^j such that

$$(\xi_i^+, \xi_k^+)(n_j) = G^j((\xi_i^-, \xi_k^-)(n_j); s, u).$$

It is obvious that the controlled simple nodes can also be put into the desired format without any further assumption. In the above derivations, we may always assume that all nodes n_j with $d_j > 2$ lie at $x = 0$ for all adjacent arcs and all nodes n_j with $d_j = 2$ lie at $x = 1$ for all adjacent arcs. This assumption can be satisfied by artificially subdividing each arc with a passive node of degree 2. Hence, we have established the following result.

Theorem 5.1 *Assume that* (5.28) *holds, that* (5.31) *is invertible for all* $j \in \mathcal{J}_\pi$, *and that* (5.32) *is invertible for all* $j \in \mathcal{J}_\alpha$. *Then, System* (5.19) *can be rewritten in standard form* (5.30).

We can verify the assumptions of Theorem 5.1 for all applications from Sect. 5.2. We consider here exemplarily the case of sewage flow. In case of gas and fresh water, similar arguments apply.

Example 5.5 For the shallow water equations, we have the Riemann invariants

$$\xi_i^\pm := \frac{Q_i}{A_i} \pm \zeta_i(A_i), \quad \zeta_i(A_i) := \int_0^{A_i} \sqrt{\frac{g}{a w_i(h_i(a))}} \, da, \quad \xi_i^3 := \frac{\mathbf{R}_i}{A_i}.$$

The diffeomorphism and its inverse are given as

$$\xi_i^+ - \xi_i^- = 2\zeta_i(A_i), \quad \xi_i^+ + \xi_i^- = 2\frac{Q_i}{A_i},$$
$$A_i = \zeta_i^{-1}\left(\frac{\xi_i^+ - \xi_i^-}{2}\right), \quad Q_i = \frac{\xi_i^+ + \xi_i^-}{2} \zeta_i^{-1}\left(\frac{\xi_i^+ - \xi_i^-}{2}\right).$$

The continuity conditions may have different formats. We choose the internal energy and the conservation of fluxes

$$E_i(A_i, Q_i) := \frac{1}{2}\left(\frac{Q_i}{A_i}\right)^2 + gh_i(A_i).$$

For details, see [56].

Theorem 5.1 can be seen as a key for the well-posedness of the abstract system (5.19) and hence of all the applications mentioned above in the following sense.

Remark 5.8 We may now use the concept of semi-global classical solutions by Li Tatsien [59] in order to show the existence of solutions of (5.30) and, hence, of (5.19), once compatibility conditions for the data of first and second order are fulfilled and these data are sufficiently small. We do not want to provide the full results, as these results can be seen from the literature as particular examples of the general result described here. See, e.g., [34, 35, 38, 40, 56, 59].

5.3.3 Discretization and Piecewise Linearization

In practical applications the switching structure, i.e., the decision driven part of the process, becomes more and more important. As there is no "sensitivity method" for discrete optimization problems, the process of linearization around an equilibrium solution may not be appropriate. To tackle a problem of the form (5.19) including switching variables, we may discretize in time and space.

For the time discretization, a typical choice is an implicit Euler scheme. To this end, we assume that $[0, T]$ is replaced by grid points $t_0 = 0 < t_1 < \cdots < t_K = T$ with time steps $\Delta t_\kappa := t_{\kappa+1} - t_\kappa$, $\kappa = 0, \ldots, K-1$. Then, the discretized state and the discretized controls can be written as $y_{i,\kappa} := y_i(t_\kappa, \cdot)$, $s_{j,\kappa} := s_j(t_\kappa)$, $u_{j,\kappa} := u_j(t_\kappa)$, and the semi-discretized dynamics become

$$\begin{aligned}
y_{i,\kappa+1} + \Delta t_\kappa \tilde{A}_i(y_{i,\kappa+1})\partial_x y_{i,\kappa+1} &= \Delta t_\kappa \tilde{S}_i(y_{i,\kappa+1}) + y_{i,\kappa}, && i \in \mathcal{I},\\
\tilde{E}_i(y_{i,\kappa+1})(n_j) &= \tilde{E}_k(y_{k,\kappa+1})(n_j), && j \in \mathcal{J}_\pi,\ i,k \in \mathcal{I}_j,\\
\sum_{i\in\mathcal{I}_j} d_{ij}\tilde{Q}_i(y_{i,\kappa+1})(n_j) &= 0, && j \in \mathcal{J} \setminus \mathcal{J}_\beta,\\
\tilde{C}_j(y_{i,\kappa+1}(n_j), y_{k,\kappa+1}(n_j), \bar{s}_{j,\kappa+1}, \bar{u}_{j,\kappa+1}) &= 0, && j \in \mathcal{J}_\alpha,\ i,k \in \mathcal{I}_j,\\
\tilde{B}_i(y_{i,\kappa+1})(n_j) &= \bar{u}_i, && j \in \mathcal{J}_\beta,\ i \in \mathcal{J}_j,\\
y_{i,0}(\cdot, 0) &= y_{i0}, && i \in \mathcal{I},
\end{aligned}$$

where $x \in (0, 1)$.

For the space discretization, various possibilities exist. For instance, in [21] an implicit Box-Scheme is used for the applications mentioned in Sect. 5.2. Such discretization schemes typically give rise to a nonlinear system of equations which then has to be solved. Given that the problem already involves discrete variables, these nonlinearities can also be approximated by piecewise linear functions (see, e.g., [60]). The idea is visualized in the left part of Fig. 5.4.

An extension of this approach covers the space of feasible states for each arc with polytopes, yielding a relaxation of the underlying nonlinear equation system; see again the left part of Fig. 5.4. These systems can then be incorporated more readily into mixed-integer optimization problems. The outlined approach was developed in [28] and used in various problems coming from gas and water network optimization (see, e.g., [30, 31, 51, 61, 67]). We discuss selected results in Sect. 5.4.3.

Remark 5.9 The idea of piecewise linear approximations can also be carried over to the abstract problem (5.19) prior to discretization. Rather than relying on the notion of linearization at some equilibrium, a piecewise linear approximation for the flux function or a piecewise constant matrix for the quasilinear form may be reasonable. To this end, we introduce a tesselation of the range space of the states y into a finite set of mutually disjoint polyhedra. On each polyhedron P_λ, we assume that the matrices $A_i(y_i)$ are constant A_i^λ. Similarly, we assume that all matrices $DE_i = E_i^\lambda$, $DQ_i = Q_i^\lambda$, $DC_j = C_j^\lambda$, $DB_i = B_i^\lambda$, and $SD_i = S_i^\lambda$ are constant on that P_λ; see Fig. 5.4 (right), where we give an illustration for a piecewise linear approximation of the source term S of Euler's momentum equation.

This turns (5.19) into a hybrid dynamical system, where the dynamics are given by a family of affine–linear PDEs along with a discrete selection rule and solutions are to be understood in the sense of characteristics. Model switching in the sense of Sect. 5.2.4 can then also be included. The quality of the approximation depends on the granularity of the tesselation. However, in continuous space and time, assuming that the solution of the piecewise-affine dynamics can be handled in each mode, the Zeno phenomenon, i.e., an eventual accumulation of discrete events, immediately becomes an issue for the global existence of solutions. We provide further information and provide some open questions in this context:

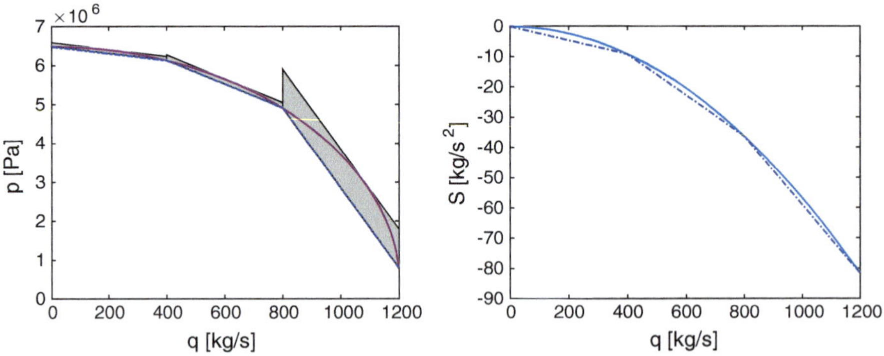

Fig. 5.4 *Left* Piecewise linear approximation (*dashed*) and relaxation (*gray boxes*) of the gas pressure p according to (5.8) in dependence of the mass flow q. *Right* Piecewise linear approximation of the source term S of Euler's momentum equation

- For scalar hyperbolic equations piecewise continuous flux functions are the base for the so-called front-tracking method [9, 50]. For systems, the piecewise linear approach is applied to the Riemann problem rather than the original PDE.
- Discontinuous flux functions have been considered by Adimurthi and Gowda [2, 3].
- Control theoretical analysis is available for the case of piecewise-affine ODEs, see e.g., [15, 71, 72].
- Hybrid dynamical PDEs in the above generality have not been studied. If piecewise linearization is understood in a lumped sense along each edge in the network, well-posedness and stability analysis is available in [6, 44, 46, 47, 68].

5.4 Control, Stabilization, and Optimization

In this section, we discuss controllability, stabilizability, and optimal control problems for the models of Sect. 5.2. We also sketch a technique that may lead to an applicable method. For this, we use the methods and results of Sect. 5.3.

5.4.1 Controllability and Stabilizability Problems

Exact controllability and observability, nodal reachability, and feedback stabilizability are crucial problems in control theory. Of course even more, the controls realizing these properties are of practical relevance. In exact controllability, one wants to reach in finite time T a prescribed full-state profile across a single element (pipe, canal, etc.) or along a network thereof at process time T using a minimum amount of boundary controls. Obviously, the control time in order to achieve this goal is limited below by the speed of propagation of information along the network. In fact, the time is twice the time a signal needs to travel from the controlled node to the farthest uncontrolled Dirichlet or Neumann node. Exact boundary observability refers to the possibility of reconstructing the initial data, and hence the entire state, from boundary measurement. As in the previous case, the speed of propagation comes in crucially. In the linear case, it is well-known that exact controllability and observability are dual concepts—they are equivalent. This is not true for the nonlinear equations discussed in this paper. A more realistic notion is that of profile nodal reachability. Here one asks whether it is possible to achieve a prescribed time function (the "profile") at a given node in the network. In terms of the application we address here, this means that one is interested whether a customer can be guaranteed to receive exactly the gas or fresh water he or she was asking for in an appropriate time window.

For a fixed switching structure, in view of Theorem 5.1 and Remark 5.8, exact controllability, exact observability, exact boundary profile nodal controllability, and uniform boundary feedback stabilizability results follow along with the lines of [13, 22, 34, 35, 56, 59]. Boundary feedback stabilization with or without time delays is typically achieved via Lyapunov-functions [7, 14, 17–19, 36].

Further, for the variable switching structure, uniform exponential stability can be addressed on the level of linearized models [4–6]. For linear switched systems also a particular Lyapunov theory is available [48, 49]. The switching mechanism may also be used for stabilization. This is demonstrated in [55] for the case when switching only changes the boundary conditions of a linear conservation law.

Remark 5.10 Despite the many individual results that are available—noting that there are many non-equivalent notions of controllability and observability—we suggest the following open questions:

- The equivalence of the problem of exact controllability and exact observability for quasilinear systems of hyperbolic balance equations is an open problem. Also the relation to nodal profile controllability is unknown.
- Exact controllability or observability for systems of nonlinear hyperbolic balance laws using switching controls is open.
- For bilinearly acting controls, as in valves, gates, compressors, or pumps, exact controllability (observability) is very unlikely to hold. In this case approximate controllability may be the right question to address. But this also remains open.
- Stability and stabilizability for switched nonlinear problems are open problems.

5.4.2 A Discrete-Continuous Optimal Control Problem

While feedback stabilizability providing closed-loop control is, of course, very significant in real applications for the operation of gas, fresh water, or sewage water networks, open-loop and hence optimal control problems are relevant for various planning purposes. To this end, we consider the formulation of a general discrete-continuous optimal control problem for non-stationary systems of nonlinear hyperbolic balance laws. Regarding our abstract model, a discrete-continuous state-control vector (y, u, s) is feasible if it satisfies the system

$$\begin{aligned} M_i^{s_i}\partial_t y_i + A_i^{s_i}(y_i)\partial_x y_i &= S_i(y_i), && i \in \mathscr{I}, \\ E_i(y_i)(n_j) &= E_k(y_k)(n_j), && j \in \mathscr{J}_\pi,\ i,k \in \mathscr{I}_j, \\ \sum_{i\in\mathscr{I}_j} d_{ij} Q_i(y_i)(n_j) &= 0, && j \in \mathscr{J} \setminus \mathscr{J}_\beta, \\ C_j(y_i(n_j), y_k(n_j), s_j, u_j) &= 0, && j \in \mathscr{J}_\alpha,\ i,k \in \mathscr{I}_j, \\ B_i(y_i)(n_j) &= u_j, && j \in \mathscr{J}_\beta, i \in \mathscr{J}_j, \\ s^i(t) \in \{1,2,3,4\}, s^j(t) &\in \{0,1\}, && i \in \mathscr{I},\ j \in \mathscr{J}_\alpha, \\ y_i(\cdot,0) &= y_{i0}, && i \in \mathscr{I}, \end{aligned} \tag{5.33}$$

for $x \in (0, 1), t \in (0, T)$. We further define the cost functional

$$I(y, u, s) := \sum_{i \in \mathscr{I}} \int_0^T \int_0^1 I_i(y_i)\,\mathrm{d}x\,\mathrm{d}t + \sum_{j \in \mathscr{I}_\alpha} \int_0^T s_j(t)\psi_j^1(u_j) + (1 - s_j(t))\psi_j^0(u_j)\,\mathrm{d}t$$
$$+ \sum_{j \in \mathscr{I}_\alpha} \int_0^T \varphi(s_j(t))\,\mathrm{d}t + \sum_{j \in \mathscr{I}_\alpha \cup \mathscr{I}_\beta} \int_0^T \|u_j(t)\|^2\,\mathrm{d}t$$

and the bounds

$$\Xi(s) := \{(y, u) : y_i^-(s) \le y_i \le y_i^+(s),\ u_j^-(s) \le u_j \le u_j^+(s),\ i \in \mathscr{I},\ j \in \mathscr{I}_\alpha \cup \mathscr{I}_\beta\}$$

on the state y and the continuous control variables u, which depends on the discrete control s. With this notation, the discrete-continuous optimal control problem reads

$$\min_{(y,u)\in\Xi(s)} \quad I(y, u, s) \quad \text{s.t.} \quad (y, u, s) \text{ satisfies (5.33).} \tag{5.34}$$

Remark 5.11 We note some related work:

- The problem belongs to the class of mixed-integer optimal control problems (MIOCP) with partial differential equations. The notion of optimal switching control problems, mixed-integer dynamic optimization problems, and hybrid optimal control problems are also used for this and related problem classes; for a discussion see [43, 47, 68].
- If the PDE model remains fixed, with e.g., $s_i \equiv 1$ or $s_i \equiv 2$, the problem reduces to optimal boundary control problems with hyperbolic PDE constraints and switched boundary data; see [44, 65, 66] for related work addressing scalar cases.
- Full discretization turns the problem into a (typically very large) mixed-integer nonlinear problem (MINLP). In the stationary case, i.e., $s_i \equiv 4$, or in the case of very coarse discretizations, these can be solved using structure exploiting algorithms; see Sect. 5.4.3. However, this approach suffers from the curse of dimensionality when discretization step sizes are reduced to fully resolve the spatiotemporal dynamics of the system. We are therefore interested in new approaches for solving such problems, possibly on the level of semi-discretizations (spatial or temporal), cf. Sect. 5.3.3, and using continuous optimality conditions for appropriate subproblems. We outline such an approach in Sect. 5.4.4 below. We note that for fixed discrete controls the problem can be approached via optimality conditions, see e.g., [73] for the scalar case.

5.4.3 Exemplary Computational Results for Special Cases

In this subsection, we discuss computational results for special cases to give an overview of what is the state of the art for the applications discussed in Sect. 5.2.

We use two examples: One from gas network optimization, where we show what state-of-the-art MINLP methods can achieve on stationary problems and one from fresh water network optimization, where we show how instationary problems can be tackled. In both cases, the solution approach is based on discretization and piecewise linearization as outlined in Sect. 5.3.3.

In the gas network setting, we discuss some of the results of [30]. Here, we consider the network given in Fig. 5.2. It is a real-world network operated by Open Grid Europe GmbH and consists of 4189 passive and boundary nodes, whereof 976 are used as boundary nodes. These nodes are connected by 3550 pipes. Additionally, the network contains roughly 1000 non-pipe elements, notably 12 groups of compressors and 401 valves.

In [30], the authors implement a piecewise linearization technique as discussed in Sect. 5.3.3 for the stationary model (M^4 and F^4 in our hierarchy) and combine it with an alternating direction method to compute accurate gas quality parameters (more precisely, the calorific value). The method was tested on 33 real-world load scenarios provided by Open Grid Europe GmbH. The results are shown in Table 5.1. Here the columns $\|\Delta_P\|_\infty$, $\|\Delta_P^{\text{rel}}\|_\infty$, $\|\Delta_\pi\|_\infty$ show different error measures to evaluate the quality of the solutions (in order: absolute error in the computed power, the relative error in the computed power, and absolute error in the squared pressures). The column N shows the number of iterations needed in the alternating direction method.

In the fresh water network example, we discuss one result of Chap. 4 of [61]. Here, the network used is shown in Fig. 5.5. It consists of 16 pipes of 10.5 km total length, 3 pumps, and 2 valves. There are also four storage tanks, which are not part of the models discussed here. The load scenario is given in Fig. 5.6. As pipe model the water hammer equations (5.11) are used, i.e., M^2 and F^2 in our model hierarchy. The optimal control problem is to be solved for a time horizon of one day with a time step size of one hour. After discretization and piecewise linearization, the resulting mixed-integer linear problem has a size of 25077 variables (10839 binary) and 25000 constraints and 19310 variables (6401 binary). The solution time for this mixed-integer linear problem is then 41 s using standard solvers. For further details on the methods used, we refer to Chap. 3 in [61]. This shows that for small networks, such methods can be used to compute solutions of discrete-continuous control problems. To achieve the goal to compute controls for larger networks in real time these methods need to be refined or other methods need to be developed to achieve a synthesis of the discrete and continuous aspects of the considered problems. The idea of such a synthesis is outlined in the following section.

Table 5.1 Computational results (taken from [30]) for the L-gas network of Open Grid Europe GmbH; see Fig. 5.2

Instance	$\|\Delta_P\|_\infty$	$\|\Delta_P^{\mathrm{rel}}\|_\infty$	$\|\Delta_\pi\|_\infty$	N	Time (s)
L-01	4.21×10^{-1}	0.0257	0.00	4	4131
L-02	3.63×10^{-2}	0.0000	0.00	4	943
L-03	6.75×10^{-2}	0.0000	0.00	4	536
L-04	3.76×10^{-1}	0.0151	0.00	3	460
L-05	6.64×10^{-2}	0.0000	0.00	3	313
L-06	6.61×10^{-2}	0.0000	0.00	3	590
L-07	6.72×10^{-2}	0.0000	0.00	3	1089
L-08	2.34×10^{-1}	0.0029	0.00	4	2774
L-09	5.12×10^{-1}	0.0022	0.00	4	3968
L-10	2.58×10^{-1}	0.0095	0.00	4	1514
L-11	2.38×10^{-1}	0.0312	0.00	3	1152
L-12	4.53×10^{-2}	0.0000	0.00	4	2752
L-13	8.38×10^{-1}	0.0110	0.00	3	2637
L-14	1.83	0.0111	0.00	3	1617
L-15	1.81×10^{-2}	0.0000	0.00	6	2671
L-16	2.49×10^{-1}	0.0028	0.00	3	1647
L-17	5.52×10^{-1}	0.0110	0.00	3	1697
L-18	4.93×10^{-2}	0.0000	0.00	5	3940
L-19	1.82	0.0472	0.00	3	2148
L-20	2.74×10^{-1}	0.0124	0.00	3	2423
L-21	8.79×10^{-1}	0.0111	0.00	3	2569
L-22	7.78×10^{-1}	0.0111	0.00	3	2127
L-23	4.03×10^{-2}	0.0000	0.00	4	1762
L-24	2.55×10^{-1}	0.0113	0.00	3	2432
L-25	2.45×10^{-1}	0.0688	0.00	3	3090
L-26	2.71×10^{-2}	0.0000	0.00	5	1705
L-27	2.27×10^{-2}	0.0000	0.00	5	1175
L-28	4.45×10^{-1}	0.0096	0.00	3	1473
L-29	3.72×10^{-1}	0.0624	0.00	3	1741
L-30	4.68×10^{-2}	0.0000	0.00	4	2215
L-31	1.17×10^{-1}	0.0061	0.00	5	3857
L-32	2.97×10^{-2}	0.0000	0.00	4	1692
L-33	3.64×10^{-1}	0.0383	0.00	3	1805

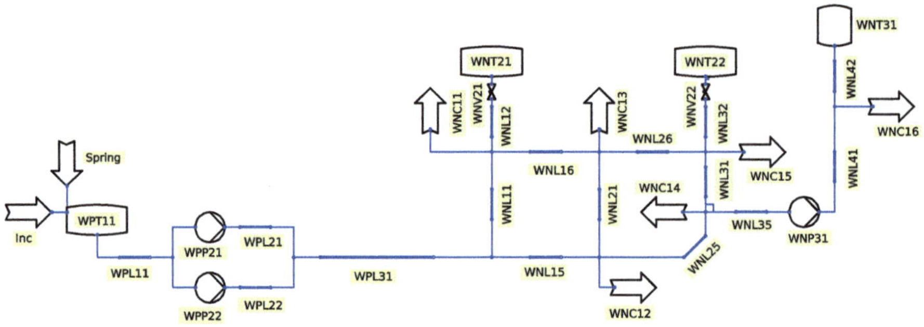

Fig. 5.5 An exemplary fresh water network (taken from [61])

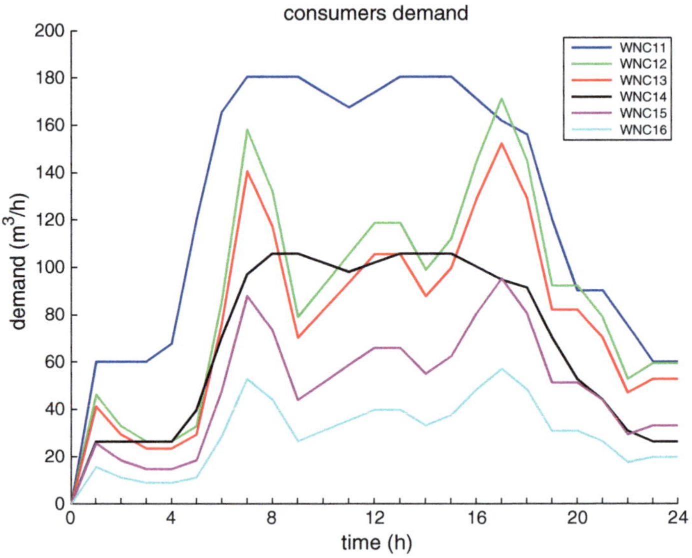

Fig. 5.6 Load scenario for the water network of Fig. 5.5 (from [61])

5.4.4 A Decomposition Approach for Discrete-Continuous Optimal Control

In what follows, we decompose Problem (5.34) along the continuous and discrete controls in order to set up an iterative framework. In addition, one may also need to decompose the network into small subnetworks, possibly consisting of single pipes. This is the approach of domain decomposition. In optimization and control for systems on metric graphs, domain decompositions should not only be applied for the sake of simulation but rather also for optimization. In the ideal case, after decomposition, we arrive at a fully parallel set of optimization problems to solve. Such strategies are known for elliptic and linear hyperbolic equations; see [54] for a general reference.

Decomposition is also possible on the level of time so that in principle small time-space units can be considered in an iterative framework. Using the abbreviation $\mathscr{K} := \{0, \ldots, K-1\}$, the optimal control problem (5.34) after time discretization reads, with $x \in (0, 1)$,

$$
\begin{aligned}
\min_{y,u,s} \quad & \sum_{i \in \mathscr{I}} \sum_{\kappa=0}^{K-1} \int_0^1 I_i(y_{i,\kappa+1})\, \mathrm{d}x \\
& + \sum_{j \in \mathscr{J}_\alpha} \sum_{\kappa=0}^{K-1} s_{j,\kappa+1} \psi_j^1(u_{j,\kappa+1}) + (1 - s_{j,\kappa+1}) \psi_j^0(u_{j,\kappa+1}) \\
& + \sum_{j \in \mathscr{J}_\alpha} \sum_{\kappa=0}^{K-1} \varphi(s_{j,\kappa+1}) + \sum_{j \in \mathscr{J}_\alpha \cup \mathscr{J}_\beta} \sum_{\kappa=0}^{K-1} \|u_{j,\kappa+1}\|^2 \\
\text{s.t.} \quad & M_i^{s_i} y_{i,\kappa+1} + \Delta t_\kappa \tilde{A}_i^{s_i}(y_{i,\kappa+1}) \partial_x y_{i,\kappa+1} = \Delta t_\kappa \tilde{S}_i(y_{i,\kappa+1}) + y_{i,\kappa}, \quad i \in \mathscr{I},\ \kappa \in \mathscr{K}, \\
& \tilde{E}_i(y_{i,\kappa+1})(n_j) = \tilde{E}_k(y_{k,\kappa+1})(n_j), \quad j \in \mathscr{J}_\pi,\ i, k \in \mathscr{I}_j,\ \kappa \in \mathscr{K}, \\
& \sum_{i \in \mathscr{I}_j} d_{ij} Q_i(y_{i,\kappa+1})(n_j) = 0, \quad j \in \mathscr{J} \setminus \mathscr{J}_\beta,\ \kappa \in \mathscr{K}, \\
& \tilde{C}_j(y_{i,\kappa+1}(n_j), y_{k,\kappa+1}(n_j), s_{j,\kappa+1}, u_{j,\kappa+1}) = 0, \quad j \in \mathscr{J}_\alpha,\ i, k \in \mathscr{I}_j,\ \kappa \in \mathscr{K}, \\
& \tilde{B}_i(y_{i,\kappa+1})(n_j) = u_i, \quad j \in \mathscr{J}_\beta,\ i \in \mathscr{J}_j,\ \kappa \in \mathscr{K}, \\
& y_{i,0}(\cdot, 0) = y_{i0}, \quad i \in \mathscr{I}, \\
& (y_{\kappa+1}, u_{\kappa+1}) \in \Xi(s_{\kappa+1}), \quad \kappa \in \mathscr{K},
\end{aligned}
$$

It is clear that the problem above involves all time steps in the cost functional. As a matter of fact, even for this discrete-time optimization problem, no published method seems to be available and the development of solution techniques for this setting is an open and great challenge. Thus, at this point in time, we can only utilize solutions for stationary problems. To this aim, we consider what has come to be known as rolling horizon control or instantaneous control. The latter amounts to reduce the sums in the cost functional of the discrete-time problem to a single time step of the discretization. Thus, for each $\kappa \in \mathscr{K}$ and given $y_{i,\kappa}$ we consider the problem

$$
\begin{aligned}
\min \quad & \sum_{i \in \mathscr{I}} \int_0^1 I_i(y_{i,\kappa+1})\,\mathrm{d}x + \sum_{j \in \mathscr{J}_\alpha} s_{j,\kappa+1}\psi_j^1(u_{j,\kappa+1}) + (1 - s_{j,\kappa+1})\psi_j^0(u_{j,\kappa+1}) \\
& + \sum_{j \in \mathscr{J}_\alpha} \varphi(s_{j,\kappa+1}) + \sum_{j \in \mathscr{J}_\alpha \cup \mathscr{J}_\beta} \|u_{j,\kappa+1}\|^2 \\
\text{s.t.} \quad & M_i^{s_i} y_{i,\kappa+1} + \Delta t_\kappa \tilde{A}_i^{s_i}(y_{i,\kappa+1})\partial_x y_{i,\kappa+1} = \Delta t_\kappa \tilde{S}_i(y_{i,\kappa+1}) + y_{i,\kappa}, \quad i \in \mathscr{I}, \\
& \tilde{E}_i(y_{i,\kappa+1})(n_j) = \tilde{E}_k(y_{k,\kappa+1})(n_j), \quad j \in \mathscr{J}_\pi,\ i,k \in \mathscr{I}_j, \\
& \sum_{i \in \mathscr{I}_j} d_{ij} Q_i(y_{i,\kappa+1})(n_j) = 0, \quad j \in \mathscr{J} \setminus \mathscr{J}_\beta, \\
& \tilde{C}_j(y_{i,\kappa+1}(n_j), y_{k,\kappa+1}(n_j), s_{j,\kappa+1}, u_{j,\kappa+1}) = 0, \quad j \in \mathscr{J}_\alpha,\ i,k \in \mathscr{I}_j, \\
& \tilde{B}_i(y_{i,\kappa+1})(n_j) = u_i, \quad j \in \mathscr{J}_\beta,\ i \in \mathscr{J}_j, \\
& (y_{\kappa+1}, u_{\kappa+1}) \in \Xi(s_{\kappa+1}),
\end{aligned}
\tag{5.35}
$$

where $x \in (0, 1)$ and where we optimize over $y_{\kappa+1}, u_{\kappa+1}, s_{\kappa+1}$. Problem (5.35) is a nonlinear optimization problem that is constrained by a system of ordinary differential equations on a graph. It contains discrete control variables $s_{\kappa+1}$ and continuous control variables $u_{\kappa+1}$. Thus, (5.35) is still in the format of a mixed-integer optimal control problem (MIOCP); cf. Remark 5.11. For the rest of this section, we give a sketch of a two-stage method that may be used to solve problems like (5.35). Our aim is to decompose the problem such that we have two problems that are easier to solve and that allow to design iterative algorithms with convergence or termination guarantees. To this end, we set up a master problem that optimizes the discrete control variables s_j, $j \in \mathscr{J}_\alpha$, for fixed continuous control variables u_j, $j \in \mathscr{J}_\alpha \cup \mathscr{J}_\beta$, and a subproblem that optimizes a continuous control u given a fixed discrete control s.

Typically, optimizing with respect to discrete controls is harder than optimizing with respect to continuous controls. This is why one often wants to simplify the physical model of the master problem. This model may be chosen as, e.g., $s_i = 4$ for all $i \in \mathscr{I}$, yielding M_i^4, $\tilde{A}_i^4$. Once this MIOCP is solved for (y, s), the optimal switching structure is delivered to the subproblem, where the more complicated physical model, i.e., $s_i < 4$ for all $i \in \mathscr{I}$, is optimized with respect to the continuous control variables u and a new state y. The optimal state of the master problem will typically be infeasible for the subproblem. Thus, there will be an error and one has to design a mechanism that drives this error to zero in the course of an iterative algorithm.

For a more detailed discussion, we now state the master and the subproblem. The master problem is obtained by (5.35) with the continuous control u fixed to $\bar{u}$. Moreover, we assume that the data $y_{i,\kappa}$ for all $i \in \mathscr{I}$ from the last time step is given. This yields the optimization problem

$$\begin{aligned}
\min \quad & \sum_{i\in\mathscr{I}} \int_0^1 I_i(y_{i,\kappa+1})\,\mathrm{d}x + \sum_{j\in\mathscr{J}_\alpha} s_{j,\kappa+1}\psi_j^1(\bar{u}_{j,\kappa+1}) + (1-s_{j,\kappa+1})\psi_j^0(\bar{u}_{j,\kappa+1}) \\
& + \sum_{j\in\mathscr{J}_\alpha} \varphi(s_{j,\kappa+1}) \\
\text{s.t.} \quad & M_i^{S_i} y_{i,\kappa+1} + \Delta t_\kappa \tilde{A}_i^{S_i}(y_{i,\kappa+1})\partial_x y_{i,\kappa+1} = \Delta t_\kappa \tilde{S}_i(y_{i,\kappa+1}) + y_{i,\kappa}, \quad i\in\mathscr{I}, \\
& \tilde{E}_i(y_{i,\kappa+1})(n_j) = \tilde{E}_k(y_{k,\kappa+1})(n_j), \quad j\in\mathscr{J}_\pi,\ i,k\in\mathscr{I}_j, \\
& \sum_{i\in\mathscr{I}_j} d_{ij} Q_i(y_{i,\kappa+1})(n_j) = 0, \quad j\in\mathscr{J}\setminus\mathscr{J}_\beta, \\
& \tilde{C}_j(y_{i,\kappa+1}(n_j), y_{k,\kappa+1}(n_j), s_{j,\kappa+1}, \bar{u}_{j,\kappa+1}) = 0, \quad j\in\mathscr{J}_\alpha,\ i,k\in\mathscr{I}_j, \\
& \tilde{B}_i(y_{i,\kappa+1})(n_j) = \bar{u}_i, \quad j\in\mathscr{J}_\beta,\ i\in\mathscr{J}_j, \\
& (y_{\kappa+1}, \bar{u}_{\kappa+1}) \in \Xi(s_{\kappa+1})
\end{aligned} \tag{5.36}$$

in $y_{\kappa+1}$ and $s_{\kappa+1}$. Let now $(\hat{y}, \hat{s})$ be an optimal pair of (5.36) for fixed $u = \bar{u}$. The subproblem (in the continuous variables $y_{\kappa+1}$ and $u_{\kappa+1}$ and for given y_κ) is then given by

$$\begin{aligned}
\min \quad & \sum_{i\in\mathscr{I}} \int_0^1 I_i(y_{i,\kappa+1})\,\mathrm{d}x + \sum_{j\in\mathscr{J}_\alpha} \hat{s}_{j,\kappa+1}\psi_j^1(u_{j,\kappa+1}) + (1-\hat{s}_{j,\kappa+1})\psi_j^0(u_{j,\kappa+1}) \\
& + \sum_{j\in\mathscr{J}_\alpha\cup\mathscr{J}_\beta} \|u_{j,\kappa+1}\|^2 \\
\text{s.t.} \quad & M_i^{S_i} y_{i,\kappa+1} + \Delta t_\kappa \tilde{A}_i^{S_i}(y_{i,\kappa+1})\partial_x y_{i,\kappa+1} = \Delta t_\kappa \tilde{S}_i(y_{i,\kappa+1}) + y_{i,\kappa}, \quad i\in\mathscr{I}, \\
& \tilde{E}_i(y_{i,\kappa+1})(n_j) = \tilde{E}_k(y_{k,\kappa+1})(n_j), \quad j\in\mathscr{J}_\pi,\ i,k\in\mathscr{I}_j, \\
& \sum_{i\in\mathscr{I}_j} d_{ij} Q_i(y_{i,\kappa+1})(n_j) = 0, \quad j\in\mathscr{J}\setminus\mathscr{J}_\beta, \\
& \tilde{C}_j(y_{i,\kappa+1}(n_j), y_{k,\kappa+1}(n_j), \hat{s}_{j,\kappa+1}, u_{j,\kappa+1}) = 0, \quad j\in\mathscr{J}_\alpha,\ i,k\in\mathscr{I}_j, \\
& \tilde{B}_i(y_{i,\kappa+1})(n_j) = u_i, \quad j\in\mathscr{J}_\beta,\ i\in\mathscr{J}_j, \\
& (y_{\kappa+1}, u_{\kappa+1}) \in \Xi(\hat{s}_{\kappa+1}),
\end{aligned} \tag{5.37}$$

where we fixed the discrete control s to $\hat{s}$.

We now receive an optimal pair (y^*, u^*) for the continuous nonlinear optimal control problem (5.37) and the errors $e_y := \|\hat{y} - y^*\|$ and $e_u := \|\bar{u} - u^*\|$. Clearly, in the next iteration we set $\bar{u} = u^*$.

If we neglect that we would like to choose different models for our hierarchy of ODEs in the master and subproblem, we mainly constructed a primal alternating direction method: We split the variables and solved the problem for one block of the variables, fixed the result, and solved the problem for the other block of the variables. Such an iterative procedure is closely related to general alternating direction methods (ADMs). ADMs have originally been proposed in the context of nonlinear variational problems in [27, 33] and have been also used recently for the optimization of large-scale real-world mixed-integer stationary gas transport problems; see, Sect. 5.4.3 and, e.g., [30, 31].

Another way to interpret the sketched iterative procedure is as a method related to generalized Benders decomposition; see [8, 32]. However, some additional assumptions must be made and some additional techniques have to be designed if one wants to embed the decomposition in a Benders-like framework. First of all, the master problem has to be a relaxation of the overall problem. This is not given if one simply chooses a coarser physics model in (5.36), since this does not translate into an embedding of the corresponding feasible sets. A possible remedy would be to use a relaxation, e.g., given by a suitably chosen outer approximation; see [23, 24]. Additionally, we also have to construct Benders-like feasibility cuts (in the case of an infeasible subproblem for a given discrete control $\hat{s}$) and optimality cuts (in case of a feasible subproblem). Since the overall problem, as well as both the master and the subproblem, are inherently nonconvex, standard Benders cuts are not globally valid and one thus has to derive problem-specific cuts; see [69].

Remark 5.12 The program outlined above is widely open. No general procedure is known, no convergence results shown on this general level. This can safely be said to be an open challenge for the discrete-continuous optimization community. More specifically, one has to answer the following questions:

- Consider a master problem that—after suitable relaxation of the ODE—is a mixed-integer linear or nonlinear problem (MIP or MINLP) and that can be solved to global optimality. Assume further that the subproblem can be solved to global optimality as well. Under which conditions is it true that the alternation between master problem and subproblem converges and if it does, is the solution globally optimal?
- What is the right way to introduce Benders-like cuts in the master problem in order to take into account (in)feasibility of the subproblem?
- Can one provide special examples for this Benders-type decomposition, where the questions above can be answered positively!

Alluding to the last point, we can provide a first result in [39], where the authors exploit MIP and MINLP techniques that have been intensively discussed in [25, 62, 63, 67] and [20, 29, 60] in the context of gas transport problems. A more general but related approach is given in the recent paper [11].

Acknowledgements The authors thank the Deutsche Forschungsgemeinschaft for their support within projects A03, A05, B07, and B08 in the Sonderforschungsbereich/Transregio 154 *Mathematical Modeling, Simulation and Optimization using the Example of Gas Networks*. In addition, parts of this research were performed as part of the Energie Campus Nürnberg and supported by funding through the "Aufbruch Bayern (Bavaria on the move)" initiative of the state of Bavaria.

References

1. Abreu, J., Cabrera, E., Izquierdo, J., García-Serra, J.: Flow modeling in pressurized systems revisited. J. Hydraul. Eng. **125**(11), 1154–1169 (1999)
2. Adimurthi, Veerappa Gowda, G.D.: Conservation law with discontinuous flux. J. Math. Kyoto Univ. **43**(1), 27–70 (2003)
3. Adimurthi, Mishra, S.: Optimal entropy solutions for conservation laws with discontinuous flux-functions. J. Hyperbolic Differ. Equ. **2**(4), 783–837 (2005)
4. Amin, S., Hante, F.M., Bayen, A.M.: On stability of switched linear hyperbolic conservation laws with reflecting boundaries. In: Hybrid Systems: Computation and Control, vol. 4981. Lecture Notes in Computer Science, pp. 602–605. Springer (2008)
5. Amin, S., Hante, F.M., Bayen, A.M.: Stability analysis of linear hyperbolic systems with switching parameters and boundary conditions. In: 47th IEEE Conference on Decision and Control, 2008, CDC 2008, pp. 2081–2086. IEEE (2008)
6. Amin, S., Hante, F.M., Bayen, A.M.: Exponential stability of switched linear hyperbolic initial-boundary value problems. IEEE Trans. Autom. Control **57**(2), 291–301 (2012)
7. Bastin, G., Coron, J.-M.: On boundary feedback stabilization of non-uniform linear 2×2 hyperbolic systems over a bounded interval. Syst. Control Lett. **60**(11), 900–906 (2011)
8. Jacques, F.: Benders. Partitioning procedures for solving mixed-variables programming problems. Numerische Mathematik **4**(1), 238–252 (1962)
9. Bressan, A., Shen, W.: Optimality conditions for solutions to hyperbolic balance laws. In: Control Methods in PDE-Dynamical Systems, vol. 426. Contemporary Mathematics, pp. 129–152. American Mathematical Society, Providence, RI (2007)
10. Brouwer, J., Gasser, I., Herty, M.: Gas pipeline models revisited: model hierarchies, nonisothermal models, and simulations of networks. Multiscale Model. Simul. **9**(2), 601–623 (2011)
11. Buchheim, C., Meyer, C., Schäfer, R.: Combinatorial optimal control of semilinear elliptic PDEs. Technical report, Fakultät für Mathematik, TU Dortmund (2015)
12. Corless, R.M., Gonnet, G.H., Hare, D.E.G., Jeffrey, D.J., Knuth, D.E.: On the Lambert *W* function. Adv. Comput. Math. **5**(4), 329–359 (1996)
13. Coron, J.-M., Glass, O., Wang, Z.: Exact boundary controllability for 1-D quasilinear hyperbolic systems with a vanishing characteristic speed. SIAM J. Control Optim. **48**(5), 3105–3122 (2009)
14. Coron, J.-M., Vazquez, R., Krstic, M., Bastin, G.: Local exponential H^2 stabilization of a 2×2 quasilinear hyperbolic system using backstepping. SIAM J. Control Optim. **51**(3), 2005–2035 (2013)
15. Daafouz, J., Di Benedetto, M.D., Blondel, V.D., Ferrari-Trecate, G., Hetel, L., Johansson, M., Juloski, A.L., Paoletti, S., Pola, G., De Santis, E., Vidal, R.: Switched and piecewise affine systems. In: Handbook of Hybrid Systems Control, pp. 87–137. Cambridge University Press, Cambridge (2009)
16. Dáger, R., Zuazua, E.: Wave Propagation, Observation and Control in 1-*d* Flexible Multi-Structures, vol. 50. Mathématiques & Applications (Berlin). Springer, Berlin (2006)
17. Dick, M., Gugat, M., Herty, M., Leugering, G., Steffensen, S., Wang, K.: Stabilization of networked hyperbolic systems with boundary feedback. In: Trends in PDE Constrained Optimization, vol. 165. International Series of Numerical Mathematics, pp. 487–504. Birkhäuser/Springer, Cham (2014)
18. Dick, M., Gugat, M., Leugering, G.: Classical solutions and feedback stabilization for the gas flow in a sequence of pipes. Netw. Heterog. Media **5**(4), 691–709 (2010)
19. Dick, M., Gugat, M., Leugering, G.: A strict H^1-Lyapunov function and feedback stabilization for the isothermal Euler equations with friction. Numer. Algebra Control Optim. **1**(2), 225–244 (2011)
20. Domschke, P., Geißler, B., Kolb, O., Lang, J., Martin, A., Morsi, A.: Combination of nonlinear and linear optimization of transient gas networks. INFORMS J. Comput. **23**(4), 605–617 (2011)
21. Domschke, P., Kolb, O., Lang, J.: Adjoint-based error control for the simulation and optimization of gas and water supply networks. Appl. Math. Comput. **259**, 1003–1018 (2015)

22. Dos Santos, V., Bastin, G., Coron, J.-M., d'Andréa Novel, B.: Boundary control with integral action for hyperbolic systems of conservation laws: stability and experiments. Autom. J. IFAC **44**(5), 1310–1318 (2008)
23. Duran, M.A., Grossmann, I.E.: An outer-approximation algorithm for a class of mixed-integer nonlinear programs. Math. Program. **36**(3), 307–339 (1986)
24. Fletcher, R., Leyffer, S.: Solving mixed integer nonlinear programs by outer approximation. Math. Program. **66**(1), 327–349 (1994)
25. Fügenschuh, A., Göttlich, S., Herty, M., Klar, A., Martin, A.: A discrete optimization approach to large scale supply networks based on partial differential equations. SIAM J. Sci. Comput. **30**(3), 1490–1507 (2008)
26. Fügenschuh, A., Geißler, B., Gollmer, R., Morsi, A., Pfetsch, M.E., Rövekamp, J., Schmidt, M., Spreckelsen, K., Steinbach, M.C.: Physical and technical fundamentals of gas networks. In: Koch, et al. [51], Chap. 2, pp. 17–44
27. Gabay, D., Mercier, B.: A dual algorithm for the solution of nonlinear variational problems via finite element approximation. Comput. Math. Appl. **2**(1), 17–40 (1976)
28. Geißler, B., Martin, A., Morsi, A., Schewe, L.: Using piecewise linear functions for solving MINLPs. In: Lee, J., Leyffer, S. (eds.) Mixed Integer Nonlinear Programming. The IMA Volumes in Mathematics and its Applications, vol. 154, pp. 287–314. Springer, New York (2012)
29. Geißler, B., Kolb, O., Lang, J., Leugering, G., Martin, A., Morsi, A.: Mixed integer linear models for the optimization of dynamical transport networks. Math. Methods Oper. Res. **73**(3), 339–362 (2011)
30. Geißler, B., Morsi, A., Schewe, L., Schmidt, M.: Solving power-constrained gas transportation problems using an MIP-based alternating direction method. Comput. Chem. Eng. **82**, 303–317 (2015)
31. Geißler, B., Morsi, A., Schewe, L., Schmidt, M.: Solving highly detailed gas transport MINLPs: block separability and penalty alternating direction methods. Technical report, 07 (2016)
32. Geoffrion, A.M.: Generalized benders decomposition. J. Optim. Theory Appl. **10**(4), 237–260 (1972)
33. Glowinski, R., Marroco, A.: Sur l'approximation, par éléments finis d'ordre un, et la résolution, par pénalisation-dualité d'une classe de problèmes de dirichlet non linéaires. ESAIM: Mathematical Modelling and Numerical Analysis - Modélisation Mathématique et Analyse Numérique, **9**(R2), 41–76 (1975)
34. Gugat, M., Leugering, G.: Global boundary controllability of the de St. Venant equations between steady states. Annales de l'Institut Henri Poincare (C) Non Linear Analysis **20**(1), 1–11 (2003)
35. Gugat, M., Leugering, G.: Global boundary controllability of the Saint-Venant system for sloped canals with friction. Annales de l'Institut Henri Poincare (C) Non Linear Analysis **26**(1), 257–270 (2009)
36. Gugat, M., Dick, M., Leugering, G.: Stabilization of the gas flow in star-shaped networks by feedback controls with varying delay. In: System Modeling and Optimization, vol. 391. IFIP Advances in Information and Communication Technology, pp. 255–265. Springer, Heidelberg (2013)
37. Gugat, M., Hante, F.M., Hirsch-Dick, M., Leugering, G.: Stationary states in gas networks. Netw. Heterog. Media **10**(2), 295–320 (2015)
38. Gugat, M., Leugering, G., Georg Schmidt, E.J.P.: Global controllability between steady supercritical flows in channel networks. Math. Methods Appl. Sci. **27**(7), 781–802 (2004)
39. Gugat, M., Leugering, G., Martin, A., Schmidt, M., Sirvent, M., Wintergerst, D.: Towards simulation based mixed-integer optimization with differential equations. Technical report, Preprint FAU (2016, submitted)
40. Gugat, M., Leugering, G., Schittkowski, K., Georg Schmidt, E.J.P.: Modelling, stabilization, and control of flow in networks of open channels. In: Online Optimization of Large Scale Systems, pp. 251–270. Springer, Berlin (2001)
41. Gugat, M., Schultz, R., Wintergerst, D.: Networks of pipelines for gas with nonconstant compressibility factor: stationary states. In: Computational and Applied Mathematics, pp. 1–32 (2016)

42. Gugat, M., Wintergerst, D.: Finite time blow-up of traveling wave solutions for the flow of real gas through pipeline networks (2016)
43. Hante, F.M.: Relaxation methods for hyperbolic PDE mixed-integer optimal control problems. ArXiv e-prints, 09 (2015)
44. Hante, F.M., Leugering, G.: Optimal boundary control of convention-reaction transport systems with binary control functions. Hybrid Systems: Computation and Control. Lecture Notes in Computer Science, vol. 5469, pp. 209–222. Springer, Berlin (2009)
45. Hante, F.M., Leugering, G., Seidman, T.I.: Modeling and analysis of modal switching in networked transport systems. Appl. Math. Optim. **59**(2), 275–292 (2009)
46. Hante, F.M., Leugering, G., Seidman, T.I.: An augmented BV setting for feedback switching control. J. Syst. Sci. Complex. **23**(3), 456–466 (2010)
47. Hante, F.M., Sager, S.: Relaxation methods for mixed-integer optimal control of partial differential equations. Comput. Optim. Appl. **55**(1), 197–225 (2013)
48. Hante, F.M., Sigalotti, M.: Existence of common Lyapunov functions for infinite-dimensional switched linear systems. In: 49th IEEE Conference on Decision and Control (CDC) 2010, CDC 2010, pp. 5668–5673. IEEE (2010)
49. Hante, F.M., Sigalotti, M.: Converse Lyapunov theorems for switched systems in Banach and Hilbert spaces. SIAM J. Control Optim. **49**(2), 752–770 (2011)
50. Holden, H., Risebro, N.H.: Front tracking for hyperbolic conservation laws. In: Applied Mathematical Sciences, vol. 152, second edn. Springer, Heidelberg (2015)
51. Koch, T., Hiller, B., Pfetsch, M.E., Schewe, L. (eds.): Evaluating Gas Network Capacities. SIAM-MOS series on Optimization. SIAM (2015)
52. Lagnese, J.E., Leugering, G., Schmidt, E.J.P.G.: On the analysis and control of hyperbolic systems associated with vibrating networks. Proc. Roy. Soc. Edinburgh Sect. A **124**(1), 77–104 (1994)
53. Lagnese, J.E., Leugering, G., Schmidt, E.J.P.G.: Modeling, analysis and control of dynamic elastic multi-link structures. In: Systems & Control: Foundations & Applications. Birkhäuser Boston, Inc., Boston (1994)
54. Lagnese, J.E., Leugering, G.: Domain decomposition methods in optimal control of partial differential equations, vol. 148. International Series of Numerical Mathematics. Birkhäuser Verlag (2004)
55. Lamare, P.-O., Girard, A., Prieur, C.: Switching rules for stabilization of linear systems of conservation laws. SIAM J. Control Optim. **53**(3), 1599–1624 (2015)
56. Leugering, G., Georg Schmidt, E.J.P.: On the modelling and stabilization of flows in networks of open canals. SIAM J. Control Optim. **41**(1), 164–180 (2002)
57. LeVeque, R.J.: Numerical Methods for Conservation Laws. Birkhäuser (1992)
58. Le Veque, R.J.: Finite Volume Methods for Hyperbolic Problems. Cambridge University Press (2002)
59. Li, T.: Controllability and observability for quasilinear hyperbolic systems, vol. 3. AIMS Series on Applied Mathematics. American Institute of Mathematical Sciences (AIMS), Springfield, MO; Higher Education Press, Beijing (2010)
60. Mahlke, D., Martin, A., Moritz, S.: A mixed integer approach for time-dependent gas network optimization. Optim. Methods Softw. **25**(4–6), 625–644 (2010)
61. Martin, A., Klamroth, K., Lang, J., Leugering, G., Morsi, A., Oberlack, M., Ostrowski, M., Rosen, R., (eds.): Mathematical Optimization of Water Networks. Birkhäuser (2012)
62. Martin, A., Möller, M., Moritz, S.: Mixed integer models for the stationary case of gas network optimization. Math. Program. **105**(2–3, Ser. B), 563–582 (2006)
63. Morsi, A., Geißler, B., Martin, A.: Mixed integer optimization of water supply networks. In: Mathematical Optimization of Water Networks, vol. 162. International Series of Numerical Mathematics, pp. 35–54. Birkhäuser/Springer Basel AG, Basel (2012)
64. Nicaise, S.: Control and stabilization of 2x2 hyperbolic systems on graphs. Technical report, Université de Valenciennes et du Hainaut Cambrésis (2016). To appear in MCRF 2017
65. Pfaff, S., Ulbrich, S.: Optimal boundary control of nonlinear hyperbolic conservation laws with switched boundary data. SIAM J. Control Optim. **53**(3), 1250–1277 (2015)

66. Pfaff, S., Ulbrich, S., Leugering, G.: Optimal control of nonlinear hyperbolic conservation laws with switching. In: Trends in PDE constrained optimization, vol. 165. International Series of Numerical Mathematics, pp. 109–131. Birkhäuser/Springer, Cham (2014)
67. Pfetsch, M.E., Fügenschuh, A., Geißler, B., Geißler, N., Gollmer, R., Hiller, B., Humpola, J., Koch, T., Lehmann, T., Martin, A., Morsi, A., Rövekamp, J., Schewe, L., Schmidt, M., Schultz, R., Schwarz, R., Schweiger, J., Stangl, C., Steinbach, M.C., Vigerske, S., Willert, B.M.: Validation of nominations in gas network optimization: models, methods, and solutions. Optim. Methods Softw. **30**(1), 15–53 (2015)
68. Rüffler, F., Hante, F.M.: Optimal switching for hybrid semilinear evolutions. Nonlinear Anal. Hybrid Syst. **22**, 215–227 (2016)
69. Sahinidis, N.V., Grossmann, I.E.: Convergence properties of generalized benders decomposition. Comput. Chem. Eng. **15**(7), 481–491 (1991)
70. Smoller, J.: Shock Waves and Reaction-Diffusion Equations, vol. 258. Grundlehren der mathematischen Wissenschaften. Springer Verlag (1983)
71. Sontag, E.: From linear to nonlinear: some complexity comparisons. In: 1995 Proceedings of the 34th IEEE Conference on Decision and Control, vol. 3, pp. 2916–2920. IEEE (1995)
72. Sontag, E.D.: Nonlinear regulation: the piecewise linear approach. IEEE Trans. Autom. Control **26**(2), 346–358 (1981)
73. Ulbrich, S.: A sensitivity and adjoint calculus for discontinuous solutions of hyperbolic conservation laws with source terms. SIAM J. Control Optim. **41**(3), 740–797 (2002). (electronic)

Chapter 6
Imaging in Acute Ischemic Stroke and Stroke Outcome Prediction

Majaz Moonis

Abstract This invited talk is mainly based on joint work of Ahmedul Kabir, Carolina Ruiz, Sergio A. Alvarez, and Majaz Moonis. There are two parts of this talk. The first part deals with imaging in acute ischemic and second is devoted to a comparison of conventional Regression with Machine Learning Methods for Stroke Outcome Prediction. The second part is based on Kabir et al., Indian J. Indust. Appl. Math. 7(2), 12, 2016 [4].

Keywords CT scans · Magnetic resonance imaging · Acute ischemic stroke Stroke outcome · Supervised machine learning

6.1 Imaging in Ischemic Stroke

Imaging modalities have continued to evolve from the time of plain uniplanar X-rays to multiplanar CT scans and more recently Magnetic Resonance Imaging (MRI) in various neurological disorders yielding diagnostic signatures. For the role of mathematics in medical imaging, we refer to [1–3].

Perhaps, the most important evolution has been in recognizing and utilizing the properties of unpaired electrons and protons in the tissue. Unpaired electrons have an inherent spin which is not synchronized and when subjected to multiplanar strong magnetic fields, there is synchronization and depending on the magnetic pulse (90–180 degree), the axis of magnetization changes. Subsequently, when the radiofrequency pulse is stopped, realignment occurs, with the release of magnetic energy, captured in appropriately placed coils. Because water and solid tissue release energy in different time frames, a differential energy is captured in 2 sequences, T1 and T2 images. This in turn using FFA and other mathematical equations transform the raw images into recognizable MRI images [1, 2]. Adding contrast agents further adds to the process of enhancement especially in T1-based sequences. Other paradigms

M. Moonis (✉)
Director, Stroke Services, UMass Memorial Medical Center, Worcester, MA, USA
e-mail: majaz.moonis@umassmemorial.org

P. Manchanda et al. (eds.), *Industrial Mathematics and Complex Systems*, Industrial and Applied Mathematics, DOI 10.1007/978-981-10-3758-0_6

are used to superimpose these images into a FLAIR and GRE images to look for ischemic damage and the presence of blood products.

In the management of acute stroke, it is crucial to visualize brain arteries, and the dead tissue (ischemic core) as well as the tissue at risk of destruction, the PENUMBRA. The implications are that if there is a viable PENUMBRA, acute interventions, to rescue tissue at risk of infarction, is possible and forms the basis of acute stroke treatment.

Parallel to the MRI technology, multiplanar CAT scans [1, 3] were developed that are capable of similar results with the advantage of less scanning time but with the disadvantage of not being to visualize the very early stroke.

The controversy of CAT versus MRI in stroke management is an ongoing hot debate. The basic physics of these techniques in acute stroke management and the advantages of either technique over the other in this life threatening condition with some basic physics and mathematical algorithms that form the concepts of the underlying basis of these techniques are investigated [5].

6.2 Stoke Outcome Prediction

In performing statistical analysis in clinical trials, the conventional method is to conduct univariate analysis for variable selection, followed by multivariate logistic regression to find more details about the variables. While this approach is widely used and does deliver useful explanations of the data, it has several shortcomings. In [4], we compare the results obtained with conventional methods with those of alternative machine learning algorithms in terms of their ability to predict stroke outcome of patients. We also examine in this paper the different models constructed by the algorithms, and compare them with the logistic regression model. We find that machine learning can be used to predict more precise outcomes and reveal more variables than previously known that may play an important role in determining stroke outcomes.

In most clinical trials, the goal is to attain information about how different independent variables X1...n affect the value of a dependent variable Y. The conventional method to perform this sort of analysis in medical trials is a two-step process. In the first step, a univariate analysis is performed that assess the effects on the dependent variable of each independent variable individually. This step filters out some variables that have no statistically significant effect on the dependent variable. The second step is multivariate logistic regression which builds a model that takes into account all the independent variables that are chosen from the univariate analysis. The results obtained from multivariate logistic regression are particularly helpful from a clinical point of view because they facilitate the computation of odds ratios. The odds ratio represents the odds that an outcome will occur given the presence of a particular variable, compared to the odds of the outcome occurring in the absence of that variable.

Most clinical trials, including those related to stroke, use multivariate logistic regression to examine the effect of one or more treatments or conditions. There have been numerous such studies that deal with factors influencing stroke outcome, of which we mention a few. Moonis et al. [5] examined the effect of statins in treating ischemic stroke patients and reports that using statins improved stroke outcome since the statins obtained an odds ratio of 1.57 in a logistic regression model predicting mRS-90 Here we demonstrate the efficacy of machine learning algorithms for prediction and variable selection on the data of stroke patients. In the context of stroke, the dependent variable is a measure of stroke outcome, and the independent variables are different factors that may affect stroke recovery. Patient's data with known stroke outcomes are fed to the machine learning algorithm to construct a model, which is used to predict the future stroke outcome of patients. Kabir et al. [4] presented the results of our experiments with several variable selection and supervised learning algorithms.

6.3 Prediction Through Supervised Learning

Figure 6.1 presents the basic idea and process of prediction through supervised learning. The data provided for the predictive algorithms to work on are called training data. From the training data, the algorithms build models that can be used to make predictions about the value of the target (or dependent) variable. The test set contains unlabeled data that can then be used to evaluate how well the model performs. Ideally, the algorithm will be able to learn general rules or methods from known data that can be applied to unknown data.

Variable Selection Algorithms Used

Very often, a larger set of predictive variables does not equate to a better predictive model. The task of variable selection is to find a smaller subset of variables that build simpler models and are likely to have better generalization performance. The basic idea behind the two variable selection methods we have used in [4] are described below.

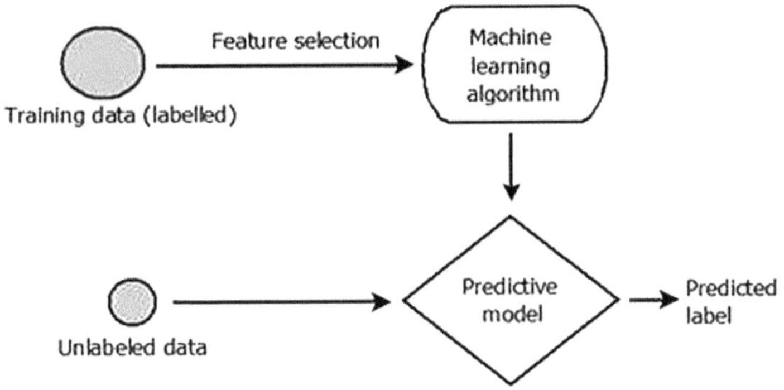

Fig. 6.1 The general process of supervised machine learning

Correlation-based Feature Selection (CFS)

This method evaluates the worth of a subset of variables by considering the individual predictive ability of each variable along with the degree of redundancy between them. The subset of variables with high correlation with the dependent variable and low internal correlation is selected. The advantage of this technique over univariate analysis is that CFS looks at the interactions between the variables, not just individual relationships with outcome.

ReliefF

The key idea of ReliefF is to estimate the quality of variables according to how well their values distinguish between the data points of the same and different classes that are close to each other. In this method, each variable is given equal initial weight and goes through an iterative process of calculation where the weights may increase or decrease. The weight of any given variable decreases if it differs from that variable in nearby instances of the same class more than nearby instances of the other class, and increases in the reverse case. After a certain number of iterations, the process stops and the variables are ranked per their weights. For variable selection, one may select the desired number of variables starting from the top of the list.

6.4 Methodology

Comparison of conventional regression with machine learning based methods in predicting stroke outcome is discussed in [4]. The comparison is done in two fronts: in assessing the predictive power of different models and in finding what the different models say about the variables involved. We also experiment with variable selection algorithms and evaluate their influence over the predictive ability of the algorithms. In our research, the statistics tool SPSS is used for conventional regression analysis, while the machine learning tool Weka is used for all the other machine learning algorithms.

We run four sets of experiments with our predictive algorithms: one with the full set of variables and the others with three separate variable selection methods. For the conventional univariate variable selection process, chi-square tests are performed for categorical variables, while Mann–Whitney U tests are performed for continuous variables in preference to a t test, after nonnormality of predictor variable distributions is determined by way of a Shapiro–Wilk test. Only the variables with a p value less than 0.05 are chosen for further analysis. For Correlation-based Feature Selection (CFS) the best subset of variables returned by the CFS algorithm is selected for use in supervised learning algorithms. ReliefF algorithm is also used, and the variables with ten largest weights are selected for further analysis.

The four separate sets of variables are then fed to the machine learning algorithms for predictive analysis. For logistic regression and Naïve Bayes, no particular parameter optimization is necessary. For C4.5 decision trees, we experiment with different values of the confidence factor and minimum number of objects in a node to try out trees of different sizes with a varied amount of pruning [6]. For Bayesian networks, we experiment with the maximum number of allowable parents of a node. We use random ordering of variables and set the network to not initialize as Naïve Bayes [7]. For k-nearest neighbors, the hold-one-out cross validation approach is used to select the best value of k. For the purpose of testing, tenfold cross validation is used. The results are evaluated based on the following two metrics: accuracy and F-measure.

6.5 Prediction Results

The results of predictive modeling with and without variable selection are shown in Table 6.1. The conventional method of univariate selection followed by multivariate regression has accuracy of 0.778 and F-measure of 0.774. Several other combinations of variable selection and prediction algorithms perform at least as good as that. However, tests to establish statistical significance did not find that there is any statistically significant improvement. The results indicate that variable selection enhances predictive ability overall, although the only statistically significant improvement is

Table 6.1 Comparison of different learning algorithms for different sets of chosen variables

Variable set	Algorithms	Accuracy	F-measure
Full set of variables	Logistic regression	0.773	0.768
	C4.5 Decision tree	0.778	0.775
	Naïve Bayes	0.769	0.768
	Bayesian network	0.752	0.750
	k-nearest neighbor	0.718	0.690
Variables selected by **Univariate analysis**	Logistic Regression	0.778	0.774
	C4.5 decision tree	0.778	0.776
	Naïve Bayes	0.769	0.768
	Bayesian network	0.763	0.762
	k-nearest neighbor	0.701	0.676
Variables selected by **CFS**	Logistic regression	0.789	0.785
	C4.5 decision tree	0.780	0.778
	Naïve Bayes	0.780	0.775
	Bayesian network	0.777	0.778
	k-nearest neighbor	0.797	0.796
Variables selected by **ReliefF**	Logistic regression	0.771	0.767
	C4.5 Decision tree	0.764	0.761
	Naïve Bayes	0.773	0.772
	Bayesian network	0.780	0.779
	k-nearest neighbor	0.707	0.699

found for k-nearest neighbor algorithm for CFS selected variables over the full set of variables (the k-nearest neighbor algorithm is very sensitive to the variable selection method used and yields significantly lower accuracy when paired with the selection methods other than CFS considered here).

6.6 Conclusions

We have described an experimental comparison of different machine learning algorithms with conventional univariate selection followed by multivariate logistic regression for the task of stroke outcome prediction in [4]. We have also examined the models to find additional insights about stroke recovery. We have worked with two different variable selection methods that consider sets of multiple variables simultaneously, as alternatives to traditional univariate selection. We have also applied several supervised learning algorithms before and after variable selection. It is found that the predictive performance of these algorithms can equal or surpass that of conventional logistic regression. Moreover, the models obtained from the supervised learning algorithm give useful additional information about the various factors affecting stroke outcome.

Regarding the significant negative effects of older age and higher NIHSS score at admission on stroke outcome, the decision tree and the Bayesian network models agree with the conventional logistic regression model. However, both these models discover the presence of atrial fibrillation as a factor for poor outcome, something the conventional model failed to do. The decision tree model also shows hemorrhagic conversion to be associated with poor stroke outcome, whereas the logistic regression model counterintuitively showed it to be associated with good outcome. In the Bayesian network model hypertension is directly linked with stroke outcome, although the conventional model did not deem it to be a significant variable. Therefore, the use of machine learning algorithms provide insights that are different from the conventional method, and thus can serve as auxiliary sources of information for decision-making in stroke treatment.

References

1. Dougherty, G.: Digital Image Processing for Medical Applications. Cambridge University Press, Cambridge (2009)
2. Epstein, C.L.: Introduction to the Mathematics of Medical Imaging, 2nd ed. SIAM, Philadelphia (2008)
3. Feernan, T.G.: The Mathematics of Medical Imaging. Springer, Berlin (2010)
4. Kabir, A., Ruiz, C., Alvarez, S.A., Moonis M.: Comparison of conventional regression with machine learning methods for stroke outcome prediction. Indian J. Indust. Appl. Math. **7**(2), 12 (2016)
5. Moonis, M., Kane, K., Schwiderski, U., Sandage, B.W., Fisher, M.: HMG-CoA reductase inhibitors improve acute ischemic stroke out come. Stroke **36**, 1298–1300 (2005)
6. Drazin, S., Montag, M.: Decision tree analysis using weka. Machine Learning-Project II, University of Miami (2012)
7. Bouckaert, R.R.: Bayesian network classifiers in weka (2004)

Chapter 7
Fourier Transforms of Multiplicative Convolutions

B.I. Golubov and S.S. Volosivets

Abstract We consider **P**-adic Fourier transform introduced by N.Y. Vilenkin that generalizes famous Walsh transform and **P**-adic convolution of functions defined on $\mathbb{R}_+$, where $\mathbf{P} = \{p_j\}_{j=1}^{\infty} \subset \mathbb{N}$ and $2 \le p_j \le N$ for all j. The order of decreasing to zero of remainder integral for **P**-adic Fourier transform is studied for convolutions of L^{q_1} and L^{q_2} functions. In the case $q_1 = q_2 = 2$ a characterization result is obtained. The sharpness of results is established.

Keywords Multiplicative Fourier transform · Multiplicative convolution Absolute integrability · Monotone function

2010 Mathematics subject classification Primary 42A38 · Secondary 42A85 · 44A35

7.1 Introduction

Let $\{p_n\}_{n=1}^{\infty}$ be a sequence of natural numbers such that $2 \le p_j \le N$. We set $p_{-j} = p_j$ for all $j \in \mathbb{N}$, $m_j = p_1 \dots p_j$ for $j \in \mathbb{N}$, $m_0 = 1$ and $m_{-l} = 1/m_l$ for $l \in \mathbb{N}$. Then each number $x \in \mathbb{R}_+$ has the expansion

$$x = \sum_{j=1}^{\infty} x_{-j} m_{j-1} + \sum_{j=1}^{\infty} \frac{x_j}{m_j}, \quad x_j \in \mathbb{Z} \cap [0, p_j), \quad |j| \in \mathbb{N}. \tag{7.1}$$

B.I. Golubov (✉)
Moscow Institute of Physics and Technology, Institutskii Per. 9, Dolgoprudnyi, Moscow Region 141700, Russia
e-mail: golubov@mail.mipt.ru

S.S. Volosivets
Saratov State University, Astrakhanskaya 83, 410012 Saratov, Russia
e-mail: volosivetsss@mail.ru

P. Manchanda et al. (eds.), *Industrial Mathematics and Complex Systems*, Industrial and Applied Mathematics, DOI 10.1007/978-981-10-3758-0_7

Here the first sum in (7.1) is finite and for $x = k/m_l$, $k, l \in \mathbb{N}$, the expansion with finite number $x_j \neq 0$ is taken. If $x, y \in \mathbb{R}_+$ are written in the form (7.1), then by definition

$$x \oplus y = z = \sum_{j=1}^{\infty} z_{-j} m_{j-1} + \sum_{j=1}^{\infty} \frac{z_j}{m_j}, \quad z_j \in \mathbb{Z} \cap [0, p_j), \quad |j| \in \mathbb{N},$$

where $z_j = x_j + y_j \pmod{p_j}$. If $z_j = p_j - 1$ for all $j \geq j_0$, then this operation is not defined, i.e. $x \oplus y$ exists for all y except countable set, where $x \in \mathbb{R}_+$ is fixed. The inverse operation $x \ominus y$ is defined in a similar way.

For $x, y \in \mathbb{R}_+$ with expansions of type (7.1) we set

$$\chi(x, y) = \exp\left(2\pi i \sum_{j=1}^{\infty} \frac{x_j y_{-j} + x_{-j} y_j}{p_j}\right).$$

It is well known that the equalities $\chi(x \oplus z, y) = \chi(x, y)\chi(z, y)$ and $\chi(x \ominus z, y) = \chi(x, y)\overline{\chi(z, y)}$ hold for almost all $z \in \mathbb{R}_+$ when $(x, y) \in \mathbb{R}_+ \times \mathbb{R}_+$ is fixed. In particular, we have $\chi(x, 0 \ominus y) = \overline{\chi(x, y)}$, $x, y \in \mathbb{R}_+$. Therefore, $\chi(x, y) = \chi(\{x\}, [y])\chi([x], \{y\})$, where $\{x\}$ be the fractional part of x and $[x]$ be the entire part of x, and the kernel $\chi(x, y)$ is constant by x on all intervals $I_j^k = [j/m_k, (j+1)/m_k)$, $j \in \mathbb{Z}_+$, for $0 \leq y < m_k$ (see [1, Sect. 1.5]).

The spaces $L^p(\mathbb{R}_+)$, $1 \leq p < \infty$, consist of Lebesgue measurable on $\mathbb{R}_+$ functions such that $\|f\|_p = \left(\int_{\mathbb{R}_+} |f(t)|^p dt\right)^{1/p} < \infty$. If $p = \infty$, then we shall use the uniform norm $\|f\|_\infty = \sup_{x \in \mathbb{R}_+} |f(x)|$ for bounded on $\mathbb{R}_+$ functions f ($f \in B(\mathbb{R}_+)$).

For $f \in L^1(\mathbb{R}_+)$ we define the multiplicative **P**-adic Fourier transform (see [1, 2]) by the formula $\widehat{f}(x) = \int_{\mathbb{R}_+} f(y)\overline{\chi(x, y)}\, dy$, where the right-hand side is the Lebesgue integral. For $f \in L^p(\mathbb{R}_+)$, $1 < p \leq 2$, we define **P**-adic Fourier transform as L^q-limit of $\int_0^a f(y)\overline{\chi(x, y)}\, dy$, $a \to +\infty$, where $1/p + 1/q = 1$. According to [1, Chap. 6, Theorem 6.1.7], a counterpart of Hausdorff–Young inequality

$$\|\widehat{f}\|_q \leq \|f\|_p, \quad f \in L^p(\mathbb{R}_+), \quad 1 \leq p \leq 2, \tag{7.2}$$

holds. For $f \in L^2(\mathbb{R}_+)$ by (7.2) we have $\widehat{f} \in L^2(\mathbb{R}_+)$ and the analogue of Parseval-Plancherel equality $\|f\|_2 = \|\widehat{f}\|_2$. The last assertion is easily deduced from [1, Theorem 6.2.4].

Finally, if f is decreasing on $(0, \infty)$, $\lim_{x \to \infty} f(x) = 0$ and f is integrable near zero, then we define $\widehat{f}(x)$ as improper integral $\int_0^\infty f(y)\overline{\chi(x, y)}\, dy$ (see [3] and Lemma 2).

Similarly, the inverse multiplicative **P**-adic Fourier transform $\check{f}(x) = \int_0^\infty f(y)\chi(x, y)\,dy$ is defined for $f \in L^1(\mathbb{R}_+)$ or $f \in L^p(\mathbb{R}_+)$, $1 < p \le 2$. We note that properties of $\check{f}$ are the same as $\widehat{f}$ ones; in particular, the analogue of (7.2) holds.

The multiplicative **P**-adic convolution of functions $f, g \in L^1_{loc}(\mathbb{R}_+)$ is defined by $f * g(x) = \int_{\mathbb{R}_+} f(x \ominus t)g(t)\,dt$, if the last integral exists. It is known that if $f \in L^p(\mathbb{R}_+)$, $g \in L^1(\mathbb{R}_+)$, then $f * g \in L^p(\mathbb{R}_+)$ and $\|f * g\|_p \le \|f\|_p\|g\|_1$ (for $p = 1$ see [1, Sect. 6.1]). Let G_n be the linear space of functions having constant value on each interval $I_k^n, k \in \mathbb{Z}_+$, for some $n \in \mathbb{Z}_+$, and $\mathscr{E}_{m_n}(f)_p := \inf\{\|f - g\|_p : g \in G_n\}$, $n \in \mathbb{Z}_+$, $1 \le p < \infty$.

Let $f \in L^p(\mathbb{R}_+)$, $1 \le p < \infty$, $D_{m_n}(x) = \int_0^{m_n} \chi(x, y)\,dy$. Then A.V. Efimov inequality [1, Sect. 10.5]

$$\mathscr{E}_{m_n}(f)_p \le \|f - f * D_{m_n}\|_p \le 2\mathscr{E}_{m_n}(f)_p \tag{7.3}$$

is valid.

In the present paper we study the conditions implying integrability of multiplicative Fourier transforms of convolutions $h = f * g$ and the asymptotic behaviour of integral norms of $\widehat{h}X_{[m_n,\infty)}$. The sharpness of these conditions in various senses is established. In the case of trigonometric series the absolute convergence and its generalizations for 2π-periodical convolutions are studied by M. Izumi and S. Izumi [4] and C.W. Onneweer [5], while in the case of multiplicative systems one can note the papers of C.W. Onneweer [6] and the second of authors [7]. The following result is contained in [5].

Theorem A. *(1) If $g, h \in L^p_{2\pi}$, $1 < p \le 2$, $1/p + 1/q = 1$, then the series of modules of their 2π-periodic convolution $(g * h)_{2\pi}$ Fourier coefficients raised to the power $q/2$ converges.*

(2) For each $1 < p \le 2$ there exist $g, h \in L^p_{2\pi}$, such that the series of modules of their 2π-periodic convolution $(g * h)_{2\pi}$ Fourier coefficients raised to the power $\beta < q/2$ diverges.

The counterpart of Theorem A for multiplicative Fourier transforms was established by authors [8], while for the multiplicative systems it was obtained in [7].

N.A. Il'yasov [9–11] considered the quantities

$$\rho_n^{(r)}(h) = \left(\sum_{|k|\ge n} |c_k(h)|^r\right)^{1/r},$$

where $\{c_k(h)\}_{k\in\mathbb{Z}}$ be the sequence of complex Fourier coefficients of the function h that is 2π-periodic convolution of functions f and g and studied relations of $\rho_n^{(r)}(h)$ with best approximations of f and g. So, he established in [9] the following theorem.

Theorem B. *(1) Let $1 < p \le 2$, $1/p + 1/p' = 1$, $f, g \in L^p(\mathbb{T})$, h be the 2π-periodic convolution of f and g, $\gamma = p'/2$. Then*

$$\rho_{n+1}^{(\gamma)}(h) \le C(p)E_n(f)_{L^p(\mathbb{T})}E_n(g)_{L^p(\mathbb{T})},$$

where $E_n(f)_{L^p(\mathbb{T})} = \inf_{t_n \in T_n} \|f - t_n\|_{L^p(\mathbb{T})}$ *is the best approximation of* f *by trigonometric polynomials* t_n *of order at most* n $(t_n \in T_n)$ *in* $L^p(\mathbb{T})$.

(2) Let $1 < p \le 2, 1/p + 1/p' = 1, \alpha, \beta > 0, \gamma = p'/2$. *Then there exist* $f, g \in L^p(\mathbb{T})$ *such that* $E_n(f)_{L^p(\mathbb{T})} \asymp n^{-\alpha}$, $E_n(g)_{L^p(\mathbb{T})} \asymp n^{-\beta}$ *and for* 2π*-periodic convolution* h *of functions* f *and* g *we have*

$$\rho_{n+1}^{(\gamma)}(h) \asymp n^{-\alpha-\beta}.$$

Here and further we write $A_n \asymp B_n$, if $A_n = O(B_n)$ and simultaneously $B_n = O(A_n)$, $n \in \mathbb{N}$. In [11] the following result is proved.

Theorem C. *Let* $\{\lambda_n\}_{n=0}^{\infty}$ *be decreasing to zero. Then the set of continuous* 2π*-periodic functions* h *with property* $\rho_{n+1}^{(1)}(h) = O(\lambda_n)$, $n \in \mathbb{Z}_+$, *coincides with the set of* 2π*-periodic convolutions of functions* $f \in L^2(\mathbb{T})$ *and* $g \in L^2(\mathbb{T})$ *such that* $E_n(f)_{L^2(\mathbb{T})}, E_n(g)_{L^2(\mathbb{T})} = O(\lambda_n^{1/2})$.

The Theorem C is a quantitative variant of the M. Riesz theorem (see [12, Chap. IX, Sect. 7]). By this theorem a function with absolutely convergent Fourier series is represented as 2π-periodic convolution of two functions from $L^2(\mathbb{T})$. The counterpart of the M. Riesz theorem for locally compact groups may be found in W. Rudin monograph [13, Sect. 1.6].

In the present paper we prove an analogue of the Hardy–Littlewood theorem concerning trigonometric series with monotone coefficients (see [12, Chap. X, Sect. 3]) in the case of multiplicative Fourier transforms. This result is applied to the proof of sharpness of some embeddings for classes of convolutions. Corresponding embeddings (Theorems 7.3 and 7.4) are analogues and generalizations of ones from [10] that in turn generalize the Theorem B. Also the counterpart of Theorem C is established.

7.2 Auxiliary Propositions

Lemma 1 *Let* $D_y(x) = \int_0^y \chi(x, t)\,dt$, $x, y \in \mathbb{R}_+$. *Then*

(1) The equality $D_{m_n}(x) = m_n X_{[0,1/m_n)}(x)$, $n \in \mathbb{Z}$, *holds, where* X_E *is the indicator of a set* E. *In particular,* $\|D_{m_n}\|_p = m_n^{1-1/p}$, $1 \le p < \infty$, $n \in \mathbb{Z}$.

(2) We have $\widehat{D_{m_n}}(x) = X_{[0,m_n)}(x)$, $x \in \mathbb{R}_+$, $n \in \mathbb{Z}$.

The Proposition (1) is well known (see [1, Sects. 1.5 and 11.1]), while the Proposition (2) follows from the definition.

Lemma 2 *Let* f *be non-increasing on* $(0, \infty)$, $f \in L^1[0, 1)$ *and* $\lim_{x\to\infty} f(x) = 0$. *Then the improper integral* $\widehat{f}(x) = \int_{\mathbb{R}_+} f(y)\overline{\chi(x, y)}\,dy$ *converges for all* $x > 0$. *If, in addition,* $f(x)x^{1-2/p} \in L^p(\mathbb{R}_+)$, $1 < p \le 2$, *then* $\widehat{f}(x) \in L^p(\mathbb{R}_+)$.

The assertion of Lemma 2 is proved in [3, Theorem 3].

Lemma 3 *Let $a(y) \in L^1_{loc}(\mathbb{R}_+)$ and the integral $\int\limits_0^\infty a(y)\overline{\chi(x, y)}\, dy$ converges everywhere on $\mathbb{R}_+$ except at most countable set to a function $f(x) \in L^1_{loc}(\mathbb{R})$. Then*

$$a(x) = \lim_{n \to +\infty} \int\limits_0^{m_n} f(y)\chi(x, y)\, dy$$

a.e. on $\mathbb{R}_+$.

Lemma 3 is established by V.A. Skvortsov [14].

Lemma 4 *Let $f \in L^p(\mathbb{R}_+)$, $g \in L^q(\mathbb{R}_+)$, $p \geq 1$, $q \geq 1$ and $1/r = 1/p + 1/q - 1 > 0$. Then the multiplicative convolution $f * g$ exists as an element of $L^r(\mathbb{R}_+)$ and $\|f * g\|_r \leq \|f\|_p \|g\|_q$.*

This lemma is a multiplicative counterpart of the Young theorem and it may be proved similarly to [15, Chap.1, Sect. 1.2] using Minkowski and Hölder inequalities and M. Riesz–Thorin theorem.

Lemma 5 *Let $1 \leq q_1, q_2 \leq 2, 3/2 \leq 1/q_1 + 1/q_2 \leq 2$, $f \in L^{q_1}(\mathbb{R}_+)$, $g \in L^{q_2}(\mathbb{R}_+)$. Then $\widehat{f * g}(x) = \widehat{f}(x)\widehat{g}(x)$ a.e. on $\mathbb{R}_+$. For $f, g \in L^1(\mathbb{R}_+)$ the equality $\widehat{f * g}(x) = \widehat{f}(x)\widehat{g}(x)$ holds for all $x \in \mathbb{R}_+$.*

Proof The second assertion of Lemma 5 easily follows from the Fubini theorem (see the proof of Theorem 6.1.4 in [1]). In the general case by Lemma 4 the function $f * g$ belongs to $L^r(\mathbb{R}_+)$, where $1/r = 1/q_1 + 1/q_2 - 1 \in [1/2, 1]$, i.e. $r \in [1, 2]$ and $\widehat{f * g}(x)$ is well defined as a function from $L^{r'}(\mathbb{R}_+)$. Let $f_t(x) = f X_{[0,t)}$, $g_t(x) = g X_{[0,t)}$. Then $f_t \in L^1(\mathbb{R}_+) \cap L^{q_1}(\mathbb{R}_+)$, $g_t \in L^1(\mathbb{R}_+) \cap L^{q_2}(\mathbb{R}_+)$ and by the second assertion of lemma we have $\widehat{f_t * g_t}(x) = \widehat{f_t}(x)\widehat{g_t}(x)$ for $x \in \mathbb{R}_+$. Since $f_t \to f$ in $L^{q_1}(\mathbb{R}_+)$ and $g_t \to g$ in $L^{q_2}(\mathbb{R}_+)$ as $t \to +\infty$, by the analogue of Hausdorff–Young -F. Riesz inequality (7.2) we obtain that $\widehat{f_t} \to \widehat{f}$ in $L^{q_1'}(\mathbb{R}_+)$ and $\widehat{g_t} \to \widehat{g}$ in $L^{q_2'}(\mathbb{R}_+)$. Therefore, by Hölder inequality taking in account that $1/q_1' + 1/q_2' = 1/r'$ we find that

$$\|\widehat{f}\widehat{g} - \widehat{f_t}\widehat{g_t}\|_{r'} = \|\widehat{f}\widehat{g} - \widehat{f_t}\widehat{g} + \widehat{f_t}\widehat{g} - \widehat{f_t}\widehat{g_t}\|_{r'} \leq \|\widehat{f} - \widehat{f_t}\|_{q_1'} \|\widehat{g}\|_{q_2'} + \|\widehat{f_t}\|_{q_1'} \|\widehat{g} - \widehat{g_t}\|_{q_2'}$$

and the right-hand side of the last inequality tends to zero as $t \to \infty$. On the other hand, $f * g - f_t * g_t = (f - f_t) * g + (g - g_t) * f_t$ and L^r-norm of the right-hand side and the left-hand side by Lemma 4 tends to zero. By (7.2) the difference $\widehat{f * g} - \widehat{f_t * g_t} = \widehat{f * g} - \widehat{f_t}\widehat{g_t}$ converges in $L^{(r')}(\mathbb{R}_+)$ to zero. These two statements give $\widehat{f * g}(x) = \widehat{f}(x)\widehat{g}(x)$ a.e. on $\mathbb{R}_+$. Lemma is proved. □

7.3 Main Results

Theorem 7.1 *Let $g(x)$ be non-increasing on $(0, \infty)$, $g \in L^1[0, 1)$ and $\lim_{x\to+\infty} g(x) = 0$, $1 < p \le 2$. Then for existence of a function $f \in L^p(\mathbb{R}_+)$ such that $\widehat{f}(x) = g(x)$ almost everywhere on $\mathbb{R}_+$ the condition $g(x)x^{1-2/p} \in L^p(\mathbb{R}_+)$ is necessary and sufficient. If this condition holds, then the following inequalities*

$$\|f(x)\|_p \le C\|g(x)x^{1-2/p}\|_p, \tag{7.4}$$

$$\mathscr{E}_{m_n}(f)_p \le C\left(m_n^{p-1}g^p(m_n) + \int_{m_n}^{\infty} g^p(x)x^{p-2}\,dx\right)^{1/p}, n \in \mathbb{Z}_+, \tag{7.5}$$

are valid and constants in (7.4) and (7.5) do not depend on g and $n \in \mathbb{Z}_+$.

Proof (a) Sufficiency. Let $g(x)x^{1-2/p} \in L^p(\mathbb{R}_+)$, $1 < p \le 2$. By Lemma 2 the function $h(x) := \widehat{g}(x)$ exists as improper integral for $x > 0$ and, due to condition, we have $h \in L^p(\mathbb{R}_+)$. By Lemma 3 we see that

$$g(x) = \lim_{n\to+\infty} \int_0^{m_n} h(y)\chi(x, y)\,dy \tag{7.6}$$

for a.e. $x \in \mathbb{R}_+$. On the other hand, similarly to the proof of Hausdorff–Young -F. Riesz inequality (7.2) we obtain

$$\left\|\check{h} - \int_0^{m_n} h(t)\chi(\cdot, t)\,dt\right\|_q \to 0, \quad n \to \infty, \quad 1/p + 1/q = 1.$$

By F. Riesz theorem about convergent a.e. subsequence (it is valid for sets of infinite measure, see [16, Sect. 13]) there exists $\{m_{n_i}\}_{i=1}^{\infty}$ such that

$$\int_0^{m_{n_i}} h(t)\chi(x, t)\,dt \to \check{h}(x) \tag{7.7}$$

a.e. on $\mathbb{R}_+$. Combining (7.6) and (7.7) we obtain $g(x) = \check{h}(x) = \widehat{h(\ominus\cdot)}(x) =: \widehat{f}(x)$ a.e. on $\mathbb{R}_+$, where $f \in L^p(\mathbb{R}_+)$.

(b) Necessity. The inverse assertion follows from [3, Theorem 2].

(c) Proof of inequalities (7.4) and (7.5). Under condition $f \in L^p(\mathbb{R}_+)$ by Theorem 5 from [17] in the case $\gamma = 2/p - 1$, $p = q = r$, we deduce the inequality (7.4).

For the proof of (7.5) let us consider the sequence of functions $\{g_n(x)\}_{n\in\mathbb{Z}_+}$ such that $g_n(x) = g(x) - g(m_n)$ if $x \in [0, m_n)$ and $g_n(x) = 0$ if $x \in [m_n, \infty)$. These functions satisfy the same conditions as the function g above, and therefore we can find $f_n \in L^p(\mathbb{R}_+)$ such that $\widehat{f_n} = g_n$. By Corollary 2 from [18] we have

$$f_n(x) = \lim_{i\to\infty} \int_{B_i} g_n(t)\chi(x,t)\,dt = \int_{B_n} g_n(t)\chi(x,t)\,dt$$

for a.e. $x \in \mathbb{R}_+$. The last expression is constant on all I_k^n, $k \in \mathbb{Z}_+$ (see Introduction). Changing f_n on a set of measure zero, we can state $f_n \in G_n$.

Further the function

$$g(x) - g_n(x) = \begin{cases} g(m_n), & x \in [0, m_n); \\ g(x), & x \in [m_n, \infty); \end{cases}$$

is non-increasing on $\mathbb{R}_+$, tends to zero when $x \to +\infty$ and also satisfies the condition $(g(x) - g_n(x))x^{1-2/p} \in L^p(\mathbb{R}_+)$. According to (7.4) we find that

$$\|f - f_n\|_p \le C_1 \left(\int_0^\infty (g(x) - g_n(x))^p x^{p-2}\,dx \right)^{1/p} \le$$

$$\le C_1 \left(\frac{m_n^{p-1}}{p-1} g^p(m_n) + \int_{m_n}^\infty g^p(x) x^{p-2}\,dx \right)^{1/p}.$$

Theorem is proved. □

Remark 1 For cosine Fourier transform close to Theorem 7.1 and Lemma 2 results were obtained by G. Hardy and J.E. Littlewood (see [19, Chap. 4, Theorems 79, 80, 82]). The estimate (7.5) is an analogue of A.A. Konjushkov inequality for best approximations of sums of trigonometric series [20].

Theorem 7.2 *Let $f \in L^1(\mathbb{R}_+)$ be such that $\widehat{f} \in L^1(\mathbb{R}_+)$, while $\{\varepsilon_n\}_{n=0}^\infty$ be decreasing to zero. Then f satisfies the relation $\int_{m_n}^\infty |\widehat{f}(t)|\,dt = O(\varepsilon_n)$, $n \in \mathbb{Z}_+$, if and only if $f = g * h$, where $g, h \in L^2(\mathbb{R}_+)$ and*

$$\mathscr{E}_{m_n}(g)_2 = O(\varepsilon_n^{1/2}), \quad \mathscr{E}_{m_n}(h)_2 = O(\varepsilon_n^{1/2}), \quad n \in \mathbb{Z}_+.$$

Proof Sufficiency. Let $f = g * h$, $g, h \in L^2(\mathbb{R}_+)$, and $\mathscr{E}_{m_n}(g)_2 = O(\varepsilon_n^{1/2})$, $\mathscr{E}_{m_n}(h)_2 = O(\varepsilon_n^{1/2})$, $n \in \mathbb{Z}_+$. We set $g_n = g * D_{m_n}$, where $D_y(x)$ is defined in Lemma 1 (h_n and f_n are defined in a similar manner). Then by A.V. Efimov inequality (7.3) we have $\|g - g_n\|_2 \le 2\mathscr{E}_{m_n}(g)_2$ and $\|h - h_n\|_2 \le 2\mathscr{E}_{m_n}(h)_2$, $n \in \mathbb{Z}_+$. We write the equality

$$(g - g_n) * (h - h_n) = g * h - g_n * h - h_n * g + g_n * h_n.$$

By Lemmas 1 and 5 we obtain $\widehat{D_{m_n} * D_{m_n}}(x) = (\widehat{D_{m_n}}(x))^2 = X_{[0,m_n)}$ and by the uniqueness theorem the equalities $D_{m_n} * D_{m_n} = D_{m_n}$ and $g_n * h = g * D_{m_n} * h = g * h_n = g_n * h_n$ are valid. Therefore, $(g - g_n) * (h - h_n) = f - f_n$ and $\widehat{f} - \widehat{f_n} = \widehat{f}(1 - X_{[0,m_n)}) = \widehat{f} X_{[m_n,\infty)}$. Thus, we obtain due to the Cauchy–Bunyakovsky–Schwarz inequality and the analogue of Plancherel theorem

$$\int_{m_n}^{\infty} |\widehat{f}(t)|\,dt = \|\widehat{f} - \widehat{f_n}\|_1 = \|(\widehat{g} - \widehat{g_n})(\widehat{h} - \widehat{h_n})\|_1 \le$$

$$\le \|\widehat{g} - \widehat{g_n}\|_2 \|\widehat{h} - \widehat{h_n}\|_2 = \|g - g_n\|_2 \|h - h_n\|_2 \le C_1 \varepsilon_n.$$

Necessity. Let us construct $g, h \in L^2(\mathbb{R}_+)$ such that $f = g * h$. We introduce the functions $\psi(x) = |\widehat{f}(x)|^{1/2}$ and $\varphi(x) = |\widehat{f}(x)|^{1/2} sign(\widehat{f}(x))$, where $sign(z) = \exp(i \arg z)$ for $z \in \mathbb{C}$. Further we define g and h as the inverse Fourier multiplicative transforms $\check{\varphi}$ and $\check{\psi}$. Then we obtain $\widehat{f} = \varphi\psi = \widehat{g}\widehat{h} = \widehat{g * h}$ by Lemma 5 and $f = g * h$ by the uniqueness theorem. Now we have again $\widehat{g} - \widehat{g_n} = \widehat{g} X_{[m_n,\infty)}$ and, by the analogue of Plancherel theorem, $\|\widehat{g} - \widehat{g_n}\|_2 = \|g - g_n\|_2$. Finally,

$$\|g - g_n\|_2 = \|h - h_n\|_2 = \left(\int_{m_n}^{\infty} |\widehat{g}(x)|^2\,dx\right)^{1/2} = \left(\int_{m_n}^{\infty} |\widehat{f}(x)|\,dx\right)^{1/2} \le C_2 \varepsilon_n^{1/2},$$

and $\mathscr{E}_{m_n}(f)_2, \mathscr{E}_{m_n}(g)_2 \le C_2 \varepsilon_n^{1/2}$. Theorem is proved. □

Theorem 7.3 *(1) Let* $1 < q_1, q_2 < 2,\ 3/2 < 1/q_1 + 1/q_2 < 2,\ 1/r = 1/q_1 + 1/q_2 - 1,\ 1/r + 1/r' = 1$. *If* $f \in L^{q_1}(\mathbb{R}_+)$, $g \in L^{q_2}(\mathbb{R}_+)$, *then* $h = f * g \in L^r(\mathbb{R}_+)$ *and the inequalities*

$$\|\widehat{h}\|_{r'} \le \|f\|_{q_1} \|g\|_{q_2}; \quad \left(\int_{m_n}^{\infty} |\widehat{h}(t)|^{r'}\,dt\right)^{1/r'} \le 4\mathscr{E}_{m_n}(f)_{q_1} \mathscr{E}_{m_n}(g)_{q_2}, \quad n \in \mathbb{Z}_+, \tag{7.8}$$

are valid.

(2) Under conditions of part (1) for $2 \ge \theta > r$ *and* $0 < \gamma < r'$ *there exist* $f_0 \in L^{q_1}(\mathbb{R}_+)$ *and* $g_0 \in L^{q_2}(\mathbb{R}_+)$ *such that* $h_0 = f_0 * g_0 \notin L^\theta(\mathbb{R}_+)$ *and* $\widehat{h_0} \notin L^\gamma(\mathbb{R}_+)$.

Proof (1) We set again $h_n = h * D_{m_n}$, f_n and g_n are defined similarly. The first inequality (7.8) follows from Lemma 4 and inequality (7.2). To prove the second inequality we write similar to the proof of Theorem 7.2 $h - h_n = (f - f_n) * (g - g_n)$ $\widehat{h} - \widehat{h_n} = \widehat{h} X_{[m_n,\infty)}$. Applying Lemma 4, (7.2) and (7.3), we obtain

$$\left(\int_{m_n}^{\infty} |\widehat{h}(t)|^{r'}\,dt\right)^{1/r'} = \|\widehat{h} - \widehat{h_n}\|_{r'} \le \|h - h_n\|_r \le$$

$$\le \|f - f_n\|_{q_1} \|g - g_n\|_{q_2} \le 4\mathscr{E}_{m_n}(f)_{q_1} \mathscr{E}_{m_n}(g)_{q_2}.$$

(2) Let $b(x) = x^{-1/q_1'}(\log_2 x + 1)^{-1}$ if $x \in [1, \infty)$ and $b(x) = 1$ if $x \in [0, 1)$. Then $b^{q_1}(x) x^{q_1 - 2} = x^{-1}(\log_2 x + 1)^{-q_1}$ for $x \ge 1$ and $b^{q_1}(x) x^{q_1 - 2}$ is equal to $x^{q_1 - 2}$ on $(0, 1)$, so $b^{q_1}(x) x^{q_1 - 2} \in L^1(\mathbb{R}_+)$ and by Theorem 7.1 there exists $f_0 \in L^{q_1}(\mathbb{R}_+)$ such that $\widehat{f_0} = b$. Also there exists $g_0 \in L^{q_2}(\mathbb{R}_+)$ such that

$$\widehat{g_0}(x) = \begin{cases} x^{-1/q_2'}(\log_2 x + 1)^{-1}, & x \ge 1; \\ 1, & x \in [0, 1). \end{cases}$$

If $2 \geq \theta > r$, we have for $h_0 = f_0 * g_0$ and $x \geq 1$

$$(\widehat{h_0}(x))^\theta x^{\theta-2} = (\widehat{f_0}(x)\widehat{g_0}(x))^\theta x^{\theta-2} = x^{-\theta(1-1/q_1+1-1/q_2)+\theta-2}(\log_2 x + 1)^{-2\theta} =$$

$$= x^{\theta/r-2}(\log_2 x + 1)^{-2\theta} \notin L^1[1, +\infty).$$

Since $\widehat{h_0}(x) = \widehat{f_0}(x)\widehat{g_0}(x)$ is decreasing and integrable on $[0, 1)$, by Theorem 7.1 we conclude that $h_0 \notin L^\theta(\mathbb{R}_+)$. If $0 < \gamma < r'$, we find that

$$\int_1^\infty (\widehat{h_0}(x))^\gamma \, dx = \int_1^\infty (\widehat{f_0}(x)\widehat{g_0}(x))^\gamma \, dx = \int_1^\infty x^{-\gamma/r'}(\log_2 x + 1)^{-2\gamma} \, dx = \infty.$$

Theorem is proved. □

Remark 2 A more general statement than existence of $h_0 = f_0 * g_0 \notin L^\theta(\mathbb{R}_+)$ may be found in Theorem 1.1 (iii) in [21] for locally compact but non-compact groups. Our proof is more simple.

Remark 3 The statements of part (1) of Theorem 7.3 are valid for $1 \leq q_1, q_2 \leq 2$ and $3/2 \leq 1/q_1 + 1/q_2 \leq 2$. In the case $r' = \infty$ the left-hand side of second inequality (7.8) must be replaced by $\|\widehat{h}X_{[m_n,\infty)}\|_\infty$. In Theorem 7.4 we establish the sharpness of this more general assertion.

Theorem 7.4 *Let $1 \leq q_1, q_2 \leq 2$, $3/2 \leq 1/q_1 + 1/q_2 \leq 2$, $1/r = 1/q_1 + 1/q_2 - 1$ and $1/r + 1/r' = 1$, the sequences $\{\nu_n\}_{n=0}^\infty$ and $\{\mu_n\}_{n=0}^\infty$ are decreasing to zero and satisfy the conditions*

$$\sum_{k=n}^\infty \nu_k^{q_1} = O(\nu_n^{q_1}), \quad \sum_{k=n}^\infty \mu_k^{q_2} = O(\mu_n^{q_2}), \quad \nu_n \leq C\nu_{n+1}, \quad \mu_n \leq C\mu_{n+1}, \quad n \in \mathbb{Z}_+.$$

Then there exist functions $f_0 \in L^{q_1}(\mathbb{R}_+)$ and $g_0 \in L^{q_2}(\mathbb{R}_+)$ such that $\mathcal{E}_{m_n}(f_0)_{q_1} \asymp \nu_n$, $\mathcal{E}_{m_n}(g_0)_{q_2} \asymp \mu_n$, $n \in \mathbb{Z}_+$, and for convolution $h_0 = f_0 * g_0 \in L^r(\mathbb{R}_+)$ we have $\widehat{h_0}(x) \geq 0$ on $\mathbb{R}_+$ and

$$\left(\int_{m_n}^\infty (\widehat{h_0(x)})^{r'} \, dx\right)^{1/r'} \asymp \nu_n\mu_n, \quad n \in \mathbb{Z}_+. \tag{7.9}$$

Proof (1) Let $q_1, q_2 > 1$. We set $b(x) = m_n^{-1/q_1'}\nu_n$ on $[m_n, m_{n+1})$, $n \in \mathbb{Z}_+$, and $b(x) = \nu_0$ for $x \in [0, 1)$. Then we obtain

$$b^{q_1}(x)x^{q_1-2} \leq \left(N^{1/q_1'}x^{-1/q_1'}\nu_n x^{1-2/q_1}\right)^{q_1} = N^{q_1-1}x^{-1}\nu_n^{q_1}$$

on $[m_n, m_{n+1})$, $n \in \mathbb{Z}_+$, and

$$\int_0^\infty b^{q_1}(x)x^{q_1-2}\,dx = \nu_0^{q_1}\int_0^1 x^{q_1-2}\,dx + \sum_{n=0}^\infty \int_{m_n}^{m_{n+1}} b^{q_1}(x)x^{q_1-2}\,dx \le$$

$$\le \frac{\nu_0^{q_1}}{q_1-1} + \sum_{n=0}^\infty \nu_n^{q_1} N^{q_1-1}\int_{m_n}^{m_{n+1}} x^{-1}\,dx \le C_1\sum_{n=0}^\infty \nu_n^{q_1} < \infty,$$

where C_1 depends on the majorant N of sequence $\{p_i\}_{i\in\mathbb{N}}$ and q_1. Since $b(x)$ is decreasing on $(0,\infty)$ and $b \in L^1[0,1)$, by Theorem 7.1 there exists $f_0 \in L^{q_1}(\mathbb{R}_+)$ such that $\widehat{f_0} = b$. For $n \in \mathbb{Z}_+$ we have

$$\mathscr{E}_{m_n}(f_0)_{q_1} \le C_2\left(m_n^{q_1-1}b^{q_1}(m_n) + \int_{m_n}^\infty b^{q_1}(x)x^{q_1-2}\,dx\right)^{1/q_1} \le$$

$$\le C_2\left(\nu_n^{q_1} + \sum_{k=n}^\infty m_k^{q_1-2}(m_{k+1}-m_k)m_k^{1-q_1}\nu_k^{q_1}\right)^{1/q_1} \le C_3\left(\nu_n^{q_1} + \sum_{k=n}^\infty \nu_k^{q_1}\right)^{1/q_1} \le C_4\nu_n.$$

On the other hand, by inequality (7.3) we obtain

$$\mathscr{E}_{m_n}(f_0)_{q_1} \ge \frac{1}{2}\|f_0 - S_{m_n}(f_0)\|_{q_1} \ge \frac{1}{2}\|\widehat{f_0} - \widehat{f_0 * D_{m_n}}\|_{q_1'} =$$

$$= \|bX_{[m_n,\infty)}\|_{q_1'} = \left(\sum_{k=n}^\infty m_k^{-1}\nu_k^{q_1'}(m_{k+1}-m_k)\right)^{1/q_1'} \ge \nu_n, \quad n \in \mathbb{Z}_+.$$

Similarly, we prove the existence of $g_0 \in L^{q_2}(\mathbb{R}_+)$ such that $\widehat{g_0}(x) = m_n^{-1/q_2'}\mu_n$ on $[m_n, m_{n+1}), n \in \mathbb{Z}_+$, $\widehat{g_0}(x) = \mu_0$ for $x \in [0,1)$ and $\mathscr{E}_{m_n}(g_0)_{q_2} \asymp \mu_n, n \in \mathbb{Z}_+$. Finally, by Lemma 5 $\widehat{h_0} = \widehat{f_0}\widehat{g_0}$ a.e. on $\mathbb{R}_+$, whence the equality

$$\widehat{h_0}(x) = \begin{cases} 1, & x \in [0,1); \\ m_n^{-2+1/q_1+1/q_2}\nu_n\mu_n, & x \in [m_n, m_{n+1}), \quad n \in \mathbb{Z}_+; \end{cases}$$

follows. Therefore $(2 - 1/q_1 - 1/q_2 = 1 - 1/r = 1/r')$,

$$\int_{m_n}^\infty (\widehat{h_0(x)})^{r'}\,dx = \sum_{k=n}^\infty m_k^{-1}\nu_k^{r'}\mu_k^{r'}(m_{k+1}-m_k) \le N\sum_{k=n}^\infty \nu_k^{r'}\mu_k^{r'}. \tag{7.10}$$

But the decreasing of $\{\nu_n\}_{n=0}^\infty$ and the condition $\sum_{k=n}^\infty \mu_k^{q_2} = O(\mu_n^{q_2})$ imply

$$\sum_{k=n}^\infty \nu_k^{r'}\mu_k^{r'} \le \nu_n^{r'}\sum_{k=n}^\infty \mu_k^{r'} \le C_5\nu_n^{r'}\mu_n^{r'}.$$

Here we use the inequality $q_2 \le 2 \le r'$ and the Jensen inequality of type

$$\left(\sum_{k=n}^{\infty} \mu_k^{r'}\right)^{1/r'} \le \left(\sum_{k=n}^{\infty} \mu_k^{q_2}\right)^{1/q_2} \le C_6 \mu_n.$$

Finally, by (7.10)

$$\int_{m_n}^{\infty} (\widehat{h_0(x)})^{r'}\, dx \ge \sum_{k=n}^{\infty} m_k^{-1} \nu_k^{r'} \mu_k^{r'} m_k \ge \nu_n^{r'} \mu_n^{r'}.$$

From obtained inequalities the relation (7.9) follows.

(2) If $q_1 = 1$, then we consider the function $f_0 = \sum_{n=0}^{\infty} \nu_n (D_{m_{n+1}} - D_{m_n})$. By Lemma 1, (7.2) and (7.3) we have

$$\mathscr{E}_{m_n}(f_0)_1 \le \sum_{k=n}^{\infty} \nu_k \| D_{m_{k+1}} - D_{m_k} \|_1 \le 2 \sum_{k=n}^{\infty} \nu_k \le C_7 \nu_n$$

and

$$\mathscr{E}_{m_n}(f_0)_1 \ge 2^{-1} \| f_0 - f_0 * D_{m_n} \|_1 \ge 2^{-1} \| \widehat{f_0} - \widehat{f_0 * D_{m_n}} \|_\infty = 2^{-1} \| \widehat{f_0} X_{[m_n, \infty)} \|_\infty. \tag{7.11}$$

Since $\widehat{f_0} = \sum_{k=0}^{\infty} \nu_k X_{[m_k, m_{k+1})}$ and $\{\nu_k\}_{k=0}^{\infty}$ is decreasing, the norm in the right-hand side of (7.11) is equal to ν_n. The function g_0 for $1 < q_2 \le 2$ is defined as in the part 1, while for $q_2 = 1$ we set $g_0 = \sum_{n=0}^{\infty} \mu_n (D_{m_{n+1}} - D_{m_n})$. We have again $\mathscr{E}_{m_n}(g_0)_{q_2} \asymp \mu_n$, $n \in \mathbb{Z}_+$, and ($1/q_2' = 0$ for $q_2 = 1$)

$$\widehat{g_0}(x) = \sum_{k=0}^{\infty} m_k^{-1/q_2'} \mu_k X_{[m_k, m_{k+1})}(x), \quad x \ge 1.$$

By Lemma 5 we obtain for $h_0 = f_0 * g_0$ the equality

$$\widehat{h_0}(x) = \widehat{f_0}(x)\widehat{g_0}(x) = \begin{cases} 0, & x \in [0, 1), \\ m_k^{-1/q_2'} \mu_k \nu_k, & x \in [m_k, m_{k+1}), \quad k \in \mathbb{Z}_+. \end{cases}$$

In the case $q_1 = 1$ we obtain $r = q_2$ and

$$\left(\int_{m_n}^{\infty} (\widehat{h_0}(x))^{r'}\, dx\right)^{1/r'} = \left(\sum_{k=n}^{\infty} m_k^{-1} (m_{k+1} - m_k) \nu_k^{q_2'} \mu_k^{q_2'}\right)^{1/q_2'}. \tag{7.12}$$

Similarly to the proof in (1) we establish that the right-hand side of (7.12) has the same order as $\nu_n \mu_n$. For $q_2 = 1$ the statement easily follows from the formula for $\widehat{h_0}(x)$. Theorem is proved. □

Acknowledgement The authors are grateful to the referee for the remarks on the first version of this paper.

References

1. Golubov, B.I., Efimov, A.V., Skvortsov, V.A.: Walsh Series and Transforms. Theory and Applications. Kluwer Academic Publishers, Dordrecht (1991)
2. Vilenkin, N.Y.: To the theory of Fourier integrals on topological groups. Mat. sbornik. **30**, 233–244 (1952). (in Russian)
3. Golubov, B.I., Volosivets, S.S.: On the integrability and uniform convergence of multiplicative Fourier transform. Georgian Math. J. **16**, 533–546 (2009)
4. Izumi, M., Izumi, S.I.: Absolute convergence of Fourier series of convolution functions. J. Approx. Theory **1**, 103–109 (1968)
5. Onneweer, C.W.: On absolutely convergent Fourier series. Arkiv Mat. **12**, 51–58 (1974)
6. Onneweer, C.W.: Absolute convergence of Fourier series on certain groups, II. Duke Math. J. **41**, 679–688 (1974)
7. Volosivets, S.S.: Convergence of series of Fourier coefficients for multiplicative convolutions. Russ. Math. (Iz. VUZ) **52**(11), 23–34 (2008)
8. Golubov, B.I., Volosivets, S.S.: Generalized weighted integrability of multiplicative Fourier transfroms. Proc. Moscow Inst. Phys. Technol. **3**, 49–56 (2011) (in Russian)
9. Ilyasov, N.A.: To the M. Riesz theorem on absolute convergence of the trigonometric Fourier series. Trans. NAS Azerbaijan **24**(1), 113–120 (2004). Series of Physics-Technology and Mathematical Sciences
10. Ilyasov, N.A.: To the M. Riesz theorem on absolute convergence of the trigonometric Fourier series (second report). Trans. NAS Azerbaijan **24**(4), 135–142 (2004). Series of Physics-Technology and Mathematical Sciences
11. Ilyasov, N.A.: The rate L_p-version of M. Riesz test on absolute convergence of trigonometric Fourier series. Proc. Inst. Math. Mech. Ural branch Russian Acad. Sci. **16**, 193–202 (2010) (in Russian)
12. Bary, N.: A Treatise on Trigonometric Series, vol. 1 and 2. Pergamon Press, New York (1964)
13. Rudin, W.: Fourier Analysis on Groups. Interscience, New York (1962)
14. Skvortsov, V.A.: Uniqueness theorem for representation of a function by multiplicative transforms. Moscow Univ. Math. Bull. No. **6**, 14–18 (1992). (in Russian)
15. Bergh, J., Löfström, J.: Interpolation Spaces. An Introduction. Springer, Berlin (1976)
16. Dyachenko, M.I., Ul'yanov, P.L.: Measure and integral. Factorial, Moscow (1998). (in Russian)
17. Volosivets, S.S., Golubov, B.I.: Weighted integrability of multiplicative Fourier transforms. Proc. Steklov Inst. Math. **269**(1), 65–75 (2010)
18. Volosivets, S.S.: Generalization of the multiplicative Fourier transform and its properties. Math. Notes **89**, 311–318 (2011)
19. Titchmarsh, E.: An Introduction to the Theory of Fourier Integrals. Oxford University Press, New-York (1937)
20. Konjushkov, A.A.: The best approximation by trigonometrical polynomials and Fourier coefficients. Mat. sbornik. **44**, 53–84 (1958). (in Russian)
21. Quek, T.S., Yap, L.Y.H.: Sharpness of Young's inequality for convolution. Math. Scand. **53**, 221–237 (1983)

Chapter 8
Tight Wavelet Frames with Matrix Dilations

Maria Skopina

Abstract Construction of compactly supported tight wavelet frames with arbitrary matrix dilation is discussed. An algorithmic method for the construction of such frames with any prescribed approximation order is proposed. The method is very suitable for practical use. Except of calculating several roots of univariate trigonometric polynomials (in the sense of Riesz's Lemma) all steps of the algorithm are described by explicit formulas. The number of generating wavelet functions depends linearly on the dimension of the space. The method is based on the polyphase approach.

Keywords Multivariate wavelets · Tight frame · Approximation order Polyphase method

A wavelet system in $L_2(\mathbb{R}^d)$ is a set of functions $\psi_{jk}^{(\nu)}$, $\nu = 1, \dots, r$, $j \in \mathbb{Z}, k \in \mathbb{Z}^d$, where $\psi_{jk}^{(\nu)}(x) = m^{j/2}\psi^{(\nu)}(M^j x + k)$, M is an integer $d \times d$ matrix such that all its eigenvalues are strictly bigger in module than 1, $m = |\det M|$. A wavelet system is a tight frame if there exist positive number $A > 0$ such that $\sum_{j,k,r} |\langle f, \psi_{jk}^{(\nu)}\rangle|^2 = A\|f\|^2$ for every $f \in L_2(\mathbb{R}^d)$. Without loss of generality, we set $A = 1$. The notion of frame was introduced by Duffin and Schaeffer in [4]. The main property of tight frames lies in the fact that any function $f \in L_2(\mathbb{R}^d)$ can be expanded into the unconditionally convergent series

$$f = \sum_{j,k,r} \langle f, \psi_{jk}^{(\nu)}\rangle \psi_{jk}^{(\nu)}. \tag{8.1}$$

A general scheme for the construction of wavelet frames is known. This scheme is based on a method called Matrix Extension Principle, which was developed by Ron and Shen in [7]. However, the implementation of this principle for the construction of compactly supported wavelet frames is difficult in practice. A difficulty arises in finding suitable trigonometric polynomials m_0 (refinable mask) and m_ν,

M. Skopina (✉)
Saint Petersburg State University, Universitetskaya Nab. 7/9,
199034 St. Petersburg, Russia
e-mail: skopina@MS1167.spb.edu

P. Manchanda et al. (eds.), *Industrial Mathematics and Complex Systems*,
Industrial and Applied Mathematics, DOI 10.1007/978-981-10-3758-0_8

$\nu = 1, \dots, r$, (wavelet masks), or equivalently, their polyphase matrices, i.e., the matrix consisting of trigonometric polynomials $\mu_{\nu k}$, $k = 0, \dots, m-1$, which are polyphase components (with respect to M) of m_ν, $\nu = 0, \dots, r$ respectively (see., e.g., [11, Sect. 6]). The columns of this matrix should be biorthonormal. Then m_ν, and hence a polyphase matrix with orthonormal columns should be constructed. Main difficulties are connected with the following. First, an analog of the Riesz Lemma does not exist in the multivariate case. Second, there is an open algebraic problem of possibility to extend any suitable row to a unitary matrix.

If a polyphase matrix is constructed, then it is easy to find all masks m_ν. Refinable mask m_0 provides refinable function φ by means of their Fourier transform defined by

$$\widehat{\varphi}(\xi) = \prod_{j=1}^{\infty} m_0(M^{*-j}\xi).$$

The wavelet functions $\psi^{(\nu)}$, $\nu = 1, \dots, r$, are defined by

$$\widehat{\psi^{(\nu)}}(\xi) = m_\nu(M^{*-1}\xi)\widehat{\varphi}(M^{*-1}\xi).$$

If the above scheme is implemented, then the wavelet system $\{\psi_{jk}^{(\nu)}\}$ is a tight frame.

Different approaches for the construction of multivariate compactly supported tight wavelet frames have been proposed in many papers, see, e.g., [1–3, 5, 6, 8, 9] and references therein. Moreover, a lot of concrete examples useful for applications can be found in the literature. We propose an approach based on the polyphase method developed in [10].

Let $\{\psi_{jk}^{(\nu)}\}$ be a tight wavelet frame. Expansion (8.1) is said to have approximation order n if there exist $C > 0$ and $\lambda > 1$ such that

$$\left\| f - \sum_{i<j} \sum_{k\in\mathbb{Z}^d} \sum_{\nu=1}^{r} \langle f, \psi_{ik}^{(\nu)} \rangle \psi_{ik}^{(\nu)} \right\|_2 \le C \frac{\|f\|_{W_2^n}}{\lambda^{jn}}. \tag{8.2}$$

for any f in the Sobolev space W_2^n.

A wavelet system $\{\psi_{jk}^{(\nu)}\}$ is said to have property $\mathbf{VM^n}$, $n \in \mathbb{Z}_+$, (vanishing moment property of order n) if $D^\beta \widehat{\psi}^{(\nu)}(\mathbf{0}) = 0$, $\nu = 1, \dots, r$, $r \ge m-1$, for all $\beta \in \mathbb{Z}_+^d$, $\|\beta\|_1 \le n$.

It is well known that the vanishing moment property is required to provide desirable approximation order of the frame expansion. This fact can be found in the literature in many different forms. We present the following statement which is a special case of Theorem 4 in [10].

Proposition 8.1 *Let $\{\psi_{jk}^{(\nu)}\}$ be a compactly supported tight wavelet frame with $\mathbf{VM^{n-1}}$ property. Then expansion (8.1) has approximation order n, and any number which is bigger than 1 and strictly smaller in modulus than each eigenvalue of M, can be taken as λ in (8.2). If $M = cI_d$, then one can take $\lambda = |c|$.*

Existence of compactly supported tight wavelet frames with vanishing moments up to any given order was proved by Han in [5]. He proved that for every matrix dilation there exist at most $\left(\frac{3}{2}\right)^d m$ smooth compactly supported wavelet functions $\psi^{(\nu)}$ generating a tight frame constructed. We provide an algorithmic method for the construction of wavelet tight frames, where the number of generating wavelet functions depends linearly on the dimension of the space d as well as on m. Namely, the following statement holds.

Theorem 8.1 *For any matrix dilation M and any natural n, there exists a tight wavelet frame generated by compactly supported functions $\psi^{(\nu)}$, $\nu = 1, \dots, r$, with vanishing moments up to order n, where $r \le d(m-1)+m$.*

The proof is based on the polyphase method developed in [10]. Combining Theorems 11, 14 in [10] and Proposition 3 in [12], we have the following statement which clarifies how to provide $\mathbf{VM^n}$ property for tight wavelet frames.

Theorem 8.2 *Let a tight wavelet frame $\{\psi_{jk}^{(\nu)}\}$ be generated by a refinable mask m_0 for which $\mu_{00}, \dots, \mu_{0m-1}$ are polyphase components. If $\{\psi_{jk}^{(\nu)}\}$ has $\mathbf{VM^n}$ property, then there exist complex numbers λ_γ, $\gamma \in \mathbb{Z}_+^d$, $\|\gamma\|_1 \le n$, $\lambda_{\mathbf{0}} = 1$, such that the relations*

$$D^\beta \mu_{0k}(\mathbf{0}) = \frac{1}{\sqrt{m}} \sum_{\mathbf{0} \le \gamma \le \beta} \lambda_\gamma \binom{\beta}{\gamma} (-2\pi i M^{-1} s_k)^{\beta-\gamma} \quad \forall \beta \in \mathbb{Z}_+^d,\ \|\beta\|_1 \le n \tag{8.3}$$

hold for all $k = 0, \dots, m-1$, and

$$\sum_{\mathbf{0} \le \gamma \le \alpha} \binom{\alpha}{\gamma} \lambda_\gamma \overline{\lambda_{\alpha-\gamma}} = 0 \quad \forall \alpha \in \mathbb{Z}_+^d, 0 < \|\alpha\|_1 \le n. \tag{8.4}$$

Theorem 8.2 gives a necessary condition for $\mathbf{VM^n}$ property, which is not sufficient. A sufficient condition is given in the following statement.

Theorem 8.3 *Let $\mu_{00}, \dots, \mu_{0m-1}$ be trigonometric polynomials. If there exist complex numbers λ_γ, $\gamma \in \mathbb{Z}_+^d$, $\|\gamma\|_1 \le n$, $\lambda_{\mathbf{0}=\mathbf{1}}$, satisfying (8.3) for all $k = 0, \dots, m-1$, and there exist trigonometric polynomials μ_{0k}, $k = m, \dots, r$, $\mu_{\nu k}$, $\nu = 1, \dots, r$, $k = 0, \dots, r$, $r \ge m-1$, such that the matrix $\mathcal{M}^{ext} := \{\mu_{\nu k}\}_{\nu,k=0}^r$ is unitary and*

$$D^\beta \mu_{0k}(\mathbf{0}) = 0,\ k = m, \dots, r, \quad \forall \beta \in \mathbb{Z}_+^d,\ \|\beta\| \le n, \tag{8.5}$$

then the matrix $\mathcal{M}$ consisting of m first columns of $\mathcal{M}^{ext}$ generates a compactly supported tight wavelet frame with $\mathbf{VM^n}$ property.

The proof follows immediately from Theorem 8 in [10].

Thus to construct a compactly supported tight wavelet frame with $\mathbf{VM^n}$ property we need a refinable mask whose polyphase row $(\mu_{00}, \dots, \mu_{0m-1})$ satisfies (8.3),

(8.4) and can be extended by trigonometric polynomials $\mu_{0m}, \ldots, \mu_{0r}$ so that (8.5) holds and $\sum_{k=0}^{r} |\mu_{0k}|^2 = 1$. After that the row $(\mu_{00}, \ldots, \mu_{0r})$ should be extended to a unitary matrix. As was mentioned above, a possibility of such extension in the general case is an open problem.

According to this method, we find trigonometric polynomials $\mu_{00}, \ldots, \mu_{0m-1}$ so that their derivatives up to order n at zero satisfy (8.3). It is possible to choose numbers λ_α such that each of these functions is a tensor product of trigonometric polynomials of one variable, and the sum of their squired magnitudes does not exceed 4. Due to the Riesz's Lemma, the difference $4 - \sum_{k=0}^{m-1} |\mu_{0k}|^2$ can be represented as the sum of squired magnitudes of trigonometric polynomials $\mu_{0m}, \ldots, \mu_{0r-1}$, where $r = d(m-1) + m$. This row cannot be extended to a unitary matrix. It is possible to fix these functions preserving the values of their derivatives at zero setting $\sigma := \sum_{k=0}^{m-1} |\mu_{0k}|^2$,

$$\mu'_{0k} := \frac{3-\sigma}{2}\mu_{0k}, \quad k = 0, \ldots, m-1, \quad \mu'_{0k} := \frac{1-\sigma}{2}\mu_{0k}, \quad k = m, \ldots, r-1.$$

Due to the identity $4 - (3-\sigma)^2\sigma = (1-\sigma)^2(4-\sigma)$, we have $\sum_{k=0}^{r-1} |\mu'_{0k}|^2 \equiv 1$. Next we add the function $\mu'_{0r} \equiv 0$. Since $\mu'_{0r} \equiv \text{const}$, the row $\mu'_{00}, \ldots, \mu'_{0r}$ can be extended to a unitary matrix using Householder transformation. Namely, the entries $\mu_{\nu k}$, $\nu = 1, \ldots, r$, of the matrix $\mathcal{M}^{ext}$ are given by

$$\mu_{\nu r} := \overline{\mu_{0,r-\nu}}, \; \mu_{\nu k} := \delta_{r-\nu,k} - \mu_{0k}\overline{\mu_{0,r-\nu}}, \; k = 0, \ldots, r-1, \nu = 1, \ldots, r.$$

All conditions of Theorem 8.3 are satisfied, and hence the matrix $\mathcal{M}$ consisting of the first m columns of the constructed unitary matrix $\{\mu_{\nu k}\}_{\nu,k=0}^{r}$ generates a compactly supported tight wavelet frame with **VM$^{\mathbf{n}}$** property.

A frame constructed according to Theorem 8.2 has approximation order $n+1$. Note also that our method is suitable for practical use. Except several calculating roots of univariate trigonometric polynomials (in the sense of Riesz's Lemma), all steps of the algorithm are described explicitly. For a large class of matrices M, slight modification of the algorithm leads to its simplification and significantly fewer number of wavelet functions.

Acknowledgements This research is supported by the RFBR grant # 15-01-05796 and the SPbGU grant # 9.38.198.2015

References

1. Charina, M., Chui, C.K.: Tight frames of compactly supported multivariate multi-wavelets. J. Comput. Appl. Math **233**, 2044–2061 (2010)
2. Charina M., Putinar M., Scheiderer, C., Stčkler. J.: An algebraic perspective on multivariate tight wavelet frames. Constr. Approx. **38**, 253–276 (2013)

3. Chui, C.K., He, W.: Construction of multivariate tight frames via Kronecker products. Appl. Comput. Harmon. Anal. **11**, 305–312 (2001)
4. Duffin, R.J., Schaeffer, A.S.: A class of nonharmonic Fourier series. Trans. Am. Math. Soc. **72**, 341–366 (1952)
5. Han, B.: Compactly supported tight wavelet frames and orthonormal wavelets of exponential decay with a general dilation matrix. J. Comput. Appl. Math. **155**, 43–67 (2003)
6. Han, B.: The projection method for multidimensional framelet and wavelet analysis. Math. Model. Nat. Phenom. **9**, 83–110 (2014)
7. Ron, A., Shen, Z.: Affine systems in $\mathbb{L}_2(R^d)$: the analysis of the analysis operator. J. Func. Anal. **148**, 408–447 (1997)
8. San Antolin, A., Zalik, R.A.: Some smooth compactly supported tight wavelet frames with vanishing moments. J. Fourier Anal. Appl. **22**, 887–909 (2016)
9. Skopina, M.: Tight wavelet frames. Dokl. Math. **77**, 182–185 (2008)
10. Skopina, M.: On construction of multivariate wavelet frames. Appl. Comput. Harmon. Anal. **27**, 55–72 (2009)
11. Skopina, M.: On construction of multivariate wavelets with vanishing moments. Appl. Comput. Harmon. Anal. **39**, 375–390 (2013)
12. Ya Novikov, I., Yu Protasov, V., Skopina, M.A.: Wavelet Theory. Translations Mathematical Monographs, vol. 239. AMS, Providence, Rhode Island (2011)

Chapter 9
First-Order and Second-Order Adjoint Methods for the Inverse Problem of Identifying Non-linear Parameters in PDEs

M. Cho, B. Jadamba, R. Kahler, A.A. Khan and M. Sama

Abstract The primary objective of this work is to develop a computational framework for the inverse problem of identifying variable parameters appearing non-linearly in a variational problem. We propose a new first-order adjoint method and two new second-order adjoint methods. All the derivative formulas are given in continuous as well as discrete setting. Detailed numerical examples are given to show the feasibility of the proposed framework.

Keywords Inverse problems · Variational problems · Regularization
Output least-squares · First-order adjoint method · Second-order adjoint method

9.1 Introduction

Let V be a Hilbert space and let B be a Banach space. Let A be a nonempty, closed and convex subset of B. Let $T : B \times V \times V \to \mathbb{R}$ be linear and symmetric with respect to the second and third arguments, but in general non-linear in the first argument, and let $m : V \to \mathbb{R}$ be a linear and continuous map. We assume that T is twice Fréchet differentiable with respect to the first argument and the partial

---

M. Cho · B. Jadamba · R. Kahler · A.A. Khan (✉)
Center for Applied and Computational Mathematics, School of Mathematical Sciences, Rochester Institute of Technology, Rochester, NY 14623, USA
e-mail: aaksma@rit.edu

M. Cho
e-mail: mxcsma1@rit.edu

B. Jadamba
e-mail: bxjsma@rit.edu

R. Kahler
e-mail: rak9698@rit.edu

M. Sama
Departamento de Matemática Aplicada, Universidad Nacional de Educación a Distancia, Calle Juan del Rosal, 12, 28040 Madrid, Spain
e-mail: msama@ind.uned.es

P. Manchanda et al. (eds.), *Industrial Mathematics and Complex Systems*,
Industrial and Applied Mathematics, DOI 10.1007/978-981-10-3758-0_9

derivative of T with respect a, denoted by $\partial_a T(a, u, v)$, is linear and symmetric with respect to the second and the third arguments, that is, for all $u, v \in V$, we have $\partial_a T\,(a, u, v) = \partial_a T\,(a, v, u)$ and $\partial_a T\,(a, \cdot, v)$ and $\partial_a T\,(a, u, \cdot)$ are linear. We assume that there are constants $\alpha > 0$, $\beta >$ such that

$$T(a, u, v) \leq \beta \|a\|_B \|u\|_V \|v\|_V, \quad \text{for all } u, v \in V, \ a \in B, \tag{9.1a}$$
$$T(a, u, u) \geq \alpha \|u\|_V^2, \quad \text{for all } u \in V, \ a \in A. \tag{9.1b}$$

Consider the following variational problem: Given $a \in A$, find $u \in V$ such that:

$$T(a, u, v) = m(v), \quad \text{for every } v \in V. \tag{9.2}$$

Our focus is on the inverse problem of identifying a parameter $a \in A$ for which a solution u of the variational problem (9.2) is closest, in some norm, to a given measurement z of u.

In view of the imposed coercivity and continuity of $T(\cdot, \cdot)$, the Lax–Milgram lemma at once ensures that for every $a \in A$, there exists a unique $u \in V$ satisfying (9.2), that is, for every $a \in A$, the map $a \to S(a) = u(a)$ is well-defined and single-valued.

9.2 Output Least-Squares Formulation for the Inverse Problem

We consider the following output least-squares (OLS) functional

$$J(a) = \frac{1}{2} \|u(a) - z\|^2 \tag{9.3}$$

where z is the measured data and $u(a)$ solves the variational problem (9.2).

Since inverse problems are known to be ill-posed, the OLS functional (9.3) needs to be regularized (see [3–6]). Consequently, we will study the following regularized optimization problem: Find $a \in A$ by solving the following:

$$\min_{a \in A} J_\kappa(a) = \frac{1}{2} \|u(a) - z\|^2 + \kappa R(a), \tag{9.4}$$

where, given a Hilbert space H, $R : H \to \mathbb{R}$ is a regularizer, $\kappa > 0$ is a regularization parameter, $u(a)$ is the unique solution of (9.2) that corresponds to the coefficient a and z is the measured data. Throughout this work, we assume that the map R is twice differentiable.

We have the following result concerning the solvability of the above optimization problem:

Theorem 9.1 *Assume that the Hilbert space H is compactly embedded into the space B, $A \subset H$ is nonempty, closed and convex, the map R is convex, lower semi-continuous and there exists $\alpha > 0$ such that $R(a) \geq \alpha\|a\|_H^2$, for every $a \in A$. Then* (9.4) *has a nonempty solution set.*

Remark 9.1 A natural choice of spaces and the regularizer involved in the above result is $H = H_2(\Omega)$, $B = L_\infty(\Omega)$ and $R(a) = \|\cdot\|_{H_2(\Omega)}^2$. Evidently, this choice is only satisfactory for smooth parameters. However, for discontinuous parameters total variation regularization can be employed and the framework given in [6] easily extends to the case of non-quadratic regularizers.

The following result will play a key role:

Theorem 9.2 *For each a in the interior of A, $u(a)$ is infinitely differentiable at a. The first derivative of u at a in the direction δa, denoted by $\delta u(a)$, is the unique solution of the variational problem:*

$$T(a, \delta u, v) = -\partial_a T(a, u, v)(\delta a), \quad \textit{for every } v \in V. \tag{9.5}$$

Proof The proof is based on standard arguments.

9.3 Derivative Formulae

In this section, our objective is to derive a first-order adjoint method to compute the first-order derivative of the regularized OLS, and two second-order adjoint methods for the computation of its second-order derivative. We note that the adjoint approach is a very commonly used idea for derivative computation in inverse and shape optimization problems. Some of the recent developments, mostly for the first-order adjoint methods, can be found in [1, 2, 7–9] and the cited references therein.

9.3.1 *First-Order Adjoint Method*

Since the regularized output least-squares functional is given by:

$$J_\kappa(a) = \frac{1}{2}\|u(a) - z\|^2 + \kappa R(a),$$

it follows, by using the chain rule, that the derivative of J_κ at $a \in A$ in any direction δa is given by

$$DJ_\kappa(a)(\delta a) = \langle Du(a)(\delta a), u(a) - z\rangle + \kappa DR(a)(\delta a),$$

where $Du(a)(\delta a)$ is the derivative of the coefficient-to-solution map u and $DR(a)(\delta a)$ is the derivative of the regularizer R, both computed at a in the direction δa.

For an arbitrary $v \in V$, we define the functional $L_\kappa : B \times V \to \mathbb{R}$ by

$$L_\kappa(a, v) = J_\kappa(a) + T(a, u, v) - m(v).$$

Since $u(a)$ is the solution of variational problem (9.2), for every $v \in V$, we have $L_\kappa(a, v) = J_\kappa(a)$, and consequently, for every $v \in V$ and for every direction δa, we have

$$\partial_a L_\kappa(a, v)\,(\delta a) = DJ_\kappa(a)\,(\delta a)\,.$$

The key idea for the first-order adjoint method is to choose v to bypass a direct computation of $\delta u = Du(a)(\delta a)$. To get an insight into such a choice of v, we use the chain rule again to obtain

$$\begin{aligned}\partial_a L_\kappa(a, v)\,(\delta a) = {} & \langle Du(a)(\delta a), u - z\rangle + \kappa DR(a)(\delta a) + \partial_a T(a, u, v)(\delta a) \\ & + T(a, Du(a)(\delta a), v). \end{aligned} \tag{9.6}$$

For $a \in A$, let $w(a)$ be the unique solution of the variational problem

$$T(a, w, v) = \langle z - u, v\rangle\,, \quad \text{for every } v \in V, \tag{9.7}$$

where the right-hand side involve the solution u of (9.2) and the data z.

By plugging $v = w$ in (9.6), using (9.7) and the symmetry of T and $\partial_a T$, we obtain

$$\begin{aligned}
&\partial_a L_\kappa(a, w)\,(\delta a) = \langle Du(a)(\delta a), u - z\rangle + \kappa DR(a)(\delta a) + \partial_a T(a, u, w)(\delta a) \\
&+ T(a, Du(a)(\delta a), w) \\
&= \langle Du(a)(\delta a), u - z\rangle + \kappa DR(a)(\delta a) + T(a, w, Du(a)(\delta a)) + \partial_a T(a, w, u)(\delta a) \\
&= \langle Du(a)(\delta a), u - z\rangle + \kappa DR(a)(\delta a) + \langle z - u, Du(a)(\delta a)\rangle + \partial_a T(a, w, u)(\delta a) \\
&= \kappa DR(a)(\delta a) + \partial_a T(a, w, u)(\delta a),
\end{aligned}$$

which yields the following formula for the first-order derivative of J_κ:

$$DJ_\kappa(a)\,(\delta a) = \kappa DR(a)(\delta a) + \partial_a T(a, w, u)(\delta a). \tag{9.8}$$

Summarizing, the following scheme computes $DJ_\kappa(a)\,(\delta a)$ for the direction δa:

1. Compute $u(a)$ by using (9.2).
2. Compute $w(a)$ by using (9.7).
3. Compute $DJ_\kappa(a)\,(\delta a)$ by using (9.8).

Remark 9.2 Note that if T is linear in the first argument, then (9.8) becomes

$$DJ_\kappa(a)\,(\delta a) = \kappa DR(a)(\delta a) + T(\delta a, w, u), \tag{9.9}$$

and the above algorithm can be used for this case by computing $DJ_\kappa(a)(\delta a)$ by (9.9).

9.3.2 Second-Order Adjoint Method

We now give a second-order adjoint method for the computation of the second-order derivative of the regularized OLS functional. The aim is to give a formula for the second-order derivative that does not involve the second-order derivative of the parameter-to-solution map u. The key idea is to compute δu directly by using Theorem 9.2 while the computation of $\delta^2 u$ is avoided by using an adjoint approach.

Given a fixed direction δa_2 and an arbitrary $v \in V$, we define

$$\begin{aligned} L_\kappa(a, v) &= DJ_\kappa(a)(\delta a_2) + T(a, Du(a)(\delta a_2), v) + \partial_a T(a, u, v)(\delta a_2) \\ &= \langle Du(a)(\delta a_2), u - z\rangle + \kappa DR(a)(\delta a_2) + T(a, Du(a)(\delta a_2), v) \\ &+ \partial_a T(a, u, v)(\delta a_2). \end{aligned}$$

Evidently, by the definition of L_κ, for every $v \in V$, and any direction δa_1, we have

$$\partial_a L_\kappa(a, v)(\delta a_1) = D^2 J_\kappa(a)(\delta a_1, \delta a_2).$$

Computing this derivative of L_κ in the direction δa_1 directly, we have

$$\begin{aligned} \partial_a L_\kappa(a, v)(\delta a_1) &= \left\langle D^2 u(a)(\delta a_1, \delta a_2), u - z\right\rangle + \langle Du(a)(\delta a_2), Du(a)(\delta a_1)\rangle \\ &+ \kappa D^2 R(a)(\delta a_1, \delta a_2) + T(a, D^2 u(a)(\delta a_1, \delta a_2), v) + \partial_a T(a, Du(a)(\delta a_2), v)(\delta a_1) \\ &+ \partial_a T(a, Du(a)(\delta a_1), v)(\delta a_2) + \partial_a^2 T(a, u, v)(\delta a_1, \delta a_2). \end{aligned}$$

Let $w(a)$ be the solution of the variational problem (9.7). By plugging $v = w$ in the above, we get

$$\begin{aligned} \partial_a L_\kappa(a, w)(\delta a_1) &= \left\langle D^2 u(a)(\delta a_1, \delta a_2), u - z\right\rangle + \langle Du(a)(\delta a_2), Du(a)(\delta a_1)\rangle \\ &+ \kappa D^2 R(a)(\delta a_1, \delta a_2) + T(a, D^2 u(a)(\delta a_1, \delta a_2), w) + \partial_a T(a, Du(a)(\delta a_2), w)(\delta a_1) \\ &+ \partial_a^2 T(a, u, w)(\delta a_1, \delta a_2) + \partial_a T(a, Du(a)(\delta a_1), w)(\delta a_2) \\ &= \left\langle D^2 u(a)(\delta a_1, \delta a_2), u - z\right\rangle + \langle Du(a)(\delta a_2), Du(a)(\delta a_1)\rangle + \kappa D^2 R(a)(\delta a_1, \delta a_2) \\ &+ \partial_a T(a, Du(a)(\delta a_2), w)(\delta a_1) + \partial_a T(a, Du(a)(\delta a_1), w)(\delta a_2) \\ &+ \partial_a^2 T(a, u, w)(\delta a_1, \delta a_2) - \left\langle D^2 u(a)(\delta a_1, \delta a_2), u - z\right\rangle \\ &= \kappa D^2 R(a)(\delta a_1, \delta a_2) + \langle Du(a)(\delta a_2), Du(a)(\delta a_1)\rangle + \partial_a T(a, Du(a)(\delta a_2), w)(\delta a_1) \\ &+ \partial_a T(a, Du(a)(\delta a_1), w)(\delta a_2) + \partial_a^2 T(a, u, w)(\delta a_1, \delta a_2). \end{aligned}$$

Consequently, it follows that:

$$\begin{aligned}D^2 J_\kappa(a)(\delta a_1, \delta a_2) &= \kappa D^2 R(a)(\delta a_1, \delta a_2) + \langle Du(a)(\delta a_2), Du(a)(\delta a_1)\rangle \\ &\quad + \partial_a T(a, Du(a)(\delta a_2), w)(\delta a_1) + \partial_a T(a, Du(a)(\delta a_1), w)(\delta a_2) \\ &\quad + \partial_a^2 T(a, u, w)(\delta a_1, \delta a_2).\end{aligned}$$

In particular, we have the following:

$$\begin{aligned}D^2 J_\kappa(a)(\delta a, \delta a) &= \kappa D^2 R(a)(\delta a, \delta a) + \langle \delta u, \delta u\rangle + 2\partial_a T(a, \delta u, w)(\delta a) \\ &\quad + \partial_a^2 T(a, u, w)(\delta a, \delta a).\end{aligned} \tag{9.10}$$

Summarizing, the following scheme computes the derivative $D^2 J_\kappa(a)(\delta a, \delta a)$ for any direction δa :

1. Compute $u(a)$ by (9.2).
2. Compute δu by (9.5).
3. Compute $w(a)$ by (9.7).
4. Compute $D^2 J_\kappa(a)(\delta a, \delta a)$ by (9.10).

Remark 9.3 Note that if T is linear in the first argument, then (9.5) reads

$$T(a, \delta u, v) = -T(\delta a, u, v), \quad \text{for every } v \in V. \tag{9.11}$$

Moreover, by using the above analogue and the linearity of T, the formula (9.10) reads:

$$D^2 J_\kappa(a)(\delta a, \delta a) = \kappa D^2 R(a)(\delta a, \delta a) + \langle \delta u, \delta u\rangle + 2T(\delta a, \delta u, \bar{w}), \tag{9.12}$$

and hence the above scheme can be modified by computing $D^2 J_\kappa(a)(\delta a, \delta a)$ by (9.12) for this case.

9.3.3 Second-Order Derivative Using the First-Order Adjoint Formula

We now give a direct second-order adjoint method for the computation of the second-order derivative of the regularized OLS functional. The goal again remains to give a formula that does not involve the second-order derivative of the map u. The key idea of the second-order adjoint approach is to use the derivative characterization Theorem 9.2 twice to avoid a direct computation of $\delta^2 u$.

We begin with defining the functional $L_\kappa : A \times V \times V \to \mathbb{R}$ by

$$\begin{aligned} L_\kappa(a,t,s) &= DJ_\kappa(a)(\delta a_2) + T(a,u,t) - m(t) + T(a,w,s) - \langle z-u,s\rangle \\ &= \kappa DR(a)(\delta a_2) + \partial_a T(a,u,w)(\delta a_2) + T(a,u,t) - m(t) + T(a,w,s) - \langle z-u,s\rangle\,, \end{aligned}$$

where δa_2 is a fixed direction, u is the solution of (9.2), w is the solution of (9.7), t and s are arbitrary elements in V, and for $DJ_\kappa(a)(\delta a_2)$ formula (9.8) was used. By the definition of the above functional, for every $t, s \in V$, we have

$$\partial_a L_\kappa(a,t,s)(\delta a_1) = D^2 J_\kappa(a)(\delta a_1, \delta a_2).$$

The derivative of L_κ at $a \in A$ in the direction δa_1 can easily be computed as follows:

$$\begin{aligned} \partial_a L_\kappa(a,t,s)(\delta a_1) &= \kappa D^2 R(a)(\delta a_1,\delta a_2) + \partial_a^2 T(a,u,w)(\delta a_1,\delta a_2) \\ &\quad + \partial_a T(a, Du(a)(\delta a_1), w)(\delta a_2) + \partial_a T(a,u,Dw(a)(\delta a_1))(\delta a_2) \\ &\quad + T(a, Du(a)(\delta a_1), t) + \partial_a T(a,u,t)(\delta a_1) + T(a, Dw(a)(\delta a_1), s) \\ &\quad + \partial_a T(a,w,s)(\delta a_1) + \langle Du(a)(\delta a_1), s\rangle\,. \end{aligned} \tag{9.13}$$

By plugging $v = Dw(a)(\delta a_1)$ in (9.5), we get

$$T(a, Dw(a)(\delta a_1), Du(a)(\delta a_2)) + \partial_a T(a,u,Dw(a)(\delta a_1))(\delta a_2) = 0. \tag{9.14}$$

Moreover, it can be shown that the derivative $Dw(a)(\delta a_2)$ of the unique solution $w(a)$ of (9.7) in any direction δa_2 is characterized as the solution of the following variational problem:

$$T(a, Dw(a)(\delta a_2), v) = -\partial_a T(a,w,v)(\delta a_2) - \langle Du(a)(\delta a_2), v\rangle\,, \text{ for all } v \in V. \tag{9.15}$$

By plugging $v = Du(a)(\delta a_1)$ into (9.15), we deduce

$$\begin{aligned} T(a, Du(a)(\delta a_1), Dw(a)(\delta a_2)) &+ \langle Du(a)(\delta a_2), Du(a)(\delta a_1)\rangle \\ &+ \partial_a T(a,w,Du(a)(\delta a_1))(\delta a_2) = 0. \end{aligned} \tag{9.16}$$

By setting $s = Du(a)(\delta a_2)$ and $t = Dw(a)(\delta a_2)$ in (9.13) and combining the resulting expression with (9.14) and (9.16), we obtain the following:

$$\begin{aligned}\partial_a L_\kappa(a,t,s)(\delta a_1) &= \kappa D^2 R(a)(\delta a_1, \delta a_2) + \partial_a^2 T(a,u,w)(\delta a_1, \delta a_2)\\ &\quad + \partial_a T(a, Du(a)(\delta a_1), w)(\delta a_2) + \partial_a T(a,u,Dw(a)(\delta a_1))(\delta a_2)\\ &\quad + T(a, Du(a)(\delta a_1), Dw(a)(\delta a_2)) + \partial_a T(a,u,Dw(a)(\delta a_2))(\delta a_1)\\ &\quad + T(a, Dw(a)(\delta a_1), Du(a)(\delta a_2)) + \partial_a T(a,w,Du(a)(\delta a_2))(\delta a_1)\\ &\quad + \langle Du(a)(\delta a_1), Du(a)(\delta a_2)\rangle\\ &= \kappa D^2 R(a)(\delta a_1, \delta a_2) + \partial_a^2 T(a,u,w)(\delta a_1, \delta a_2)\\ &\quad + \partial_a T(a, Du(a)(\delta a_2), w)(\delta a_1) + \partial_a T(a,u,Dw(a)(\delta a_2))(\delta a_1),\end{aligned}$$

and consequently we obtain the following formula for the second-order derivative of the regularized OLS that has no involvement of the second-order derivatives of the solution map:

$$\begin{aligned}D^2 J_\kappa(a)(\delta a_1, \delta a_2) &= \kappa D^2 R(a)(\delta a_1, \delta a_2) + \partial_a^2 T(a,u,w)(\delta a_1, \delta a_2)\\ &\quad + \partial_a T(a, Du(a)(\delta a_2), w)(\delta a_1) + \partial_a T(a,u,Dw(a)(\delta a_2))(\delta a_1).\end{aligned}$$

In particular,

$$\begin{aligned}D^2 J_\kappa(a)(\delta a, \delta a) &= \kappa D^2 R(a)(\delta a, \delta a) + \partial_a^2 T(a,u,w)(\delta a, \delta a)\\ &\quad + \partial_a T(a, \delta u, w)(\delta a) + \partial_a T(a,u,\delta w).\end{aligned} \tag{9.17}$$

Summarizing, the following scheme computes $D^2 J_\kappa(a)(\delta a, \delta a)$ given $a \in A$ and a direction δa:

1. Compute u by (9.2).
2. Compute δu by (9.5).
3. Compute w by (9.7).
4. Compute δw by (9.15).
5. Compute $D^2 J_\kappa(a)(\delta a, \delta a)$ by (9.17).

Remark 9.4 Note that if T is linear in the first argument, then (9.15) becomes as follows:

$$T(a, Dw(a)(\delta a), v) = -T(\delta a, w, v) - \langle Du(a)(\delta a), v\rangle\,, \quad \text{for all } v \in V. \tag{9.18}$$

Moreover, by using (9.11) and the above analogue, the formula (9.17) reads as follows:

$$\begin{aligned}D^2 J_\kappa(a)(\delta a, \delta a) &= \kappa D^2 R(a)(\delta a, \delta a) + T(\delta a, u, Dw(a)(\delta a))\\ &\quad + T(\delta a, w, Du(a)(\delta a)),\end{aligned} \tag{9.19}$$

and hence the above algorithm can be modified by computing $D^2 J_\kappa(a)(\delta a, \delta a)$ by (9.19) for this case.

9.4 Computational Framework

We begin with a triangulation T_h on Ω, L_h is the space of all piecewise continuous polynomials of degree d_a relative to T_h and U_h is the space of all piecewise continuous polynomials of degree d_u relative to T_h. Let the basis for A_h and U_h be given by $\{\varphi_1, \varphi_2, \ldots, \varphi_m\}$, and $\{\psi_1, \psi_2, \ldots, \psi_k\}$, respectively. The space A_h is then isomorphic to $\mathbb{R}^m$ and for any $a \in L_h$, we define $A \in \mathbb{R}^m$ by $A_i = a(x_i)$, for $i = 1, 2, \ldots, m$, where the nodal basis $\{\varphi_1, \varphi_2, \ldots, \varphi_m\}$ corresponds to the nodes $\{x_1, x_2, \ldots, x_m\}$. Conversely, each $A \in \mathbb{R}^m$ corresponds to $a \in A_h$ defined by $a = \sum_{i=1}^m A_i\varphi_i$. Similarly, $u \in U_h$ will correspond to $U \in \mathbb{R}^k$, where $U_i = u(z_i)$, $i = 1, 2, \ldots, k$, and $u = \sum_{i=1}^k U_i\psi_i$, where $z_1, z_2, \ldots, z_k$ are the nodes of the mesh defining U_h.

The discrete variational problem seeks, for each a_h, the unique $u_h \in V_h$ such that:

$$T(a_h, u_h, v) = m(v), \quad \text{for every } v \in V_h. \tag{9.20}$$

We define $S : \mathbb{R}^m \to \mathbb{R}^k$ to be the finite element solution operator that assigns to each $a_h \in A_h$, the unique discrete solution $u_h \in U_h$. Then $S(A) = U$, where U is given by the following:

$$K(A)U = F, \tag{9.21}$$

where $K(A)_{i,j} = T(a, \psi_j, \psi_i), \quad i, j = 1, 2, \ldots, k$.

We discretize T and the derivatives $\partial_a T$ and $\partial_a^2 T$ directly to get

$$\begin{aligned}
T(a, u, v) &= \sum_{i,j=1}^{k} T(a, U_i\psi_i, V_j\psi_j) = V^T K(A)U \\
\partial_a T(a, u, v)(\delta a) &= \sum_{i,j=1}^{k} \partial_a T(a, U_i\psi_i, V_j\psi_j)(\delta a) = V^T \widetilde{K}_A(\delta A)U, \\
\partial_a^2 T(a, u, v)(\delta a, \delta a) &= \sum_{i,j=1}^{k} \partial_a^2 T(a, U_i\psi_i, V_j\psi_j)(\delta a, \delta a) = V^T \widetilde{K}_A^2(\delta A, \delta A)U,
\end{aligned}$$

with

$$\begin{aligned}
\delta a &= \sum_{s=1}^{m} (\delta A)_s \varphi_s \\
\widetilde{K}_A(\delta A)_{i,j} &= \partial_a T(a, \psi_i, \psi_j)(\delta a) \\
\widetilde{K}_A^2(\delta A, \delta A)_{i,j} &= \partial_a^2 T(a, \psi_i, \psi_j)(\delta a, \delta a).
\end{aligned}$$

9.4.1 Gradient Computation by a Direct Approach

Recall that the regularized OLS functional is given by

$$J_\kappa(a) = \frac{1}{2} \|u(a) - z\|^2 + \kappa R(a),$$

where z is the measured data and $u(a)$ solves (9.2). The discrete analogue of the above functional is given by

$$J_\kappa(A) := \frac{1}{2}(U - Z)^T \mathbb{M}(U - Z) + \kappa R(A),$$

where U solves (9.21), the matrix $\mathbb{M}$ is given by $\langle u_1, u_2\rangle_V = U_2^T \mathbb{M} U_1$, for any $u_1, u_2 \in V_h$, and Z is the discrete data.

The first-order derivative of the above functional

$$DJ_\kappa(a)(\delta a) = \langle \delta u, u - \bar{z}\rangle + DR(a)(\delta a),$$

involves $\delta u(a)$, which by Theorem 9.2, solves the following variational problem:

$$T(a, \delta u, v) = -\partial_a T(a, u, v)(\delta a), \quad \text{for every } v \in V.$$

By standard arguments, the discrete version of the above system reads as follows:

$$K(A)\delta U = -\widetilde{K}_A(\delta A)U = -\mathbb{K}_A(U)\delta A, \tag{9.22}$$

where $\mathbb{K}_A$ is the *directional stiffness matrix* given by $\widetilde{K}_A(\delta A)U = \mathbb{K}_A(U)\delta A$, for every $U \in \mathbb{R}^k$, $\delta A \in \mathbb{R}^m$. Let $\{E_1, E_2, \ldots, E_m\}$ be the basis of $\mathbb{R}^m$. Since $\widetilde{K}_A(\delta A)U = \left(\sum_{s=1}^m \delta A_s \widetilde{K}_A(E_s)\right) U$, the matrix $\mathbb{K}_A(U)$, consists of columns $\widetilde{K}_A(E_s)U$, that is, $\mathbb{K}_A(U) := \left[\widetilde{K}_A(E_1)U \quad \widetilde{K}_A(E_2)U \quad \cdots \quad \widetilde{K}_A(E_m)U\right]$.

Nonetheless, the directional stiffness matrices has the following explicit expression:

$$\mathbb{K}_A(U)_{i,j} = \partial_a T(a, u, \varphi_j)(\chi_i) \text{ for every } i = 1, \ldots, k, \; j = 1, \ldots, m.$$

The Jacobian ∇U is computed by solving m equations:

$$K(A)\nabla_i U = -\mathbb{K}_A(U)_i, \quad i = 1, \ldots, m, \tag{9.23}$$

where $\mathbb{K}_A(U)_i$ denotes ith column. A discrete gradient formula is then given by the following:

$$DJ_\kappa(A)(\delta A) = \delta U^{\mathrm{T}} \mathbb{M}(U - Z) + \kappa \delta A^{\mathrm{T}} \nabla R(A) = \delta A^{\mathrm{T}} \nabla U^{\mathrm{T}} \mathbb{M}(U - Z) + \kappa \delta A^{\mathrm{T}} \nabla R(A),$$

which at once leads to the following expression for the gradient as follows:

$$\nabla J_\kappa(A) = \nabla U^{\mathrm{T}}\mathbb{M}\,(U - Z) + \kappa\nabla R\,(A)\,. \tag{9.24}$$

Summarizing, we propose the following scheme for the gradient computation:

Step 1. Compute U by solving linear system (9.21).
Step 2. Compute ∇U by solving m linear systems (9.23).
Step 3. Compute $\nabla J_\kappa(A)$ by using formula (9.24).

9.4.2 *Computation of the Gradient Using the First-Order Adjoint Method*

We shall now give a scheme for computing the gradient of the first-order adjoint approach. Recall that the first-order adjoint approach led to the following formula for the first-order derivative:

$$DJ_\kappa(a)(\delta a) \;= \kappa DR(a)(\delta a) + \partial_a T(a, u, w)(\delta a) \tag{9.25}$$

where u and w are the solutions of (9.2) and (9.7), respectively. The discrete counterparts of these elements are U, which solves (9.21), and W which solves the system:

$$K(A)W = \mathbb{M}(Z - U). \tag{9.26}$$

Therefore, the discrete derivative formula reads as follows:

$$DJ_\kappa(A)(\delta A) = \kappa\nabla R(A)(\delta A) + \delta A^{\mathrm{T}}\mathbb{K}_A(U)^{\mathrm{T}}W,$$

which at once leads to an explicit formula for the gradient as follows:

$$\nabla J(A) = \kappa\nabla R(A) + \mathbb{K}_A(U)^{\mathrm{T}}W. \tag{9.27}$$

We have the following scheme for the derivative:

Step 1. Compute U by solving linear system (9.21).
Step 2. Compute W by solving linear system (9.26).
Step 3. Compute $\nabla J(A)$ by using formula (9.27).

9.4.3 *Hessian Computation by the Second-Order Adjoint Approach*

We recall that the second-order adjoint approach led to the following formula

$$D_a^2 J_\kappa(a)(\delta a) = \kappa D^2 R(a)(\delta a, \delta a) + \langle \delta u, \delta u \rangle + 2\partial_a T(a, \delta u, w)(\delta a) + \partial_a^2 T(a, u, w)(\delta a, \delta a). \tag{9.28}$$

To give a discretization of the above formula, we note that for the first term, we have

$$D^2 R(a)(\delta a, \delta a) = \delta A^T \nabla^2 R(A) \delta A$$

where $\nabla^2 R(A) \in \mathbb{R}^{m\times m}$ is the Hessian matrix of the regularization functional.

For the second term, we have $\langle \delta u, \delta u \rangle = \delta A^T \nabla U^T \mathbb{M} \nabla U \delta A$, and for the third term, we note that

$$\partial_a T(a, \delta u, w)(\delta a) = \delta U^T \widetilde{K}_A(\delta A) W = \delta A^T \nabla U^T \mathbb{K}_A(W) \delta A,$$

and for the last term, we have

$$\partial_A^2 K(A, u, w)(\delta A, \delta A) = \delta A^T \widetilde{K}_A^2(A, U, W) \delta A,$$

where $\widetilde{K}_A^2(A, U, W) \in \mathbb{R}^{m\times m}$ is the matrix defined by the following:

$$\widetilde{K}_A^2(A, U, W)_{i,j} = \partial_a^2 T(a, u, w)(\varphi_i, \varphi_j), \quad \text{for every } i, j = 1, \ldots, m.$$

In view of the above formulae, we deduce an explicit formula for the Hessian as follows:

$$\nabla^2 J_\kappa(A) = \kappa \nabla^2 R(A) + \nabla U^T \mathbb{M} \nabla U + 2\nabla U^T \mathbb{K}_A(W) + \widetilde{K}_A^2(A, U, W). \tag{9.29}$$

We have the following scheme for the computation of the Hessian of the regularized OLS:

Step 1. Compute U by solving linear system (9.22).
Step 2. Compute W by solving linear system (9.26).
Step 3. Compute ∇U by solving m linear systems (9.23).
Step 4. Compute $\nabla^2 J_\kappa(A)$ by using formula (9.29).

9.5 Numerical Results

In this section we present some numerical results. We consider the following elliptic boundary value problem (Example 1)

$$\begin{aligned} -\nabla \cdot (a \nabla u) &= f \text{ in } \Omega \\ u &= 0 \text{ on } \partial\Omega \end{aligned} \tag{9.30}$$

where the domain is $\Omega = [0, 1]^2$. The exact coefficient a and the load function f are given by the following:

$$a(x, y) = 1 + \frac{0.5}{1 + e^{50||p_1-(x,y)||^2-3}} + \frac{0.3}{1 + e^{100||p_2-(x,y)||^2-3}}$$
$$f(x, y) = 1 + 4||(x, y)||$$

where $p_1 = (0.6, 0.3)$ and $p_2 = (0.4, 0.75)$. Additionally, we consider an example (Example 2) again with the same parameter as Example 1 but that appears in the following:

$$-\nabla \cdot (e^a \nabla u) = f \text{ in } \Omega \tag{9.31}$$

We note that in Example 2, T is non-linear in the first argument. For these examples, the exact solution u is not known. The measured solution z is computed by solving the problem with the exact parameter a and f. Thereafter the noisy data is created from a uniform distribution on the interval $[-\alpha, \alpha]$ with $\alpha = 0.001$, where α is the noise level, then added to the measured solution z (Fig. 9.1). The optimization was performed using the Newton method for the second-order algorithm. All coefficients were identified in a finite dimensional space of dimension of 1129 on a mesh with 2136 triangles. The H^1 semi-norm regularization was used with $\kappa = 7 \cdot 10^{-8}$. For Example 2, the optimization problem was solved using the non-linear Conjugate Gradient algorithm for the first-order method and the Newton method for the second-order method, respectively. While the first-order method converges to the solution in 155 iterations, the second-order method needs 97 iterations when both are started from the same initial points and performed under the similar stopping criteria ($\nabla J < 10^{-10}$). Subfigure (a) in Figs. 9.2 and 9.3 shows how the comparison of the computed parameter at several algorithm steps can be seen. We consider (Example 3) the boundary value problem (9.30), but where the exact coefficient a is

$$a(x, y) = 1 + xy^2 \tag{9.32}$$

and the exact solution u is given by

$$u(x, y) = xy(1 - x)(1 - y) \tag{9.33}$$

The function f in (9.30) can be found from this information. In Figs. 9.4 and 9.5, we compare two algorithms to Example 3 where T is linear in all arguments.

Finally, we consider a similar example (Example 4) with identifying parameter a as Example 1 but the differential equation becomes as follows:

$$-\nabla \cdot (a^3 \nabla u) = f \text{ in } \Omega \tag{9.34}$$

where the exact parameter function a and the load function f are given by the following:

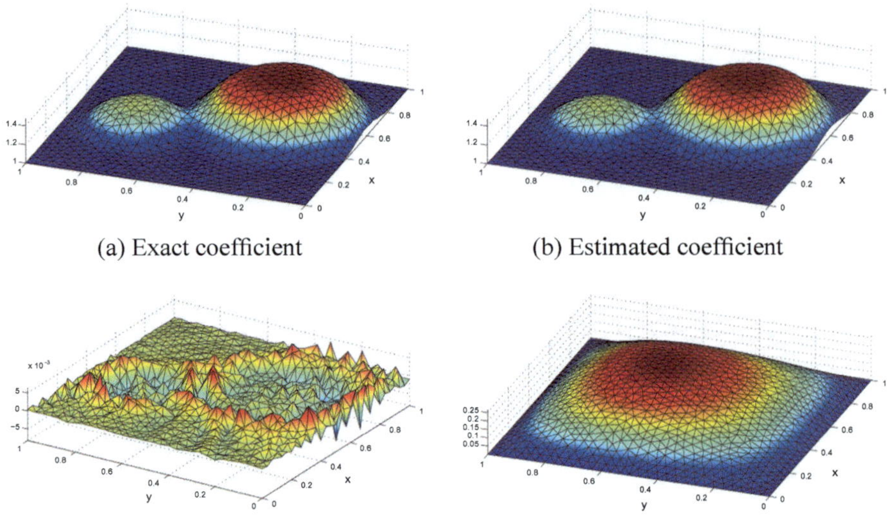

(a) Exact coefficient (b) Estimated coefficient

(c) Error in coefficient (d) Estimated solution

Fig. 9.1 Numerical results of Example 1 using the second-order algorithm (total 47 iterations)

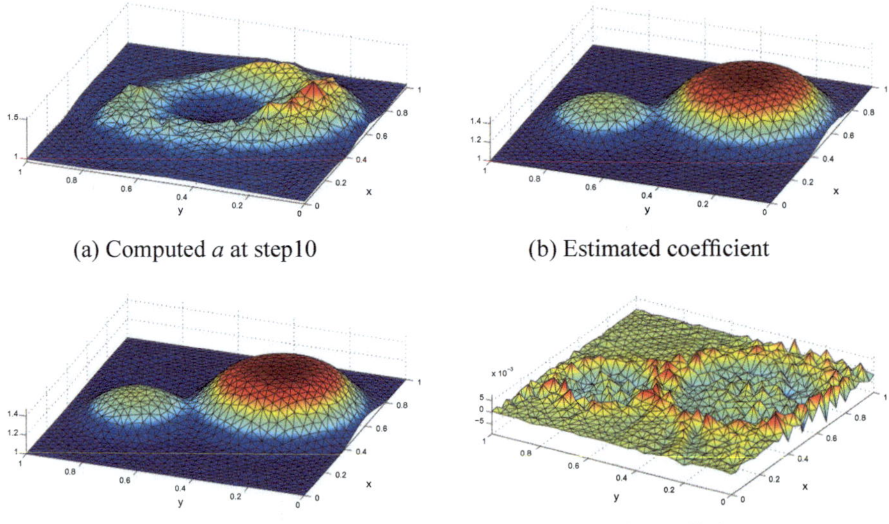

(a) Computed a at step10 (b) Estimated coefficient

(c) Exact coefficient (d) Error in coefficient

Fig. 9.2 Numerical results of Example 2 using the first-order algorithm (total 155 iterations)

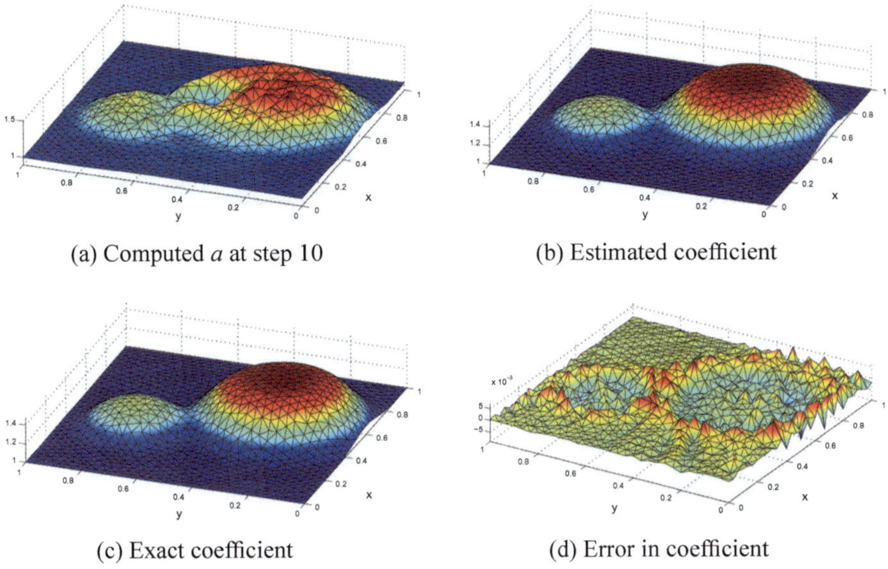

(a) Computed a at step 10

(b) Estimated coefficient

(c) Exact coefficient

(d) Error in coefficient

Fig. 9.3 Numerical results of Example 2 using the second-order algorithm (total 97 iterations)

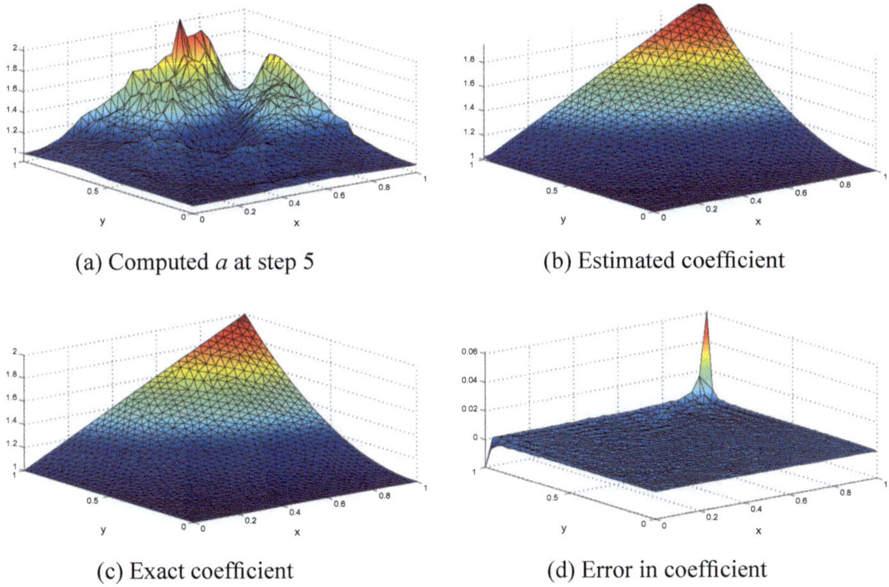

(a) Computed a at step 5

(b) Estimated coefficient

(c) Exact coefficient

(d) Error in coefficient

Fig. 9.4 Numerical results of Example 3 using the first-order algorithm (total 89 iterations)

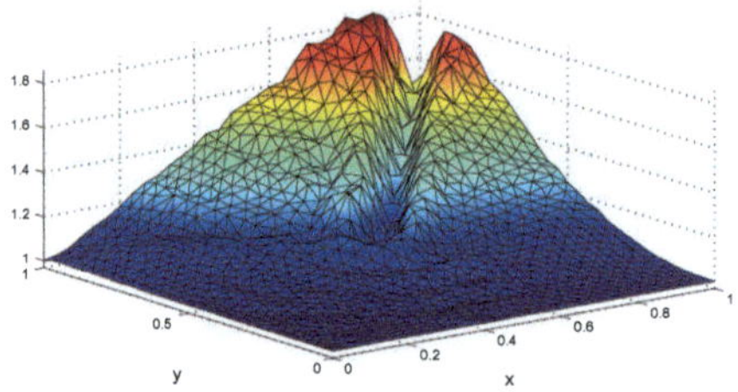

(a) Computed a at step 5

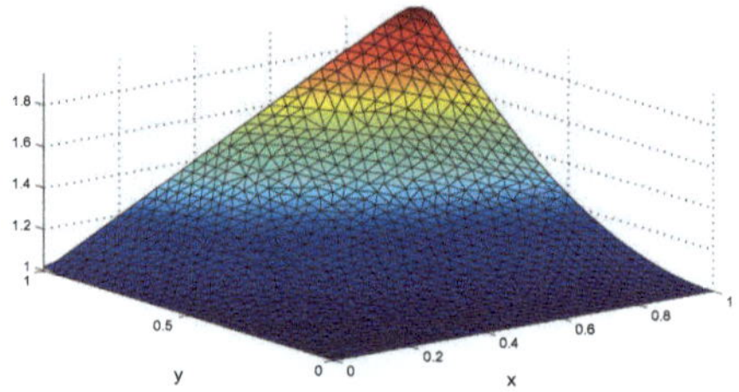

(b) Estimated coefficient

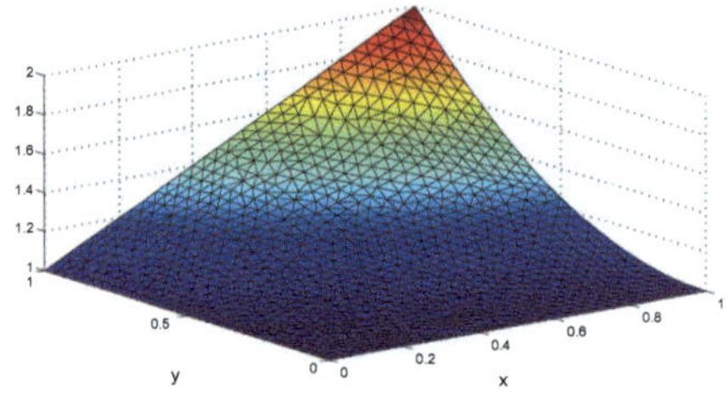

(c) Exact coefficient

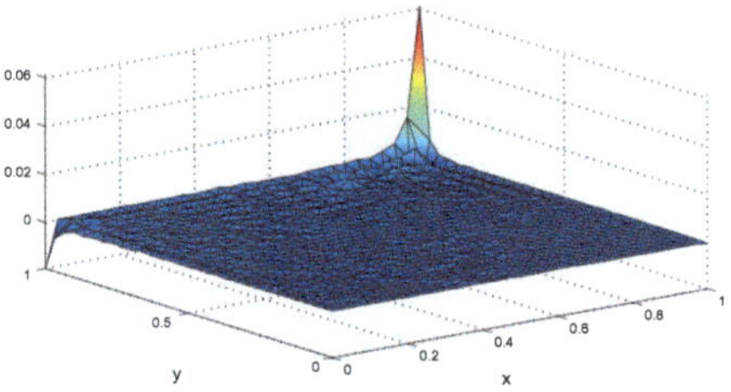

(d) Error in coefficient

Fig. 9.5 Numerical results of Example 3 using the second-order algorithm (total 29 iterations)

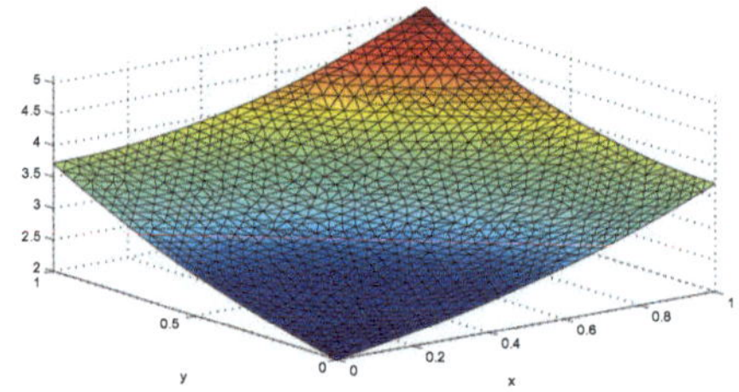

(a) Exact coefficient

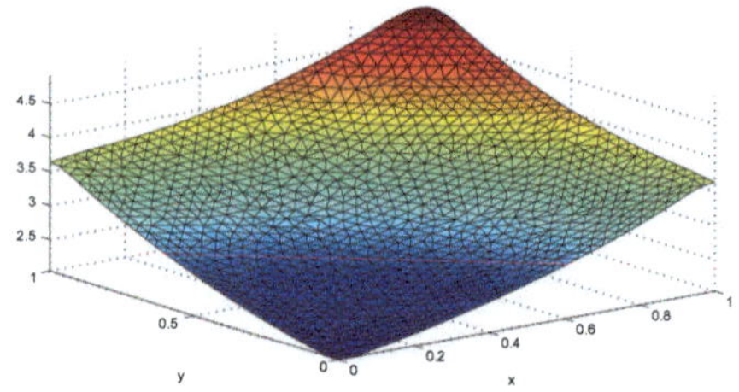

(b) Estimated coefficient

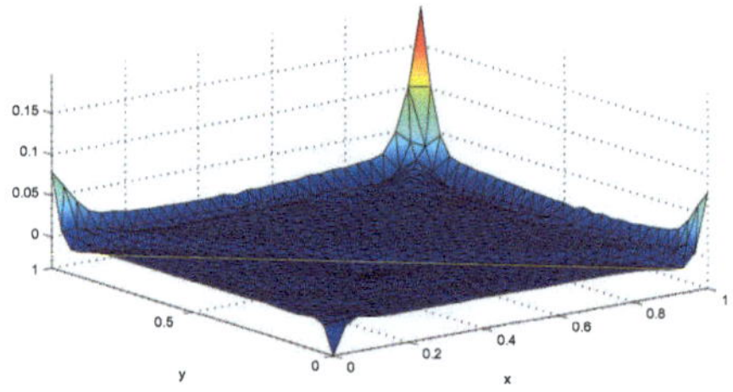

(c) Error in coefficient

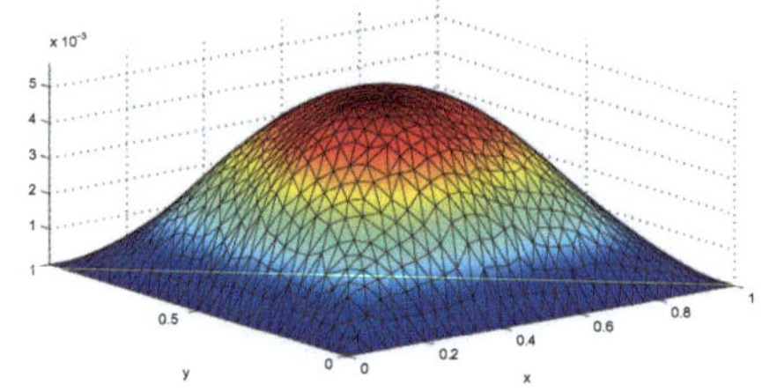

(d) Estimated solution

Fig. 9.6 Numerical results of Example 4 using the second-order algorithm (total 26 iterations)

$$a(x, y) = 1 + e^{||(x,y)||}$$

$$f(x, y) = 0.5 + ||(x, y)||^2 + 1.3[\sin(20||(x, y)||) + 1]$$

In this example, the regularization value is taken as $\kappa = 7 \times 10^{-9}$ (Fig. 9.6).

Acknowledgements Baasansuren Jadamba's work is supported by RIT's COS FEAD grant for 2016-2017. Akhtar Khan is supported by a grant from the Simons Foundation (#210443) and RIT's COS FEAD grant for 2016-2017. Miguel Sama's work is partially supported by Ministerio de Economa y Competitividad (Spain), project MTM2015-68103-P.

References

1. Cioaca, A., Sandu, A.: An optimization framework to improve 4D-Var data assimilation system performance. J. Comput. Phys. **275**, 377–389 (2014)
2. Dominguez, N., Gibiat, V., Esquerre, Y.: Time domain topological gradient and time reversal analogy: an inverse method for ultrasonic target detection. Wave Motion **42**(1), 31–52 (2005)
3. Gockenbach, M.S., Jadamba, B., Khan, A.A.: Numerical estimation of discontinuous coefficients by the method of equation error. Int. J. Math. Comput. Sci. **1**(3), 343–359 (2006)
4. Gockenbach, M.S., Khan, A.A.: Identification of Lamé parameters in linear elasticity: a fixed point approach. J. Ind. Manag. Optim. **1**(4), 487–497 (2005)
5. Gockenbach, M.S., Khan, A.A.: An abstract framework for elliptic inverse problems. II. An augmented Lagrangian approach. Math. Mech. Solids **14**(6), 517–539 (2009)
6. Jadamba, B., Khan, A.A., Rus, G., Sama, M., Winkler, B.: A new convex inversion framework for parameter identification in saddle point problems with an application to the elasticity imaging inverse problem of predicting tumor location. SIAM J. Appl. Math. **74**(5), 1486–1510 (2014)
7. Knopoff, D.A., Fernández, D.R., Torres, G.A., Turner, C.V.: Adjoint method for a tumor growth PDE-constrained optimization problem. Comput. Math. Appl. **66**(6), 1104–1119 (2013)
8. Reséndiz, E., Pinnau, R.: Adjoint-based optimization of particle trajectories in laminar flows. Appl. Math. Comput. **248**, 567–583 (2014)
9. Tottorelli, D.A., Michaleris, P.: Design sensitivity analysis: overview and review. Inverse Probl. Eng. **1**, 71–105 (1994)

Chapter 10
The Solution of the Hierarchy of Quantum Kinetic Equations with Delta Potential

Martin Brokate and Mukhayo Rasulova

Abstract The existence of a unique solution, in terms of initial data of the hierarchy of quantum kinetic equations with delta potential, has been proven. The proof is based on nonrelativistic quantum mechanics and application of semigroup theory methods.

10.1 The Dynamics of a One-Dimensional System Bosons Interacting via a Delta Function Potential

In 1931, Hans Bethe used an ansatz, now known as the Bethe ansatz, to find the exact eigenvalues and eigenvectors of the one-dimensional Heisenberg model [2]. This model describes a chain of spin-$\frac{1}{2}$ particles with nearest-neighbor interactions. Since then, the Bethe ansatz has been used to find a number of exactly solvable quantum many-body models in one dimension. In 1963, Lieb and Liniger used the Bethe ansatz to determine the exact solution of a one-dimensional model of interacting spinless particles with bosonic exchange symmetry [8]. In this model, now known as the Lieb–Liniger model, N bosons interact on a line of length L via a repulsive contact potential. Unlike the one-dimensional Heisenberg model, in which spins are fixed at discrete lattice sites, the Lieb–Liniger model is a continuum model in which the particles are free to move along a line. The Hamiltonian operator describing this model is given by [1]

$$H = -\sum_{i=0}^{N} \frac{\partial^2}{\partial x_i^2} + 2c \sum_{i<j} \delta(x_i - x_j). \tag{10.1}$$

M. Brokate (✉)
Technical University Munich, 85747 Garching, Germany
e-mail: brokate@ma.tum.de

M. Rasulova
National University Uzbekistan, Tashkent 100214, Uzbekistan
e-mail: rasulova@live.com

P. Manchanda et al. (eds.), *Industrial Mathematics and Complex Systems*,
Industrial and Applied Mathematics, DOI 10.1007/978-981-10-3758-0_10

Here Hamiltonian operator is total energy

$$H = -\frac{\hbar}{2m}\sum_{i=0}^{N}\frac{\partial^2}{\partial x_i^2} + V(x_1, \ldots, x_N) \tag{10.2}$$

and where the constant $\hbar$ is called the reduced Planck constant, m is the mass and V is the potential energy, $c > 0$ (the repulsive case) and $R : \quad all \quad 0 \leq x_i \leq L$. In (10.1) we suppose that $\hbar = 2m = 1$.

Function that is often used in physics is the Dirac delta function, designated $\delta(x - x_0)$. It is a generalized function that is defined as

$$\delta(x - x_0) = \begin{cases} \infty, & if \quad x = x_0, \\ 0 & if \quad x \neq x_0 \end{cases}$$

and has the property

$$\int_a^b f(x)\delta(x - x_0)dx = \begin{cases} f(x_0) & if \quad a < x_0 < b, \\ 0 & otherwise. \end{cases}$$

Let the particles be confined between 0 and L on the real line and let S_N denote the symmetric group of all N! permutations of the numbers (1;2;:::;N). The problem of solving the differential problem related to the Schrödinger equation is reduced by the Bethe ansatz to a much simpler system of algebraic equations, called the Bethe ansatz equations:

$$\psi|_{x_j = x_k + 0} = \psi|_{x_j = x_k - 0},$$

$$\left(\frac{\partial\psi}{\partial x_j} - \frac{\partial\psi}{\partial x_k}\right)|_{x_j = x_k + 0} - \left(\frac{\partial\psi}{\partial x_j} - \frac{\partial\psi}{\partial x_k}\right)|_{x_j = x_k - 0} = 2c\psi|_{x_j = x_k},$$

for all $x_j = x_k$ for all j, k=1,2, ...,N and $j \neq k$. The solution of the Schrödinger equation in $R_1 : 0 \leq x_1 \leq x_2 \leq \ldots \leq x_N \leq L$ in this case will have Bethe ansatz form:

$$\psi_B(x_1, \ldots x_N) = \sum_P a(P)Pexp(i\sum_{j=1}^{N} k_j x_j),$$

where the summation extends over all permutations P of an ordered of N numbers $k = k_1, \ldots, k_N$, and $a(P)$ are certain coefficients depending on P:

$$a(P) = -\frac{c - i(k_\alpha - k_\beta)}{c + i(k_\alpha - k_\beta)}.$$

In [5] using the ideas of Bethe ansatz given a method to solve the time-dependent Schrödinger equation for a system of one-dimensional bosons interacting via a repulsive delta function potential. Authors of [5] considered the problem of the solution of time-dependent Schrödinger equation:

$$H\psi = i\frac{\partial \psi}{\partial t},$$

with initial condition

$$\psi(x;0) = \psi(x_1, x_2, \dots, x_N; 0) = \psi_0(x_1, x_2, \dots, x_N).$$

The problem of solving reduced to the solution of equation

$$i\frac{\partial \psi}{\partial t} = -\sum_i \frac{\partial^2 \psi}{\partial x_j}$$

in the $\mathscr{R}^0 : x_1 < x_2, \dots < x_N$ with the initial condition Schrödinger equation

$$\psi(x;0) = \prod_{j=1}^{N} \delta(x_j - y_j) \tag{10.3}$$

in $\mathscr{R} : -\infty < x_1 \leq x_2 \leq \dots \leq x_N < \infty$ and the boundary condition:

$$(\frac{\partial}{\partial x_{j+1}} - \frac{\partial}{\partial x_j})\psi|_{x_{j+1}=x_j} = c\psi|_{x_{j+1}=x_j}. \tag{10.4}$$

for a system of one-dimensional bosons interacting via a delta function potential. In (10.3) $y_j \in \mathscr{R}$ are fixed and $y_1 < y_2, \dots < y_N$. Equation (10.4) is the effect of the δ function which is confined to the boundary of $\mathscr{R}$, e.g., on the hyperplanes $x_{j+1} = x_j$. The interest of authors [5] in the Lieb–Liniger model has arisen because of its connection to ultracold gases confined in a quasi-one-dimensional trap [5, 16].

In present paper, we give approach to solve the time-dependent BBGKY [3, 4] chain of quantum kinetic equations for a system of one-dimensional bosons interacting via a delta function potential.

We will consider a system N one-dimensional particles contained in a finite region (vessel) $\Lambda = L^3$ with volume $V = |\Lambda|$. The operators ρ_N^Λ and Hamiltonian H_N^Λ act in the space H with zero boundary condition [9]. Finally, we get the equation

$$i\frac{\partial \rho_s^\Lambda(t, x_1, \dots, x_s; x_1', \dots, x_s')}{\partial t} = [H_s^\Lambda, \rho_s^\Lambda](t, x_1, \dots, x_s; x_1', \dots, x_s')$$

$$+\frac{N}{V}\left(1 - \frac{s}{N}\right) Tr_{x_{s+1}} \sum_{1 \leq i \leq s} \left(\phi_{i,s+1}(|x_i - x_{s+1}|) - \phi_{i,s+1}(|x_i' - x_{s+1}|)\right) \times$$

$$\times\ \rho_{s+1}^\Lambda(t, x_1, \dots, x_s, x_{s+1}; x_1', \dots, x_s', x_{s+1}) \tag{10.5}$$

with initial date

$$\rho_s^\Lambda(t, x_1, \ldots, x_s; x_1', \ldots, x_s')|_{t=0} = \rho_s^\Lambda(0, x_1, \ldots, x_s; x_1', \ldots, x_s').$$

for $1 \leq s < N$. For $s = N$, we have

$$i\frac{\partial \rho_N^\Lambda(t, x_1, \ldots, x_N; x_1', \ldots, x_N')}{\partial t} = [H_N^\Lambda, \rho_N^\Lambda](t, x_1, \ldots, x_N; x_1', \ldots, x_N').$$

For the case when potential $\phi_{i,j}(x_i - x_j)$ has form delta function $\delta(x_i - x_j) = \delta(x_i - x_{i+1}) = 0 \quad \text{for} \quad x_i \neq x_j, \quad (i, j = 1, \ldots s), \quad 1 \leq s < N$ we can reduce problem (10.5) to solution of system equations:

$$i\frac{\partial \rho_s^\Lambda(t, x_1, \ldots, x_s; x_1', \ldots, x_s')}{\partial t} = [-\sum_{j=1}^{s} \frac{\partial^2}{\partial x_j^2}, \rho_s^\Lambda](t, x_1, \ldots, x_s; x_1', \ldots, x_s')$$

with initial date

$$\rho_s^\Lambda(t, x_1, \ldots, x_s; x_1', \ldots, x_s')|_{t=0} = \rho_s^\Lambda(0, x_1, \ldots, x_s; x_1', \ldots, x_s'),$$

where

$$-\sum_{j=1}^{s}(\frac{\partial^2}{\partial x_j^2})\rho_s^\Lambda(0) = E\rho_s^\Lambda(0)$$

inside R_1, and

$$\Big(\frac{\partial}{\partial x_{j+1}} - \frac{\partial}{\partial x_j}\Big)\rho_s^\Lambda(0)|_{x_{j+1}=x_j} = c\rho_s^\Lambda(0)|_{x_{j+1}=x_j}$$

on the boundary of R_1.

According to the theory group [9, 11, 12] the solutions of these equations have the following form:

$$\rho_s^\Lambda(t, x_1, \ldots, x_s; x_1', \ldots, x_s') = U^\Lambda(t)\rho_s^\Lambda(0, x_1, \ldots, x_s; x_1', \ldots, x_s') =$$

$$= \Big(e^{\Omega(\Lambda)} e^{-i(-\sum_{j=1}^{s} \frac{\partial^2}{\partial x_j^2})t} e^{-\Omega(\Lambda)} \rho^\Lambda e^{i(-\sum_{j=1}^{s} \frac{\partial^2}{\partial x_j^2})t}\Big)_s (0, x_1, \ldots, x_s; x_1', \ldots, x_s') =$$

$$\Big(e^{-i(-\sum_{j=1}^{s} \frac{\partial^2}{\partial x_j^2})t} \rho^\Lambda e^{i(-\sum_{j=1}^{s} \frac{\partial^2}{\partial x_j^2})t}\Big)_s (0, x_1, \ldots, x_s; x_1', \ldots, x_s'),$$

where [9]

$$(\Omega(\Lambda)\rho^\Lambda)_s(0, x_1, \ldots, x_s; x_1', \ldots, x_s') =$$

$$= \frac{N}{V}(1 - \frac{s}{N}) \int_\Lambda \sum_i \rho_{s+1}^\Lambda(0, x_1, \ldots, x_s, x_{s+1}; x_1', \ldots, x_s', x_{s+1}) g_i^1(x_{s+1}) \tilde{g}_i^1(x_{s+1}) dx_{s+1},$$

$g_i^1(x_{s+1})$ is a complete orthonormal system of vectors in the one-particle space $L_2(\Lambda)$ and [8]

$$\rho_s^{\Lambda}(0, x_1, \ldots, x_s; x_1', \ldots, x_s') = \sum_{\alpha=1} w_\alpha \psi_\alpha(x_1, \ldots, x_s)\psi_\alpha^*(x_1', \ldots, x_s').$$

Here $\psi_\alpha(x_1, \ldots, x_s)$ is a symmetric function

$$\psi_\alpha(x_1, \ldots x_s) = \frac{1}{s!}\sum_P (-1)^{|P|} exp\Big(i\sum_{j=1}^{s} x_j k_{Pj}^{\alpha}\Big) exp\Big[\frac{i}{2}\sum_{j>i}\theta(k_{P_j}^{\alpha} - k_{P_i}^{\alpha})\Big],$$

in fundamental domain $x_1 < x_2 < \cdots < x_s, \quad (s = 1, \ldots, N)$ with eigenvalues $E_s = \sum_{i=1}^{s} k_i^2$.

In the R^s, $(-\infty < x_j < +\infty, j = 1, 2, \ldots, s)$ function ψ_α have the following form:

$$\psi_\alpha(x_1, \ldots x_s) = \frac{1}{s!}\prod_{j>i}(\varepsilon(x_j - x_i)\sum_P (-1)^{|P|} exp(i\sum_{j=1}^{s} x_j k_{Pj}^{\alpha})\times$$

$$\times exp[\frac{i}{2}\sum_{j>i}(\varepsilon(x_j - x_i)\theta(k_{P_j}^{\alpha} - k_{P_i}^{\alpha})],$$

where

$$\theta(k) = i \log\frac{ic + k}{ic - k},$$

$k = k_{P_j} - k_{P_i}$ and the branch of the logarithm is chosen so that $\theta(k)$ is continuous antisymmetric function if k is real ($\theta(k) = 2arctan(\frac{k}{c})$, $Imk = 0$). This form of the wave function is quite typical of models solvable by mean of Bethe ansatz. We are interested in the repulsive case, so that $c \geq 0$. In this case in the domain R^N the corresponding k's entering Ψ should be all real, $Imk_j = 0$, and spectrum has form $E_s = \sum_{j=1}^{s} k_j^2$ with $-\infty < k_j < +\infty$, so the spectrum consists of the elementary particles only [6, 7].

It should be noted that the dynamics of an infinite number of bosons can also explore on the basis of the nonlinear Schrödinger equation [10–14] or Gross–Pitaevskii equation. Using the method proposed by H.G. Spohn [17] for the derivation of the Hartree equation from BBGKI, B. Schlein derived Gross–Pitaevskii equation for the time evolution of the Bose–Einstein condensates [15].

Acknowledgements Authors are grateful to Prof. H. Spohn for the discussion of results and useful comments.

References

1. Albertsson Martin: Analysis of the Many-Body Problem in One Dimension with Repulsive Delta-Function Interaction. Uppsala Universitet, TVE 14 025 juni Examensarbete 15 hp Juni (2014)
2. Bethe H.A.: On the theory of metals, I. Eigenvalues and eigenfunctions of a linear chain of atoms. Zeits. Phys. 205–226 (1931); Bethe H.A.: Selected Works of Hans A. Bethe With Commentary. World Scientific, Singapour (1996)
3. Bogolyubov N.N.: Lectures on Quantum Statistics, London (1970); Selected Works, **2** Naukova Dumka, Kiev (1970). [in Russian]
4. Bogolyubov, N.N.: (Jr): Introduction to Quantum Statitic Mechanics. Nauka, Moscow (1984)
5. Tracy, C.A., Harold, W.: The dynamics of one-dimentional delta-function Bose gas. J. Phys. A: Math. Theor. **41**, 485204 (2008)
6. Izergin, A.G.: Introduction to the Bethe Ansatz Solvible Models. University di Firenze (2000)
7. Korepin, V.E., Bogoliubov, N.M., Izergin, A.G.: Quantum Inverse Scattering Method and Correlation Functions. Cambridge University Press, Cambridge (1993)
8. Lieb, E.H., Liniger, W.: 1963 Exact analysis of an interacting Bose gas. I: the general solution and the ground state. Phys. Rev. **130**, 1605–1616 (1963)
9. Petrina, D.Y.: Mathematical Foundation of Quantum Statistical Mechanics, Continuous Systems. Kluwer Academic Publishers, Dordrecht (1995)
10. Petrina, D.Y., Enolsky, V.Z.: About the vibrations of one-dimensional systems. Rep. AS UkrSSR A. **8**, 756–760 (1976)
11. Petrina D.Y., Vidybida, A.K.: Couchy problem for kinetic equation of Bogolubov, 370–378, IM AS SSSR, **136**, Part 2, 370–378 (1975)
12. Rasulova, M.Y.: Couchy problem for kinetic equation of Bogolubov. Quantum case. Rep. AS UzSSR **2**, 6–9 (1976)
13. Rasulova, MYu.: The soliton solution of BBGKY'S chain of quantum kinetic equations for bose systems, interacting by delta potential. Rep. Math. Phys. **40**, 551–556 (1997)
14. Rasulova, MYu.: The soliton solution of BBGKY'S chain of quantum kinetic equations for system of particles, interacting by delta potential. Phys. A. **315**, 72–78 (2002)
15. Schlein, B.: Derivation of effective evolution equations from many body quantum dynamics. In: Exner, P. (ed.) Sixteenth International Congress on Mathematical Physics, pp. 406–416. World Scientific, Singapore (2010)
16. Seringer, R., Yin, J.: The Lieb-Liniger model as a limit of dilute bosons in three dimentions. Commun. Math Phys. **284**, 459 (2008). doi:10.1007/s00220-008-0521-6
17. Spohn, H.: Kinetic equations from Hamiltonian dynamics. Rev. Mod. Phys. **52**, 569–615 (1980)
18. Zakharov, V.E., Shabat, A.B.: Exact theory of two-dimensional self-focusing and one-dimensional self-modulation of waves in nonlinear media. Sov. Phys. JETP **34**(1), 62–69 (1972)
19. Zakharov, V.E., Shabat, A.B.: Interaction between solitons in a stable medium. Sov. Phys. JETP **37**(5), 823–828 (1973)

Chapter 11
1D Wavelet and Partial Correlation Application for MS Subgroup Diagnostic Classification

Yeliz Karaca, Zafer Aslan and Abul Hasan Siddiqi

Abstract In this study, 1 D wavelet and Partial correlation analyses were applied to a data set obtained from patients with Multiple Sclerosis along with a control group of healthy individuals. The analysis is limited to a sample of 139 individuals, 76 being with Relapsing-Remitting Multiple Sclerosis, 38 with Secondary Progressive Multiple Sclerosis, 6 with Primary Progressive Multiple Sclerosis, and 19 being Healthy individuals. It is the main objective of the study to develop a clinical decision support system in order to classify the patients' diagnostic data based on features gathered from Magnetic Resonance Imaging. The 1-D Continuous Wavelet Transforms are developed to measure the health status of the patients based on features gathered from Magnetic Resonance Imaging and Expanded Disability Status Scale (EDSS). Classification of the Multiple Sclerosis (MS) diagnosis level indicates that it can be used as an important indicator for making decisions to identify MS health status of patients. Our results of relative distribution of three indicators help to identify some differences of "Remitting Relapsing Multiple Sclerosis", "Secondary Progressive Multiple Sclerosis", and "Primary Progressive Multiple Sclerosis". Features like sex, the maximum and minimum lesion sizes, and maximum and minimum values of EDSS scores are widely known and applied in medical studies. This study has fulfilled what lacked in terms of mathematical explanation concerning the significance of such features.

Y. Karaca (✉)
Visiting Engineering School (DEIM), Tuscia University, Viterbo, Italy
e-mail: yelzkaraca@gmail.com; yeliz.karaca@ieee.org

Z. Aslan
Faculty of Engineering, Computer Engineering, Istanbul Aydin University, Istanbul, Turkey
e-mail: zaferaslan@aydin.edu.tr

A.H. Siddiqi
School of Basic Sciences and Research, Sharda University, Greater Noida, India
e-mail: siddiqi.abulhasan@gmail.com

P. Manchanda et al. (eds.), *Industrial Mathematics and Complex Systems*, Industrial and Applied Mathematics, DOI 10.1007/978-981-10-3758-0_11

11.1 Introduction

Recently, studies on the interaction of mathematics with different fields have evoked interest and started to be used in medicine. In this respect, researchers have attempted to design smart interfaces to ensure the interaction between mathematics and multiple sclerosis in an effective way. Among such interfaces are Magnetic Resonance Imaging (MRI), Expanded Disability Status Scale (EDSS), and Cerebrospinal Fluid which helps diseases be diagnosed. Using magnetic resonance (MR) images and lesion numbers, following the assessment of their magnitudes, through EDSS scale, the limit of the patients movements can be detected. Based on the variables obtained from the devices and mathematical models, the correlation of diagnosis along with its subtypes type has been examined [1–3].

Multiple Sclerosis is a disorder that affects the spinal cord and brain in the central nervous system. Determining the causes of MS is difficult since the clinical manifestations and its course may change significantly [1–3]. Multiple Sclerosis has three subgroups examined in the study [1–4]:

1 Relapsing-Remitting Multiple Sclerosis (RRMS): RRMS is the typical form of multiple sclerosis which often has an onset in the late teens or twenties of individuals. At the beginning, there is a severe attack which is followed by either full recovery or partial recovery. Further attacks may be seen in intervals that are not predictable, which are followed by increasing disability. In the late thirties, the relapsing-remitting pattern tends to change into the secondary progressive type [1–4].
2 Secondary Progressive Multiple Sclerosis (SPMS): The relapsing-remitting form of MS frequently develops into secondary progressive multiple sclerosis mentioned above following a variable period of time usually in the late thirties [1–4].
3 Primary Progressive Multiple Sclerosis (PPMS): The disease shows a steady worsening course. This course is interrupted by periods of dormancy without improvement. The progression is variable. In the worst case, this situation may end up with death within a few years [1–4].

Linear model has been formed in the study of Karaca et al. with the same data group used in this present study. In Fig. 11.1, a general block diagram of the model is presented [2, 3]. The researchers used linear models for each bifurcation. Distinction between Healthy/Patient was made in the first node. Upon determination that the subject was with MS, a distinction was made whether it was RRMS–SPMS or PPMS. The final step included the differentiation of RRMS/SPMS if it was determined that the patient had RRMS or SPMS. RRMS, SPMS, and PPMS subgroups were determined as a result through three bifurcations [2, 3].

A linear model is designed based on the number of lesions. 94.17% accuracy rate is attained for the distinction between patients. In the division between RRMS/SPMS, 73.68% accuracy is attained. Based on EDSS scores, an accuracy rate of 99.28% is achieved for the division between MS patients and healthy individuals. 94.17% accuracy rate is obtained for RRMS–SPMS/PPMS and the rate is 61.67% for

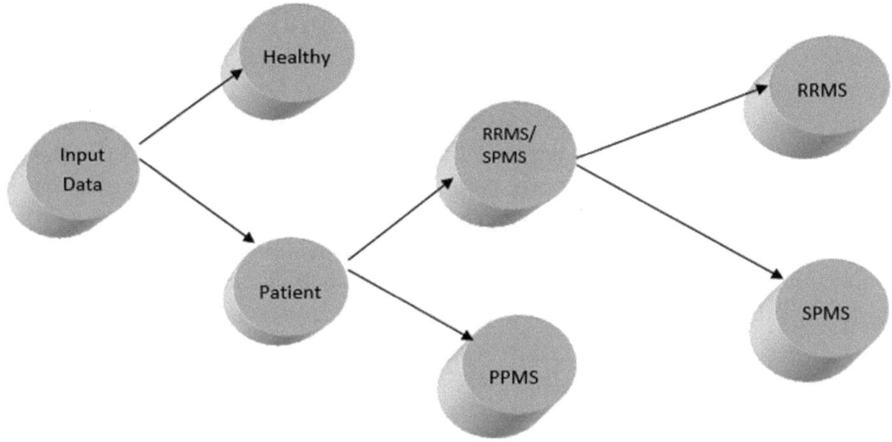

Fig. 11.1 Healthy, Patient, RRMS, SPMS, and PPMS distinction [2, 3]

RRMS/SPMS. As can be seen, it is possible to attain highly accurate results so as to identify MS and subgroups. These results are parallel to clinical findings [2, 3].

In the literature, Shahvar et al. have introduced a new approach to form synthetic models of sonic by logs using wavelet coefficients and artificial neural network. They obtained results that ascertain the applicability of this model for sonic log prediction [5]. In another study by Safarinezhadian et al., a wavelet transform was applied in a single coherent analysis to decompose data in spatial nature into a series of independent components at different scales [6]. A study by Zope-Chaudhari provides distortion assessment of geospatial data. The study used wavelet-based invisible watermarking. The study used eight wavelets at different levels of wavelet decomposition levels. Error measures like maximum and mean square errors were used for accuracy assessment [7]. Safarinezhadian et al. proposed a new method for development of Digital Elevation Model (DEM) of soil surface. Data collection basis and wavelet modeling were used in the study. The multiresolution nature of the method has proven to have many advantages compared to other methods. The power and versatility of wavelets are shown upon analysis of the laser scanner data [6]. In an article by Aloe, synthesization of two indices under different situations is evaluated for accuracy. In such situations partial correlation and the semi-partial correlation both appeared to behave as expected with regard to bias and root-mean-squared error (RMSE) [8]. Algina and Olejnik studied applications, in partial correlation and multiple regression analyses, of the tables. Other than those used in forming the tables, sample size for levels of probabilities, accuracy and parameter values, and for Type I error rates could be selected by researchers using SAS and SPSS computer programs. It has been revealed that planning correlation studies in order to obtain an adequate level of estimation accuracy is important. Another important goal revealed is to obtain adequate power. It is also possible to calculate from the distribution of r the sample size necessary to accomplish a target for power [9].

Karaca et al. analyzed 1-D Continuous Wavelet Analysis, 1-D Wavelet Coefficient Method and Partial Correlation Method on various Wechsler Adult Intelligence Scale-Revised parameters including School Education, Gender, Age, Information Verbal and Performance Full-Scale Intelligence Quotient. Particularly, gender variable has been shown to have a negative yet important role on age and Performance Information Verbal factors. The age parameters, too, have a significant relation in Full-Scale Intelligence Quotient change and Performance Information Verbal [10].

The literature review has shown that as to the application and comparison of 1D continuous wavelet analyses and partial correlation method, no study related to MS exists comprising three parts of the brain using the analyses under discussion.

11.2 Material and Method

11.2.1 Material

11.2.1.1 MS Data

In the present study, the monitoring was conducted in Hacettepe University (Ankara, Turkey), Faculty of Medicine, Neurology Department and Radiology Department as well as Primary Magnetic Resonance Imaging center. 120 patients (88 females, 32 males) who were definitely diagnosed to have MS with RRMS, SPMS, or PPMS based on the McDonald criteria were taken. MS patients were aged 20–65, and those 19 ones were healthy subjects who did not have the disease (as shown by MRI scans). They were also enrolled as the control subjects. The disability level of MS patients was determined using the EDSS through different devices. Magnetic resonance was read for three regions lesion covering data of different years. Comparison of data was made based on the criteria mentioned above [2, 3, 11].

11.2.1.2 Expanded Disability Status Scale

Neurologists make use of this scale in their diagnostic classification of MS for their patients. EDSS scale is used by neurologists and physicians for MS causes to measure disability. The details of the scale can be found in reference [11, 12].

11.2.1.3 Magnetic Resonance Imaging

Magnetic Resonance Imaging (MRI) is one of the most beneficial tools for the diagnosis of MS [11–13].

The number of lesions was used in this study. The locations of the brain were considered along with EDSS score for modeling. The first, second, and third regions were designated as brain stem, periventricular corpus callosum, and upper cervical regions, respectively. The first, second, and third MRI brain regions of each patient were taken into consideration.

For MRI scans of the first, second, and third regions, the numbers of lesions in the three regions were designated as features. The labels for such features are provided below:
Max. MRI1: The maximum lesion number in the first region of MRI scan.
Min. MRI1: The minimum lesion number in the first region of MRI scan.
Max. MRI2: The maximum lesion number in the second region of MRI scan.
Min. MRI2: The minimum lesion number in the second region of MRI scan.
Max. MRI3: The maximum lesion number in the third region of MRI scan.
Min. MRI3: The minimum lesion number in the third region of MRI scan.

11.2.2 Methodology

11.2.2.1 1-D Continuous Wavelet Transforms

Wavelet theory is the result of a multidisciplinary approach in science, through which different scholars have started to work together. For example, mathematicians, physicists, and engineers perform studies on wavelet theory and applications. Likewise, wavelet analysis has involved a great deal of interest recently in signal processing. Many applications in image analysis, transient signal analysis, communications systems, and other signal processing have been performed effectively by means of this. Despite not being a new type of analysis, it has an innovative aspect in that the development of recent results on wavelet mathematical foundations has provided a unified framework for the scope [10, 14–17].

A common link is formed between a lot of varied problems which are of interest for various fields. In order for further development in mathematical understanding of wavelet, there exist opportunities in the future regarding its applications in the scopes of natural sciences as well as in engineering [10, 14–17].

The following part outlines the main formulae for wavelet analyses [10, 14–17].

With a complex-valued function ψ, the conditions below are met:

$$\int_{-\infty}^{\infty} |\psi\,(\mathrm{t})|^2 dx < +\infty \tag{11.1}$$

$$c_{\psi} = 2\pi \int_{-\infty}^{\infty} \frac{|\psi\,(\omega)\,|^2}{\omega} d\omega < +\infty, \tag{11.2}$$

where ψ is the Fourier transform of ψ. The first condition hints finite energy of the function ψ, whereas the other, being admissibility, shows that if $\psi(\omega)$ is smooth, then $\psi(0) = 0$. The mother wavelet is the function ψ. For a continuous wavelet transform, see [10, 14–17].

If ψ meets the aforementioned conditions, then the wavelet transform of a real signal $s(t)$ with respect to the wavelet function $\psi(t)$ is termed as in [10, 14–17]:

$$S(b,a) = \frac{1}{\sqrt{a}} \int_{-\infty}^{\infty} \psi' \left(\frac{t-b}{a}, \right) s(t)dt \tag{11.3}$$

where ψ' displays the complex conjugate of ψ, which defines the open (b,a) half-plane ($b \in R, a$). The parameter b refers to the time shift, whereas the parameter a refers to the scale of analyzing wavelet. If $\psi_{a,b}(t)$ is defined as

$$\psi_{a,b}(t) = a^{-1/2} \psi \left(\frac{\mathrm{t}-\mathrm{b}}{\mathrm{a}} \right). \tag{11.4}$$

This means doing rescaling by a and doing the shifting by b. Given this Eq. 11.3 can be formulated as either a scalar or the inner product of the real signal $s(t)$ with the function of $\psi_{a,b}(t)$:

$$S(b,a) = \int_{-\infty}^{\infty} \psi'_{a,b}(t)\, s(t)\, s(t)dt \tag{11.5}$$

when function $\psi(t)$ fulfills the condition of admissibility, Eq. 11.2, the original signal $s(t)$ can be received from the wavelet transform $S(b,a)$ through the inverse formula as given below [10, 14–17]:

$$s(t) = \frac{1}{C_\psi} \int_{-\infty}^{\infty} \int_{-\infty}^{\infty} S(b,a)\, \psi_{a,b}(t) \frac{dadb}{a^2}. \tag{11.6}$$

11.2.3 Partial Correlation

The Pearson correlation coefficient (r) defines linear association among continuous random variables. It has been extended in the literature for several variable interaction mapping problems [10, 18]. However, the direct or indirect relationship between variables cannot be distinguished by the correlation coefficient per se. Table 11.1 below provides the general terminology for partial correlation coefficients [10, 18]:

When we consider two variables, e.g., A and B, the association between them can be seen in different ways like a direct correlation, $A \to B$, both are co-regulated by a third variable, being C, (i.e., $C \to B$ and $C \to A$) or an indirect one $A \to C \to B$. r, the regular correlation coefficient, is defined on two variables, A and B not distinguished between the relation types. It shows that A and B have direct relationship or they do not have relation [10, 18–20].

Table 11.1 Partial correlation coefficients: general terminology [10, 18]

A, B, C, D, X, Z selected variable in a given system	
r correlation coefficient	i, j, k subscripts used to identify the variables
n number of observations	N data matrices used in training
p number of variables	M correlation coefficient matrix
k number of classes	l number of samples in a class
x order of partial correlation	d number of correlation defined in the system

The partial correlation coefficient illustrates such difference by separating indirect relations namely path relations. The correlation between two variables is conditioned upon the third or a filtered form, A and B prior to computing the coefficient. Hence, partial correlation $r_{AB/C}$ underscores that correlation exists between A and B if the effect of the conditioned variable C is removed. With no conditioning upon any variable, the order of the partial correlation coefficients is zero where one defines the correlation directly between A and B. The order is x when the correlation is calculated subsequent to the conditioning on x number of different variables apart from A and B. Equations 11.1–11.3 provide an overall definition as to the first three orders of partial correlations [10, 18–20].
Zeroth-order correlation is presented as follows:

$$r_{AB} = \frac{COV\,(A, B)}{\sqrt{var\,(A) * var\,(B)}}. \tag{11.7}$$

First-order partial correlation is presented as follows:

$$r_{AB/Z} = \frac{[r_{AB} - (r_{AZ} * r_{BZ})]}{\sqrt{\left(1 - r_{AZ}^2\right) * \left(1 - r_{BZ}^2\right)}}. \tag{11.8}$$

Second-order partial correlation can be seen as follows:

$$r_{AB/XZ} = \frac{\left[r_{AB} - \left(r_{AZ/X} * r_{BZ/X}\right)\right]}{\sqrt{\left(1 - r_{AZ/X}^2\right) * \left(1 - r_{BZ/X}^2\right)}}. \tag{11.9}$$

Correlation measures r_{AB}, partial correlation measures $r_{AB/Z}$ and $r_{AB/XZ}$ show symmetric property (i.e., $r_{AB} = r_{BA}, r_{AB/Z} = r_{BA/Z}$, etc.). Such coefficients are limited to a range of values from -1 to 1 [10, 18–20].

For this reason, instead of making the estimation of a full complete correlation matrix that has redundant entries, the inter-variable association pattern may be considered a single array which contains unique values of correlation coefficients. This depicts a defined order of variable combinations. Such vector coefficients are referred to as PCCM in this study. PCCM stores the pattern defined as well as the strong aspects of interval-variable associations for a particular system. In a system with p number of variables, zeroth-order PCCM will have $[p(p-1)/2]$ elements, and first-order PCCM will have $[p(p-1)(p-2)(p-3)/4]$ elements in the vector [18].

Partial correlation coefficient infers direct and indirect associations between random measurements, which is stated in the literature [10, 18–20].

The following could be expressed to generally state the classification problem; consider a system $N[nxp; k]$ where the measurement of p variables yields n observations that belong to k different classes in the system. Discriminant analysis aims at improving a classifier that uses the observations in N. Each of the k classes is modeled accordingly. The classifier capability is afterward tested based on its estimation ability as to the classes of samples in N, which may be regarded as resubstitution test or self-consistency. Those of new set of samples $N_{test}[mxp; k]$, not used during the modeling process (independent sample test) [10, 18–20].

11.3 Results

11.3.1 Application of Wavelet Analysis MRI Lesion Extreme Values

MRI1 Lesion Extreme Values

The first region means brain stem region. 1D wavelet has been applied taking maximum and minimum lesion numbers in brain stem into consideration. The illustrations can be seen in Figs. 11.2 and 11.3, respectively.

Maximum MRI1 Lesion Numbers

Maximum data number from 1 to 88 belongs to maximum lesion number of female group for the first region.

If we divide females' group into three subsections:

Patients 0–35 constitute the first group; they have small and large lesions size.

Patients 35–60 constitute the second group. It is found out in this study that large lesions are dominant in this group. Finally, the last group (patients between 60 and 120) seemed to have medium and large lesion sizes.

If we analyze the smaller data sets for male groups between the patient 89–110, small- and medium-size lesions have been accrued but for the last group (for the last 10 female) large-size lesions are dominant.

Minimum MRI1 Lesion Numbers

Minimum numbers of lesions for females are associated with small-, medium-, and large-size lesions. For the male group mainly small- and medium-size lesions are dominant.

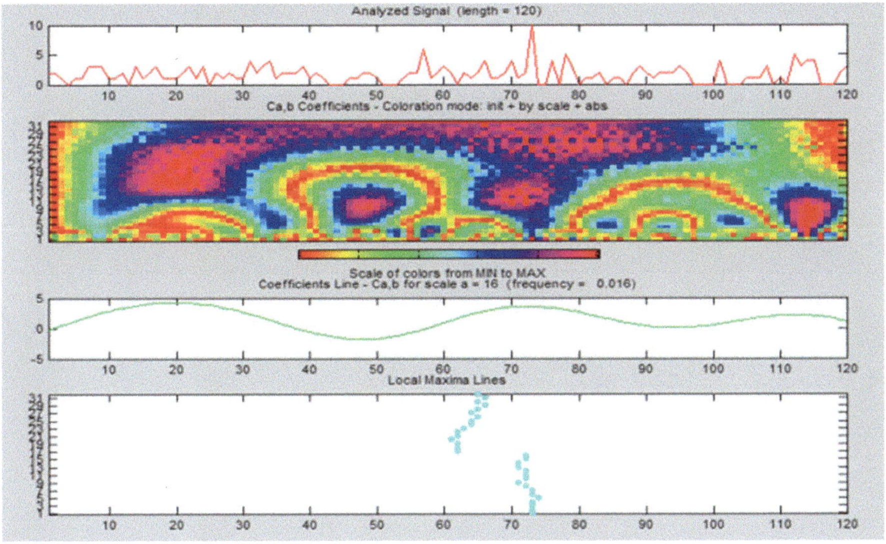

Fig. 11.2 Maximum MRI1 lesion numbers

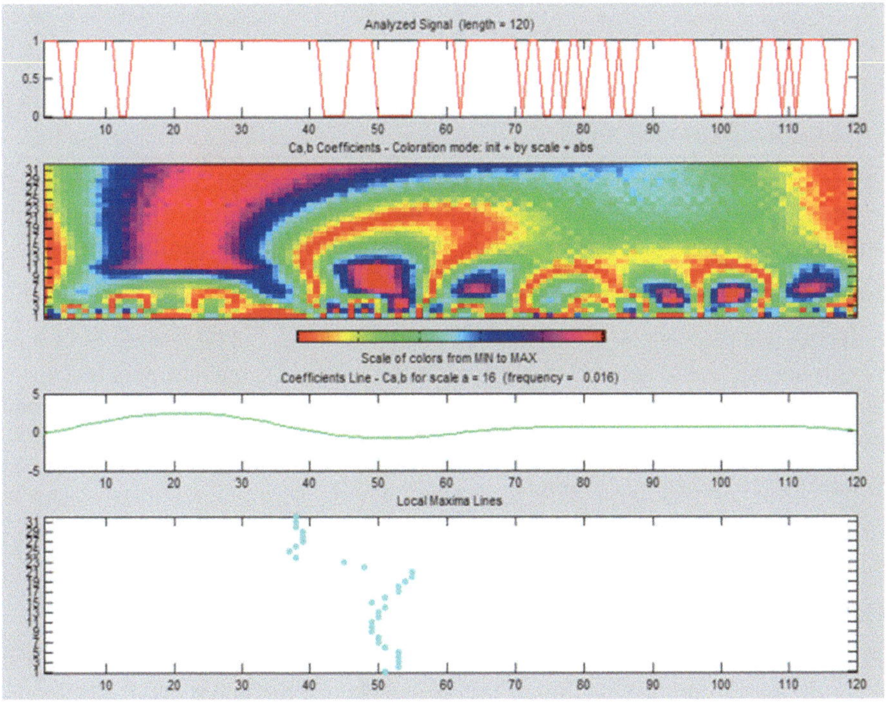

Fig. 11.3 Minimum MRI1 lesion numbers

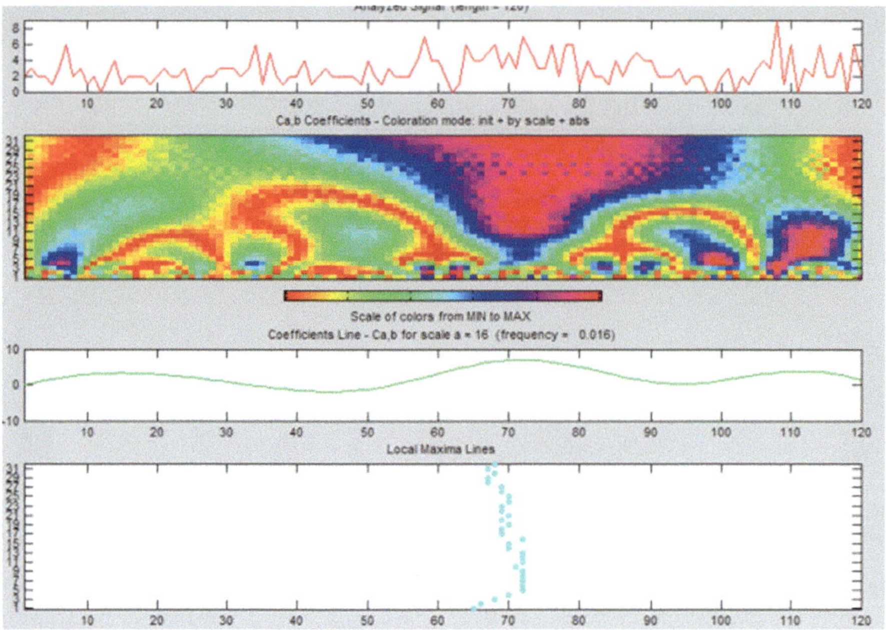

Fig. 11.4 Maximum MRI2 lesion numbers

MRI2 Lesion Extreme Values

Periventricular corpus callosum region is shown as the second region. 1D wavelet has been applied using maximum and minimum number of lesions in periventricular corpus callosum region as can be seen in Figs. 11.4 and 11.5, respectively.

Maximum MRI2 Lesion Numbers

For the female patients, the last group has small, medium, and large sizes of lesions (mixed) but the first of half of female data set has small- and medium-size lesions in this region. For the male group, small, medium, and large sizes of lesions have been observed.

Minimum MRI2 Lesion Numbers

Variations of lesion size are similar for all the patients in this region (small-, medium-, and large-size lesions could be recorded).

MRI3 Lesion Extreme Values

The third region illustrates the upper cervical regions. 1D wavelet has been applied taking the maximum and minimum lesion numbers in the upper cervical regions, which can be seen in Figs. 11.6 and 11.7, respectively.

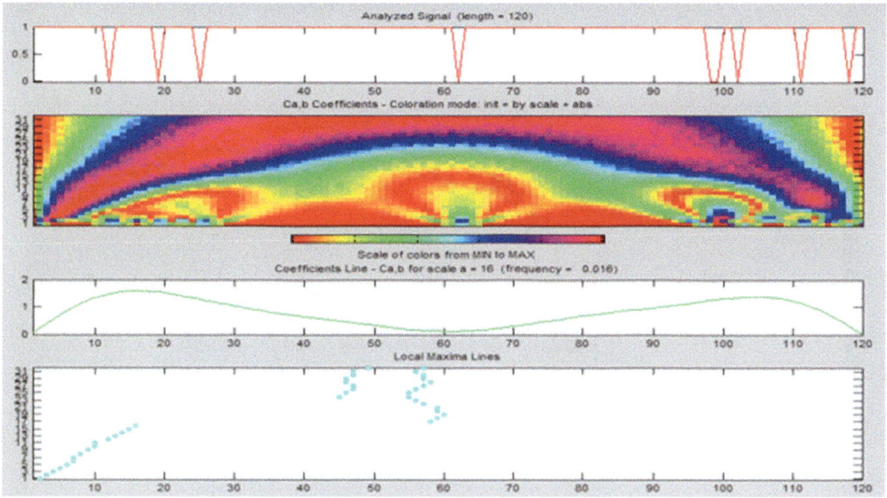

Fig. 11.5 Minimum MRI2 lesion numbers

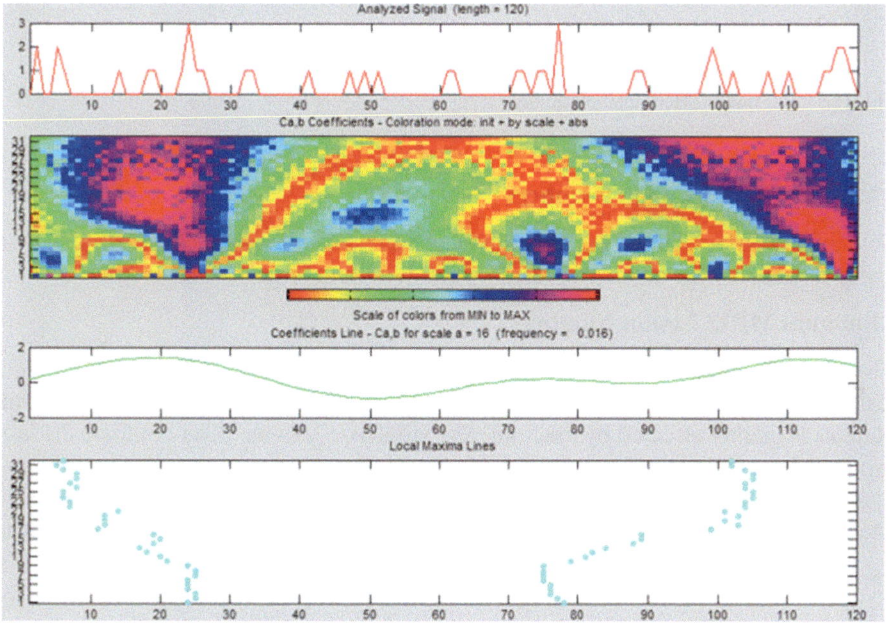

Fig. 11.6 Maximum MRI3 lesion numbers

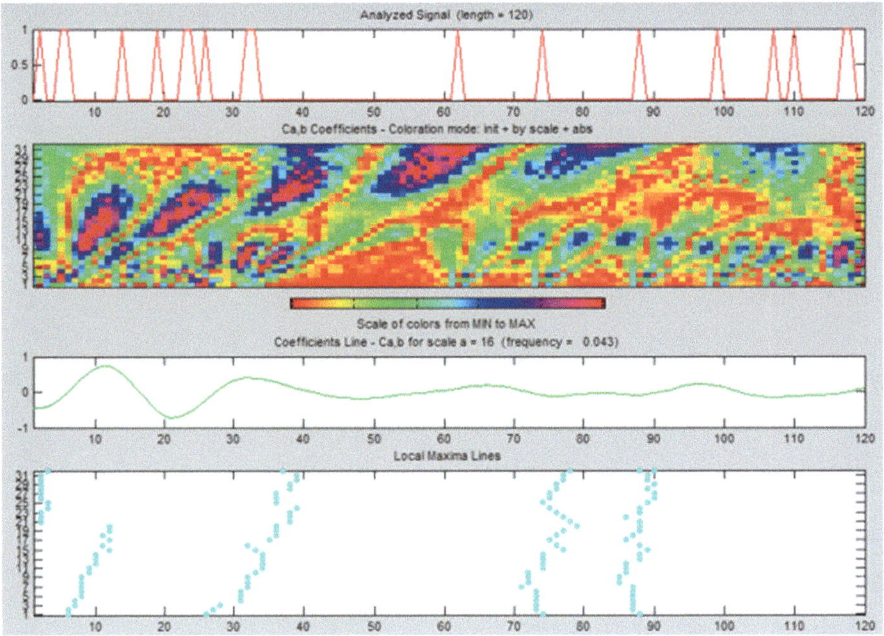

Fig. 11.7 Minimum MRI3 lesion numbers

Maximum MRI3 Lesion Numbers

For the male group, small, medium, and large sizes of lesions are observed. The first group of patients are mainly affected by medium- and large-size lesions.

Minimum MRI3 Lesion Numbers

The sizes of minimum lesions for the first 60 patients have been compared. Minimum lesion characters are totally different in the male and female group. The first group of patients is mainly affected by medium- and large-size lesions. If we compare the last group of female and the first group of male data, they are very similar and minimum lesions are generally small and medium sizes. As for the last group of male patients, it can be said that they have basically medium or large sizes for both groups of lesions.

11.3.2 Application of Partial Correlation

In each MRI scan of each patient, the three parts of brain, brain stems, corpus callosum–periventricular region, and the upper cervical spine, have been chosen; and different variables (max. MRI1, min. MRI1, EDSS) have been used in Tanagra program [21]. Partial correlation is put into application (for $\alpha = 0.05$). Control

Table 11.2 Table for control variables regarding first region

Variables	
1	Max. MRI1
2	Min. MRI1
3	EDSS

Table 11.3 Partial correlation coefficients

No	Att. *Y*	Att. *X*	*r*	r^2	*T*	*p*-value
1	SPMS	Max. MRI1	0.00712	0.00005	0.07636	0.93927
2	SPMS	Min. MRI1	0.00643	0.00004	0.06897	0.94513
3	SPMS	EDSS	0.00618	0.00004	0.06633	0.94723
4	PPMS	Max. MRI1	0.00318	0.00001	0.03406	0.97289
5	PPMS	Min. MRI1	−0.00051	0.00000	−0.00546	0.99565
6	PPMS	EDSS	−0.00454	0.00002	−0.04866	0.96128
7	RRMS	Max. MRI1	0.00898	0.00008	0.0963	0.92345
8	RRMS	Min. MRI1	0.00648	0.00004	0.06953	0.94469
9	RRMS	EDSS	0.00428	0.00002	0.04595	0.96343

Table 11.4 Control variables

Variables	
1	Max. MRI2
2	Min. MRI2
3	EDSS

variables for the first region (namely brain stems region) can be seen in the table provided below:

Table 11.2 shows the partial correlation coefficients for the first region. Partial correlation coefficients are concluded in the first region and EDSS relation of maximum lesions at first region, with a minimum lesion in the first region and EDSS is positive. There is a positive correlation for the minimum lesion numbers at the first region. EDSS is higher than the other parameters.

For the second region (corpus callosum–periventricular region), control variables can be seen in the table provided below:

Table 11.2 shows partial correlation coefficients for the second region.

Table 11.3 shows the role of partial relations between maximum and minimum lesion numbers and EDSS. Partial relations are negative in general.

For PPMS, the highest partial role of maximum lesions is 0.025 (alpha = 0.22).

For the third region (the upper cervical spine) control variables are provided below:

Table 11.4 shows the partial correlation coefficients for the third region.

Table 11.5 shows partial correlation coefficients among maximum lesion numbers in the third region. It also provides their impact on variations of the minimum lesion

Table 11.5 Partial correlation coefficients

No	Att. Y	Att. X	r	r^2	t	p-value
1	SPMS	Max. MRI2	−0.01521	0.00023	−0.16317	0.87067
2	SPMS	Min. MRI2	−0.00303	0.00001	−0.03253	0.97411
3	SPMS	EDSS	−0.00162	0.00000	−0.01739	0.98615
4	PPMS	Max. MRI2	0.02546	0.00065	0.27313	0.78525
5	PPMS	Min. MRI2	−0.01064	0.00011	−0.1141	0.90036
6	PPMS	EDSS	0.00136	0.00000	0.01456	0.98841
7	RRMS	Max. MRI2	−0.00343	0.00001	−0.03673	0.97076
8	RRMS	Min. MRI2	0.00834	0.00007	−0.08943	0.9289
9	RRMS	EDSS	0.00103	0	−0.011	0.99124

Table 11.6 Control variables

Variables	
1	Max. MRI3
2	Min. MRI3
3	EDSS

Table 11.7 Partial correlation coefficients

No	Att. Y	Att. X	r	r^2	t	p-value
1	SPMS	Max. MRI3	0.00652	0.00004	0.06987	0.94442
2	SPMS	Min. MRI3	−0.01701	0.00029	−0.18241	0.85558
3	SPMS	EDSS	−0.00756	0.00006	−0.08106	0.93554
4	PPMS	Max. MRI3	−0.02403	0.00058	−0.25772	0.79708
5	PPMS	Min. MRI.3	0.02584	0.00067	0.27718	0.78214
6	PPMS	EDSS	−0.00738	0.00005	−0.07912	0.93708
7	RRMS	Max. MRI3	−0.00514	0.00003	−0.0551	0.95615
8	RRMS	Min. MRI3	−0.0048	0.00002	−0.05151	0.95901
9	RRMS	EDSS	−0.01145	0.00013	−0.1228	0.90248

numbers and EDSS for the same region concerning SPMS and PPMS. There are only two positive correlations between the maximum and minimum numbers of lesions. The variation of minimum lesion numbers for the third region has some significant features pertaining to SPMS (r $= 0.025$, $\alpha = 0.22$) (Tables 11.6 and 11.7).

11.4 Conclusion

The first part of the present paper is concerned with the 1D wavelet applications on lesion sizes and numbers belonging to MS patients' MRI images. For the wavelet analyses specific results have been yielded. For maximum MRI1 Lesion Numbers, small- and medium-size lesions have been accrued but for the last group (for the last 10 females). For the second region, based on Minimum MRI2, lesion numbers are seen to be small-, medium-, and large-size lesions for females. For the second region, Maximum MRI2 lesion numbers are seen to be small- and medium-size lesions. It has been observed that lesion characters are totally different in the male and female group. The first of group of patients are mainly affected by medium- and large-size lesions. The second part of the present paper is concerned with partial correlation analyses. Lesion sizes and numbers of male and female test groups have been compared by partial correlation. Results describe only the existences of positive or negative relationships. The highest partial correlation coefficient is $r = 0.026$ ($n = 120$, $\alpha = 0.05$). These analyses should be extended for an additional data set.

Acknowledgements The authors are thankful to Prof. Dr. Rana Karabudak and her team along with Hacettepe University Neurology Department and Radiology Units. Dr. Yeliz Karaca is particularly grateful to Turkish Neurology Association.

References

1. Gilroy, J.: Basic Neurology, 3rd edn, pp. 225–278. The McGraw-Hill Companies, New York (2000)
2. Karaca, Y., Osman, O., Karabudak, R.: Linear modeling of multiple sclerosis and its subgroups. Turk. J. Neurol. **21**(1), 7–12 (2015)
3. Karaca, Y.: Constituting an optimum mathematical model for the diagnosis of multiple sclerosis. Ph.D. thesis, Institute of Science and Technology, Marmara University, İstanbul, Turkey (2012)
4. Karabudak, R., Işık, N., Siva, A.: Multiple Sklerozda Tanı ve Tedavi Kılavuzu, Bilimsel Tıp yayınevi, pp. 5–20 (2009)
5. Shahvar, M.B., Badounak, N.D., Kharrat, R.: A new approach for compressional slowness modeling using wavelet coefficients, energy sources, part a: recovery. Util. Environ. Eff. **36-19**, 2106–2112 (2014)
6. Safarinezhadian, B., Karimaghaee, P., Prof, A., Safavi, A.A., Abedini, M.J.: Analysis of laser-scanned topographic data using wavelet methods. Instrum. Sci. Technol. **36**(3), 323–336 (2008)
7. Zope-Chaudhari, S., Venkatachalam, P., Buddhiraju, K.M.: Assessment of distortion in water-marked geospatial vector data using different wavelets. Geo-Spat. Inf. Sci. **18**(2–3), 124–133 (2015)
8. Aloe, A.M.: An empirical investigation of partial effect sizes in meta-analysis of correlational data. J. Gen. Psychol. **141-1**, 47–64 (2014)
9. Algina, J., Olejnik, S.: Sample size tables for correlation analysis with applications in partial correlation and multiple regression analysis. Multivar. Behav. Res. **38-3**, 309–323 (2003)
10. Karaca, Y., Aslan, Z., Cattani, C., Galletta, D., Zhang, Y.: Rank determination of mental functions by 1D wavelets and partial correlation, Springer. J. Med. Syst. **41-1**, 1–10 (2017)
11. Karaca, Y., Zhang, Y., Cattani, C., Ayan, U.: The differential diagnosis of multiple sclerosis using convex combination of infinite kernels. CNS Neurol. Disord. Drug Targets **16-1**, 36–43 (2016)

12. Expanded disability status scale (EDSS). https://www.mstrust.org.uk/a-z/expanded-disability-status-scale-edss (14.11.2016)
13. Joy, J.E., Jonston, R.B., Jr. Editors: Multiple Sclerosis Current Status and Strategies for the Future. National Academy Press, Washington (2001)
14. Karaca, Y., Aslan, Z.: Wavelet analysis of anxiety and mathematics. Indian J. Ind. Appl. Math. **4-2**, 118–130 (2013)
15. Abramovich, F., Bailey, T.C., Sapatinas, T.: Wavelet analysis and its statistical applications. Stat. Part 1, 1–29 (2000)
16. Lee, D.T.L., Yamamoto, A.: Wavelet analysis: theory and applications. Hewlett Packard J., 44–52 (1994)
17. Siddiqi, A.H.: Applied Functional Analyses: Numerical Methods, Wavelet Methods, and Image Processing, vol. 258, pp. 356–425. CRS Press, Boca Raton (2003)
18. Melissa, A.S., Raghuraj, K.R., Lakshminarayanan, S.: Partial correlation metric based classifier for food product characterization. J. Food Eng. **90**(2), 146–152 (2009)
19. Cramer, D.: A cautionary tale of two statistics: partial correlation and standardized partial regression. J. Psychol. **137-5**, 507–511 (2003)
20. Marreles, G., Horwitz, B., Kim, J., Pelegrini-Issac, M., Benali, H., Doyan, J.: Using partial correlation to enhance structural equation modeling of functional MRI data. Magn. Reson. Imaging **25**, 1181–1189 (2007)
21. http://eric.univ-lyon2.fr/~ricco/tanagra/en/tanagra.html

Chapter 12
Numerical Methods for Nonlinear System of Hyperbolic Equations Arising in Oil Reservoir Simulation

G.D. Veerappa Gowda

Abstract Here, we propose a higher order finite volume scheme by using the idea of discontinuous flux for the numerical study of two-phase flow in a heterogeneous porous media, arising in oil reservoir simulation. To enhance the oil recovery, chemical components called polymers are dissolved in the aqueous phase. This results in studying the Buckley–Leverett model with multicomponent polymer flooding, which is a coupled non-strictly hyperbolic system of conservation laws in the absence of capillary pressure. In the presence of gravity, the flux function is non-monotone and the construction of Godunov type upwind scheme for this system becomes difficult and computationally expensive. To overcome this difficulty, the coupled system is reduced to an uncoupled system of scalar conservation laws with discontinuous coefficients and applied the idea of discontinuous flux to solve these scalar equations.

Keywords Conservation laws · Finite volume · Riemann problems
Multi-component · Polymer flooding · Buckley–Levrett model
Enhanced oil recovery

12.1 Introduction

In this article briefly I review the works done with Adimurthi, Jerome Jaffre, C. Praveen, and K. Sudarshan Kumar on applications of discontinuous flux and how it can be used to solve numerically a certain class of systems of hyperbolic conservation laws such as systems modeling polymer flooding in the heterogeneous oil reservoir engineering.

It is well known that in a heterogeneous media fingering instability will be developed when the water (aqueous phase) with less viscous displaces the oil with more viscous, due to which water can reach the recovery well before all the oil reaches and this end up with a poor oil recovery. To overcome this difficulty, in an enhanced oil

G.D. Veerappa Gowda (✉)
TIFR Centre for Applicable Mathematics, Sharada Nagar,
Yelahanka New Town, Bangalore 560065, India
e-mail: gowda@math.tifrbng.res.in

P. Manchanda et al. (eds.), *Industrial Mathematics and Complex Systems*,
Industrial and Applied Mathematics, DOI 10.1007/978-981-10-3758-0_12

recovery (EOR) process different polymers of varying concentrations are injected along with the water. As the concentration of polymers increases, viscosity of water increases that increases the oil and water mobility ratio. Improving in the oil and water mobility ratio reduces the fingering effect which in turn enhances the oil production.

In the absence of capillary pressure, flow is described by the Buckley–Leverett equation with multicomponent polymer flooding, see [3–5, 9]. In two dimensions, this corresponds to a system of $(m+1)\times(m+1)$, non-strictly, nonlinear hyperbolic conservation laws given by

$$\begin{aligned} s_t + \nabla\cdot F(s, c_1, c_2, \dots c_m, x) &= 0 \\ (sc_l + a_l(c_l))_t + \nabla\cdot(c_l F(s, c_1, c_2, \dots c_m, x)) &= 0, \quad l = 1, 2, \dots, m \end{aligned} \tag{12.1}$$

where $s = s(x,t)\,,\, c_l = c_l(x,t)\,,\,(x,t) \in I\!R^2 \times (0,\infty)$, are saturation of water and concentration of the polymers, respectively, and the flux function, $F(s,c,x) = (F_1, F_2) \in I\!R^2$ is given by

$$\begin{aligned} F_1(s,c,x) &= v_1(x) f(s,c) \\ F_2(s,c,x) &= [v_2(x) - (\rho_w - \rho_o) g \lambda_o(s) K(x)]\, f(s,c), \end{aligned} \tag{12.2}$$

where $c = (c_1, c_2, \dots, c_m)$, ρ_w and ρ_o are the densities of water and oil, g is the acceleration due to gravity, $a_l,\ l = 1, \dots, m$ are adsorption functions, $K(x)$ is the absolute permeability of the rock, $\lambda_w(s,c)$ and $\lambda_o(s,c)$ are mobilities of water and oil, respectively, and $v = (v_1, v_2) \in I\!R^2$ is the total velocity given by Darcy's law:

$$v = -((\lambda_w + \lambda_o) K(x) \frac{\partial p}{\partial x_1}, (\lambda_w + \lambda_o) K(x) \frac{\partial p}{\partial x_2} + (\lambda_w \rho_w + \lambda_o \rho_o) g K(x)),$$

where p is a pressure. The velocity v is governed by the incompressibility of the flow

$$\nabla\cdot v = 0,$$

which leads to a Poisson equation for the pressure p.

For this system, developing a Godunov type upwind numerical scheme is difficult as it needs the exact or approximate Riemann solvers. The presence of gravity makes the flux in s no longer monotone and can also change the sign. All this combined with heterogeneity of the porous medium makes the computation of Reimann problem solution complicated and expensive. Hence, the construction of Godunov type scheme becomes complicated. Most often these numerical methods require the eigenstructure of the system. Here by using the idea of discontinuous flux, coupled system is reduced to an uncoupled scalar equations with discontinuous coefficients. Next we study each scalar equation by using the idea of discontinuous flux (DFLU), developed in [1, 2]. This approach does not require detailed information about the eigenstructure of the full system but one has to handle discontinuous coefficients properly.

12.1.1 One-Dimensional Problem with Single Polymer Component

In one space dimension and for $m = 1$, system (12.1) reduces to the following 2×2 system of conservation laws:

$$\begin{aligned} s_t + f(s, c, x)_x &= 0 \\ (sc + a(c))_t + (cf(s, c, x))_x &= 0, \end{aligned} \tag{12.3}$$

where

$$f(s, c, x) == \frac{\lambda_1(s, c)}{\lambda_1(s, c) + \lambda_2(s, c)}[v + (\rho_o - \rho_w)gK(x)\lambda_2(s, c)]. \tag{12.4}$$

12.1.2 Finite Volume Method to Solve the System (12.3)

We define the space grid points as $x_{i+\frac{1}{2}} = ih, \ h > 0$ and $i \in Z$ and for $\Delta t > 0$ define the time discretization points $t_n = n\Delta t$ for all nonnegative integer n, and $\lambda = \frac{\Delta t}{h}$. The finite volume scheme for the system (12.3) is given by

$$\begin{aligned} s_i^{n+1} &= s_i^n - \lambda(F_{i+\frac{1}{2}}^n - F_{i-\frac{1}{2}}^n) \\ c_i^{n+1}s_i^{n+1} + a(c_i^{n+1}) &= c_i^n s_i^n + a(c_i^n) - \lambda(G_{i+\frac{1}{2}}^n - G_{i-\frac{1}{2}}^n), \end{aligned} \tag{12.5}$$

where numerical fluxes $F_{i+\frac{1}{2}}^n$ and $G_{i+\frac{1}{2}}^n$ are associated with the flux functions $F(S, c, x)$ and $G(S, c, x) = cF(S, c, x)$ respectively and they are functions of the left and right values of the saturation s and the concentration c at $x_{i+\frac{1}{2}}$:

$$F_{i+\frac{1}{2}}^n = \bar{F}(s_i^n, c_i^n, s_{i+1}^n, c_{i+1}^n, x_{i+\frac{1}{2}}), \quad G_{i+\frac{1}{2}}^n = \bar{G}(s_i^n, c_i^n, s_{i+1}^n, c_{i+1}^n, x_{i+\frac{1}{2}}).$$

The choice of the numerical flux functions $\bar{F}$ and $\bar{G}$ determines the numerical scheme. Once we compute s_i^{n+1} from the first equation of (12.5), then we recover c_i^{n+1} from second equation using an iterative method, like Newton-Raphson method.

Now we briefly explain the discontinuous flux given in [1, 2].

12.1.3 The DFLU Numerical Flux

The DFLU flux is an extension of the Godunov scheme that was proposed and analyzed in [2] for scalar conservations laws with a flux function discontinuous in space.

We define

$$G^n_{i+\frac{1}{2}} = \begin{cases} c^n{}_i F^n_{i+\frac{1}{2}} & \text{if } F^n_{i+\frac{1}{2}} > 0 \\ c^n{}_{i+1} F^n_{i+\frac{1}{2}} & \text{if } F^n_{i+\frac{1}{2}} \leq 0. \end{cases} \tag{12.6}$$

Now the choice of the numerical scheme depends on the choice of $F^n_{i+1/2}$. To do so we treat $c(x, t)$ in $F(s, c, x)$ as a known function which may be discontinuous at the space discretization points and F is allowed to be discontinuous in the x variable at the same space discretization points. Therefore on each rectangle $(x_{i-\frac{1}{2}}, x_{i+\frac{1}{2}}) \times (t_n, t_{n+1})$, we consider the conservation law:

$$s_t + F(s, c^n_i, x)_x = 0,$$

with initial condition $s(x, 0) = s^n_i$ for $x_{i-\frac{1}{2}} < x < x_{i+\frac{1}{2}}$ (see the figure below).

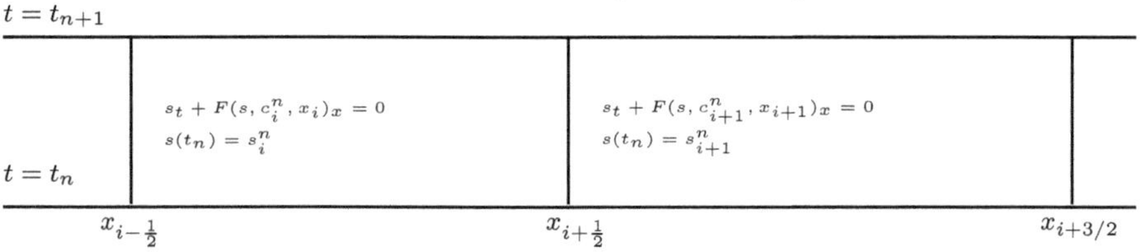

The above problem can be considered as a conservation law with flux function discontinuous in x for which DFLU flux can be used. Then the DFLU flux is given as

$$\begin{aligned} F^n_{i+\frac{1}{2}} &= F^{DFLU}(s^n_i, c^n_i, s^n_{i+1}, c^n_{i+1}) \\ &= \max\{F(\max\{s^n_i, \theta^n_i\}, c^n_i, x_i), F(\min\{s^n_{i+1}, \theta^n_{i+1}\}, c^n_{i+1}, x_{i+1})\}, \end{aligned} \tag{12.7}$$

where $\theta^n_i = \operatorname{argmin} F(., c^n_i, x_i)$.

12.2 Multicomponent Polymer Flooding in Two Dimensions

The ideas of previous sections can be extended to two dimensions with multicomponent polymer flooding and higher order accurate schemes can be constructed by introducing slope limiter in space variable and a strong stability preserving Runge–Kutta scheme in the time variable, see [7, 8]. The resulting schemes are shown to respect a maximum principle and other desired properties. These schemes are easy to implement on a computer and they are less expensive from the point of computation. Numerical experiments clearly show that (see [7, 8]) by adding polymers (in the presence/absence of gravity) to water improves the efficiency of oil recovery even in a highly heterogeneous medium.

References

1. Adimurthi, Jaffre, J., Veerappa Gowda, G.D.: Godunov-type methods for conservation laws with a flux function discontinuous in space. SIAM. J. Numer. Anal. **42**(1), 179–208 (2004)
2. Adimurthi, Veerappa Gowda, G.D., Jaffre, J.: The DFLU flux for systems of conservation laws. J. Comput. Appl. Math. **247**, 102–123 (2013)
3. Chavent, G., Jaffre, J.: Mathematical Models and Finite Elements for Reservoir Simulation. North-Holland, Amsterdam (1978)
4. Johansen, T., Winther, R.: The solution of the Riemann problem for a hyperbolic system of conservation laws modeling polymer flodding. SIAM J. Math. Anal. **19**(3), 541–566 (1988)
5. Johansen, T., Tveito, A., Winther, R.: A Riemann solver for a two phase multicomponent process. SIAM. J. Sci. Statist. Comput. **10**(5), 846–879 (1989)
6. Kaasschieter, E.F.: Solving the Buckley-Levrettequation with gravity in a hetrogeneous porous medium. Comput. Geosci. **3**(1), 23–48 (1999)
7. Sudarshan Kumar, K.: A finite volume method for non-linear system of hyperbolic conservation laws arising in oil reservoir simulations, Ph.D. thesis, TIFR (2014)
8. Sudarshan Kumar, K., Praveen, C., Veerappa Gowda, G.D.: A finite volume method for a two-phase multicomponent polymer flooding, J. Comput. Phys. **275**, 667–695 (2014)
9. Tveito, A., Winther, R.: Existence, uniqueness, and continuous dependence for a system of hyperbolic conservation laws modeling polymer flooding. SIAM J. Mat. Anal. **22**(4), 905–933 (1991)

Chapter 13
Construction and Properties of Haar-Vilenkin Wavelets

Meenakshi and P. Manchanda

Abstract The Haar wavelet based on Haar system, introduced by the Hungarian mathematician Alfred Haar [5], is the simplest example of wavelets. Recently, we have studied the concept of Haar-Vilenkin wavelet in [8] which is a generalization of Haar wavelet. We have introduced a Special Type of Multiresolution Analysis [10] generated by Haar-Vilenkin wavelet which is a special case of matrix multiresolution analysis studied in [18]. In this paper we represent Haar-Vilenkin wavelets in discrete form by introducing Haar-Vilenkin matrices and expand a function in Haar-Vilenkin wavelet series. We have applied this method for solving ordinary differential equations in this paper.

Keywords Wavelets · Haar-Vilenkin system multiresolution analysis · Matrices

AMS classification 42A38 · 42A55 · 42C15 · 42C40 · 43A70

13.1 Introduction

We have introduced the concept of Haar-Vilenkin wavelet and Haar-Vilenkin scaling function and have studied the basic properties of Haar-Vilenkin wavelet series and coefficients in [8]. Haar-Vilenkin wavelet is a generalization of Haar wavelet. Haar wavelet basis is the simplest and historically the first example of an orthonormal wavelet basis. Haar basis functions are step functions with jump discontinuities. Haar wavelet basis provides a very efficient representation of functions that consist of smooth, slowly varying segments punctuated by sharp peaks and discontinuities. Haar system is an orthonormal system such that each continuous function on [0, 1] has a uniformly convergent Fourier series with respect to this system.

Meenakshi (✉)
Department of Mathematics, Lajpat Rai D.A.V. College, Jagraon, India
e-mail: meenakshi_wavelets@yahoo.com

P. Manchanda
Department of Mathematics, Guru Nanak Dev University, Amritsar, India
e-mail: pmanch2k1@yahoo.co.in

P. Manchanda et al. (eds.), *Industrial Mathematics and Complex Systems*,
Industrial and Applied Mathematics, DOI 10.1007/978-981-10-3758-0_13

The Haar wavelet is the function defined on the real line $\mathbb{R}$ as

$$h(x) = \begin{cases} 1 & x \in [0, \frac{1}{2}) \\ -1 & x \in [\frac{1}{2}, 1) \\ 0 & \text{otherwise} \end{cases}.$$

It can be expressed in the form

$$h(x) = \chi_{[0,\frac{1}{2})}(x) - \chi_{[\frac{1}{2},1)}(x).$$

By taking the translations and dilations of $h(x)$, the system $\{h_{j,k}(x)\}_{j,k\in\mathbb{Z}}$ is referred as the Haar system on $\mathbb{R}$ where $h_{j,k}(x) = 2^{j/2}h(2^j x - k),\ \ j, k \in \mathbb{Z}$. The Haar scaling function on the real line is $p(x) = \chi_{[0,1)}(x)$. The collection $\{p_{j,k}(x)\}_{j,k\in\mathbb{Z}}$ is referred to as the system of Haar scaling functions:

$$\operatorname{supp} h_{j,k} = \Big[\frac{k}{2^j}, \frac{k+1}{2^j}\Big),$$

where the intervals $\big[\frac{k}{2^j}, \frac{k+1}{2^j}\big)$ for $j, k \in \mathbb{Z}$ form the family of dyadic intervals. The various properties of Haar system have been extensively studied. It has been shown that the system $\{h_{j,k}(x)\}_{j,k\in\mathbb{Z}}$ is an orthonormal system in $L^2(\mathbb{R})$ [2, 16, 18]. The family $\{h_{j,k}\}_{j,k\in\mathbb{Z}}$ is also associated with multiresolution analysis, for example:
Let $S_n = \ \operatorname{span}\{h_{j,k}\}_{j<n,k\in\mathbb{Z}}$, and

$$L_n = \{\text{all functions in } L^2(\mathbb{R}) \text{ constant on all intervals } [k2^{-n}, (k+1)2^{-n}) \text{ for } k \in \mathbb{Z}\}.$$

Both the families have the following properties:

$$\begin{aligned} \ldots S_{-1} \subset S_0 \subset S_1 \subset \ldots, \\ f(t) \in S_n \Longleftrightarrow f(2t) \in S_{n+1}, \\ f(t) \in S_0 \Longleftrightarrow f(t+k) \in S_0 \text{ for } k \in \mathbb{Z}. \end{aligned}$$

It can be proved that $L_n = S_n$ for all $n \in \mathbb{Z}$, then the family $\{L_n\}_{n=-\infty}^{\infty}$ form a multiresolution analysis.

Theorem 13.1 ([18]) *The system $\{h_{j,k}(x)\}_{j,k\in\mathbb{Z}}$ forms an orthonormal basis in $L^2(\mathbb{R})$.*

Theorem 13.2 [18] *If $f \in L^p(\mathbb{R})$ with $1 < p < \infty$ or f is C^0 on $\mathbb{R}$, then $\lim_{r\to\infty} P_r(f) = f$ and for each $r \in \mathbb{Z}$, $\lim_{\mu\to\infty} P_r(f) + Q_r^{\mu}(f) = P_{r+1}(f)$ where $P_r(f) = \sum_{j<r}\sum_{k\in\mathbb{Z}} < f, h_{j,k} > h_{j,k}$ and $Q_j^{\mu}(f) = \sum_{k\le\mu} < f, h_{j,k} > h_{j,k}$. The convergence is in the norm of the space.*

The convergence in $L^p[0, 1]$ for $1 \le p < \infty$ has been shown by Schauder in [11]. Comparison of Fourier series of a function $f \in L^2(\mathbb{R})$ and its expansion with respect to the Haar system has been investigated. Behavior of Haar coefficients is also studied.

13.2 Behavior of Haar Coefficients Near Jump Discontinuities

The following estimates are obtained in [16].

Suppose $f(x)$ is a function defined on [0, 1] with a jump discontinuity at $x_0 \in (0, 1)$ and continuous at all other points in [0, 1].

Let us assume that the function $f(x)$ is C^2 on the intervals $[0, x_0]$ and $[x_0, 1]$. Now we have two possibilities, either $x_0 \in \left[\frac{k}{2^j}, \frac{k+1}{2^j}\right)$ or $x_0 \notin \left[\frac{k}{2^j}, \frac{k+1}{2^j}\right)$. Let $x_{j,k}$ is the mid point of the interval $\left[\frac{k}{2^j}, \frac{k+1}{2^j}\right)$, i.e., $x_{j,k} = 2^{-j}(k + 1/2)$.

Case I If $x_0 \notin \left[\frac{k}{2^j}, \frac{k+1}{2^j}\right)$, then

$$| < f, h_{j,k} > | \approx \frac{1}{4}|f'(x_{j,k})|2^{-3j/2}.$$

Case II If $x_0 \in \left[\frac{k}{2^j}, \frac{k+1}{2^j}\right)$, then

$$| < f, h_{j,k} > | \approx \frac{1}{4}|f(x_0^-) - f(x_0^+)|2^{-j/2}.$$

Thus we see that the decay of $| < f, h_{j,k} > |$ for large j is considerably slower if $x_0 \in \left[\frac{k}{2^j}, \frac{k+1}{2^j}\right)$ than if $x_0 \notin \left[\frac{k}{2^j}, \frac{k+1}{2^j}\right)$. That is, large coefficients in the Haar expansion of a function $f(x)$ that persist for all scales suggest the presence of jump discontinuity in the intervals $\left[\frac{k}{2^j}, \frac{k+1}{2^j}\right)$ corresponding to the large coefficients.

It may be observed that Haar function was introduced in 1910 [5], Walsh function in 1923 [17] and Haar type Vilenkin system in 1947, see for e.g., [12, 14, 15]. Certain properties of multidimensional generalized Haar type Fourier series have been investigated [13].

In the recent years various extensions and concepts related to Haar wavelet have been studied, see e.g., [1, 3, 4, 7, 13]. The matrix form of Haar wavelets, the integrals related to it, and the solution of ODE's using Haar Wavelet coefficients are studied in [6]. In this paper we have introduced matrix form of Haar-Vilenkin wavelets. System of ODE's is also solved using Haar-Vilenkin wavelet coefficients.

This paper is organized as follows: In section 2 we have recalled the concept of Haar-Vilenkin wavelets and a special type of multiresolution analysis. Integrals related to Haar-Vilenkin wavelets have been evaluated in section 3 and wavelets have been represented in matrix form and procedure for expanding a function or a signal in Haar-Vilenkin wavelet series is given. In section 4 the procedure for solving ODE's using Haar-Vilenkin wavelets is introduced.

13.3 Haar-Vilenkin Wavelet

Let us recall the system of Haar-Vilenkin wavelets studied in [8]:
The following system which is a generalization of Haar system is connected with the name of Vilenkin. Very often it is termed as a generalized Haar system or a Haar type Vilenkin system.

Let $m = (m_k, k \in \mathbf{N})$ be a sequence of natural numbers such that $m_k \geq 2$, $\mathbf{N}$ denotes the set of nonnegative integers. Let $M_0 = 1$ and $M_k = m_{k-1}M_{k-1}$, $k \in \mathbf{P}$.

Let $\mathbf{P}$ denotes the set of positive integers and let $k \in \mathbf{P}$ can be written as

$$k = M_n + r(m_n - 1) + s - 1, \tag{13.1}$$

where $n \in \mathbf{N}$, $r = 0, 1, \ldots, M_n - 1$ and $s = 1, 2, \ldots, m_n - 1$. This expression is unique for each $k \in \mathbf{P}$. Let us write an arbitrary element $t \in [0, 1)$ in the form

$$t = \sum_{k=0}^{\infty} \frac{t_k}{M_{k+1}}, \ (0 \leq t_k < m_k). \tag{13.2}$$

It may be noted that there exists two such expressions (13.2), for so-called m-adic rational numbers. In such cases we use the expression which contains only a finite number of terms different from zero.

Define the function system $(h_k, k \in \mathbf{N})$ by $h_0 = 1$ and

$$h_k(t) = \begin{cases} \sqrt{M_n} e^{\frac{2\pi i s t_n}{m_n}} & \frac{r}{M_n} \leq t < \frac{r+1}{M_n} \\ 0 & otherwise \end{cases}. \tag{13.3}$$

This system can be extended to $\mathbb{R}$ by periodicity of period 1: $h_k(t + 1) = h_k(t)$, $t \in [0, 1)$. It can be checked that $\{h_k(t)\}$ is a complete orthonormal system in $L^2(\mathbb{R})$. This system is called Haar-Vilenkin system. It is clear that

$$h_k(t) = \chi_{[\frac{r}{M_n}, \frac{r+1}{M_n}]}(t)\sqrt{M_n} e \frac{2\pi i s t_n}{m_n}.$$

Certain properties of this system have been recently studied [8].

For $k \in \mathbf{P}$ and $t \in [0, 1)$ as defined in (13.1) and (13.2) the **Haar-Vilenkin scaling function** is defined as

$$\begin{aligned} p_k(t) &= \sqrt{M_n} \chi_{[\frac{r}{M_n}, \frac{r+1}{M_n})} \\ &= \begin{cases} \sqrt{M_n}, & \frac{r}{M_n} \leq t < \frac{r+1}{M_n} \\ 0 & otherwise \end{cases}. \end{aligned} \tag{13.4}$$

The collection $\{\phi_{a,b}(t)\}_{a,b \in \mathbb{Z}}$ is referred to as the system of Haar-Vilenkin scaling functions where $\phi_{a,b}(t) = m_n^{a/2} p_k(m_n^a t - b)$.

We have studied the basic properties of Haar-Vilenkin system in [8]. We have proved the orthogonality of Haar-Vilenkin wavelet, convergence of Haar-Vilenkin wavelet series, and properties of Haar-Vilenkin wavelet coefficients. We have introduced a multiresolution analysis where translation and dilation are taken by $\frac{b}{M_n}$ $(b \in \mathbb{Z})$ and m_n, respectively.

Define

$$\psi_{a,b}(t) = m_n^{a/2} h_k(m_n^a t - b). \tag{13.5}$$

The collection $\{\psi_{a,b}(t)\}_{a,b\in\mathbb{Z}}$ is referred to as the Haar-Vilenkin system.
$\psi_{a,b}(t)$ is supported on the interval $I_{a,b}$ where
$I_{a,b} = \left[\frac{r}{m_n^a M_n} + \frac{b}{m_n^a}, \frac{r+1}{m_n^a M_n} + \frac{b}{m_n^a}\right), a, b \in \mathbb{Z}$.
The system $\psi_{a,b}(t)$ can also be written as $\{m_n^{\frac{a}{2}} h_k(m_n^a t - b)\} = D_{m_n^a} T_b h_k(t)$.

Theorem 13.3 *The system* $\{m_n^{\frac{a}{2}} h_k(m_n^a t - b)\} = \{\psi_{a,b}\}, a, b \in \mathbb{Z}$ *is an orthonormal system in* $L^2(\mathbb{R})$.

Behavior of Haar-Vilenkin Coefficients Near Jump Discontinuities.

Suppose that $f(x)$ is defined on interval $\left[\frac{r}{M_n}, \frac{r+1}{M_n}\right]$ with a jump discontinuity at $x_0 \in \left(\frac{r}{M_n}, \frac{r+1}{M_n}\right)$ and continuous at all other points in $\left[\frac{r}{M_n}, \frac{r+1}{M_n}\right]$. We have to check whether Haar-Vilenkin coefficients $< f, \psi_{a,b} >$ such that $x_0 \in I_{a,b}$ behave differently than do the Haar-Vilenkin coefficients such that $x_0 \notin I_{a,b}$.

Let us assume that given function $f(x)$ is C^2 on the intervals $[\frac{r}{M_n}, x_0]$ and $[x_0, \frac{r+1}{M_n}]$. This means that both $f'(x)$ and $f''(x)$ exist, are continuous functions and hence are bounded on these intervals. Fix integers $a \geq 0$ and $0 \leq b \leq m_n^a - 1$ and let $x_{a,b}$ be the midpoint of the interval $I_{a,b}$, i.e., $x_{a,b} = \frac{r+1/2}{m_n^a M_n} + \frac{b}{m_n^a}$.

Case I If $x_0 \notin I_{a,b}$, then we find that for the large values of a

$$|\langle f, \psi_{a,b}\rangle| \approx \frac{1}{4} m_n^{-3a/2} M_n^{-3/2} |f'(x_{a,b})|.$$

Case II If $x_0 \in I_{a,b}$,

Thus for the large values of a

$$\begin{aligned} | < f, \psi_{a,b} > | &\approx m_n^{a/2}\sqrt{M_n}\frac{1}{2m_n^a M_{n+1}}|f(x_0^-) - f(x_0^+)| \\ &= \frac{m_n^{-a/2} M_n^{1/2}}{2M_{n+1}}|f(x_0^-) - f(x_0^+)|. \end{aligned}$$

Comparing the two cases, we see that the decay of $| < f, \psi_{a,b} > |$ for the large a is considerably slower if $x_0 \in I_{a,b}$ than if $x_0 \notin I_{a,b}$.

The large coefficient in the Haar-Vilenkin expansion of the coefficient f(x) that persist for all scales suggests the presence of jump discontinuity in the intervals $I_{a,b}$ corresponding to the large coefficient.

13.3.1 A Special Type of Multiresolution Analysis

Definition 13.1 For k as in (13.1), a special type of multiresolution analysis is a sequence $\{V_j\}_{j\in\mathbb{Z}}$ of closed subspaces of $L^2(\mathbb{R})$ such that

1. $V_j \subset V_{j+1}$ for all $j \in \mathbb{Z}$.
2. $\cup_{j\in\mathbb{Z}} V_j$ is dense in $L^2(\mathbb{R})$.
3. $\cap_{j\in\mathbb{Z}} V_j = \{0\}$.
4. $f(x) \in V_j$ iff $f(m_n^{-j}x) \in V_0$ for all $j \in \mathbb{Z}$.
5. There exists a function $g_k(x)$ in $L^2(\mathbb{R})$, called the scaling function such that the collection $\{g_k(t - \frac{b}{M_n})\}_{b\in\mathbb{Z}}$ is an orthonormal system of translates and

$$V_0 = \overline{span}\{T_{\frac{b}{M_n}} g_k(x)\}.$$

For details on wavelet generated by a special type of multiresolution analysis see [9].

Remark 13.1 A special type of multiresolution analysis is defined by first identifying the space V_0, defining V_j by letting

$$V_j = \{f(x) : f(x) = D_{m_n^j} g(x), g(x) \in V_0\}$$

so that the Definition (13.1)(4) is satisfied and then proving that Definition (13.1)(1), (2), (3), and (5) hold. V_0 can be defined by just identifying the function $g_k(x)$ such that $\{T_{\frac{b}{M_n}} g_k(x)\}_{b\in\mathbb{Z}}$ is an orthonormal system of translates and then defining

$$V_0 = \overline{span}\{T_{\frac{b}{M_n}} g_k(x)\}.$$

Example 13.3.1 (Haar-Vilenkin Multiresolution Analysis) Let V_0 consist of all step functions $f(x)$ such that
(i) $f(x) \in L^2(\mathbb{R})$.
(ii) $f(x)$ is constant in the intervals $I_{0,\frac{b}{M_n}} \equiv \left[\frac{r+b}{M_n}, \frac{r+b+1}{M_n}\right)$ for all $b \in \mathbb{Z}$.
It can be verified that for $l \in \mathbb{Z}$

$$V_0 = \overline{span}\{T_{\frac{l}{M_n}} p_k(x)\}$$

, where $p_k(x) = \sqrt{M_n}\chi_{[\frac{r}{M_n}, \frac{r+1}{M_n})}(x)$.

13.4 Haar-Vilenkin Wavelets and Their Integrals

The Haar-Vilenkin system $(h_n,\ n \in \mathbf{N})$ over the interval $[A, B]$ of length 1 is defined by $h_0 = 1$ and

$$h_k(t) = \begin{cases} \sqrt{M_n} e^{\frac{2\pi i s t_n}{m_n}} & A + \frac{r}{M_n} \leq t < A + \frac{r+1}{M_n} \\ 0 & otherwise \end{cases}. \tag{13.6}$$

We have proved the orthogonality of Haar-Vilenkin wavelets in [8].
We need the following integrals of Haar-Vilenkin functions:

$$\begin{aligned} P_{v,i}(x) &= \int_A^x \int_A^x \cdots \int_A^x h_i(t)\, dt^v \\ &= \frac{1}{(v-1)!} \int_A^x (x-t)^{v-1} h_i(t)\, dt. \end{aligned}$$

For $i \neq 1$,we have
If $\frac{r}{M_n} \leq x < \frac{r}{M_n} + \frac{1}{M_{n-1}}$, then we have

$$\begin{aligned} P_{\alpha,i}(x) &= \frac{1}{(\alpha-1)!} \int_{r/M_n} x(x-t)^{\alpha-1} \sqrt{M_n}\, dt \\ &= \frac{\sqrt{M_n}}{(\alpha-1)!} \int_{r/M_n} x(x-t)^{\alpha-1}\, dt \\ &= \frac{\sqrt{M_n}}{(\alpha-1)!} . \frac{1}{\alpha} \left(x - \frac{r}{M_n}\right)^\alpha \\ &= \frac{\sqrt{M_n}}{\alpha!} \left(x - \frac{r}{M_n}\right)^\alpha . \end{aligned}$$

If $\frac{r}{M_n} + \frac{1}{M_{n-1}} \leq x < \frac{r}{M_n} + \frac{2}{M_{n-1}}$, then on solving as above, we have

$$P_{\alpha,i}(x) = \frac{\sqrt{M_n}}{\alpha!} \left[x - \left(\frac{r}{M_n} + \frac{1}{M_{n-1}}\right)\right]^\alpha .e\frac{2\pi i s}{m_n}.$$

...

...

If $\frac{r}{M_n} + \frac{m_n-1}{M_{n-1}} \leq x < \frac{r+1}{M_n}$, we have

$$P_{\alpha,i}(x) = \frac{\sqrt{M_n}}{\alpha!} \left[x - \left(\frac{r}{M_n} + \frac{m_n - 1}{M_{n-1}}\right)\right]^\alpha .e\frac{2\pi i s(m_n - 1)}{m_n}.$$

Therefore,

$$P_{\alpha,i}(x) = \begin{cases} 0 & x < \frac{r}{M_n} \\ \frac{\sqrt{M_n}}{\alpha!}\left(x - \frac{r}{M_n}\right)^{\alpha} & \frac{r}{M_n} \le x < \frac{r}{M_n} + \frac{1}{M_{n-1}} \\ \frac{\sqrt{M_n}}{\alpha!}\left[x - \left(\frac{r}{M_n} + \frac{1}{M_{n-1}}\right)\right]^{\alpha} .e^{\frac{2\pi i s}{m_n}} & \frac{r}{M_n} + \frac{1}{M_{n-1}} \le x < \frac{r}{M_n} + \frac{2}{M_{n-1}} \\ \cdots & \\ \frac{\sqrt{M_n}}{\alpha!}\left[x - \left(\frac{r}{M_n} + \frac{m_n-1}{M_{n-1}}\right)\right]^{\alpha} .e^{\frac{2\pi i s(m_n-1)}{m_n}} & \frac{r}{M_n} + \frac{m_n-1}{M_{n-1}} \le x < \frac{r+1}{M_n} \\ 0 & otherwise \end{cases}. \tag{13.7}$$

If $i = 0$, we have $h_i(t) = 0$ and

$$\begin{aligned} P_{\alpha,1}(x) &= \frac{1}{(\alpha-1)!}\int_A^x (x-t)^{\alpha-1}\,dt \\ &= \frac{1}{(\alpha-1)!}.\frac{1}{\alpha}(x-A)^{\alpha} \\ &= \frac{(x-A)^{\alpha}}{\alpha!} \end{aligned} \tag{13.8}$$

Equation (13.7) holds for $i > 1$.
For $i = 1$, we have Eq. (13.8).

13.4.1 Matrix Form of Haar-Vilenkin Wavelets

Consider the case where $A = 0$ and $B = 1$.
We have formulated the Haar-Vilenkin wavelets in the discrete form:
Denote the grid points by

$$\tilde{x}_l = A + l\delta x, \ l = 0, 1, \ldots, m_0. \tag{13.9}$$

We have considered

$$x_l = \frac{1}{2}(\tilde{x_{l-1}} + \tilde{x}_l), \ l = 1, 2, \ldots, m_0. \tag{13.10}$$

On replacing x by x_l in Eqs. (13.1), (13.7) and (13.8), we will obtain Haar-Vilenkin wavelet matrices. We introduce the square matrices $H, P_1, P_2, \ldots, P_v$. The elements of these matrices are

$$H(i,l) = h_i(x_l), \quad P_v(i,l) = P_{v_i}(x_l), \quad v = 1, 2, 3, \ldots.$$

$A = 0$, $B = 1$ and

$$\delta x = \frac{B-A}{m_0} = \frac{1}{m_0}.$$

Example 13.1 Consider the sequence $(m_k,\ k \in \mathbf{N}) = (2, 2, 2, 2, \ldots)$.
Then $x_1 = \frac{1}{4}, x_2 = \frac{3}{4}$.
Therefore,

$$h_1(t) = \begin{cases} 1 & 0 \le t < 1/2 \\ -1 & 1/2 \le t < 1 \\ 0 & otherwise \end{cases}$$

$$h_2(t) = \begin{cases} \sqrt{2} & 0 \le t < 1/4 \\ -\sqrt{2} & 1/4 \le t < 1/2 \\ 0 & otherwise \end{cases}.$$

The Haar-Vilenkin matrix H is

$$H = \begin{bmatrix} h_1(x_1) & h_1(x_2) \\ h_2(x_1) & h_2(x_2) \end{bmatrix}. = \begin{bmatrix} 1 & -1 \\ -\sqrt{2} & 0 \end{bmatrix}.$$

Similarly, the other Haar-Vilenkin matrices are

$$P_1 = \begin{bmatrix} P_{11}(x_1) & P_{11}(x_2) \\ P_{12}(x_1) & P_{12}(x_2) \end{bmatrix}, \quad P_2 = \begin{bmatrix} P_{21}(x_1) & P_{21}(x_2) \\ P_{22}(x_1) & P_{22}(x_2) \end{bmatrix} \quad \ldots.$$

Using Eqs. (13.7) and (13.8), we obtain

$$P_1 = \begin{bmatrix} \frac{1}{4} & \frac{3}{4} \\ \frac{1}{32} & \frac{9}{32} \end{bmatrix}.$$

Example 13.2 Consider the sequence $(m_k,\ k \in \mathbf{N}) = (4, 3, 2, 2, \ldots)$.
Then $x_1 = \frac{1}{8}, x_2 = \frac{3}{8}, x_3 = \frac{5}{8}, x_4 = \frac{7}{8}$.
Therefore,

$$h_1(t) = \begin{cases} e^{\frac{\pi i t_0}{2}} & 0 \le t < 1 \\ 0 & otherwise \end{cases}$$

$$h_2(t) = \begin{cases} e\pi i t_0 & 0 \le t < 1 \\ 0 & otherwise \end{cases}$$

$$h_3(t) = \begin{cases} e^{\frac{3\pi i t_0}{2}} & 0 \le t < 1 \\ 0 & otherwise \end{cases}$$

$$h_4(t) = \begin{cases} 2e2\pi i t_0 & 0 \le t < 1/4 \\ 0 & otherwise \end{cases}.$$

The Haar-Vilenkin matrix H is

$$H = \begin{bmatrix} h_1(x_1) & h_1(x_2) & h_1(x_3) & h_1(x_4) \\ h_2(x_1) & h_2(x_2) & h_2(x_3) & h_2(x_4) \\ h_3(x_1) & h_3(x_2) & h_3(x_3) & h_3(x_4) \\ h_4(x_1) & h_4(x_2) & h_4(x_3) & h_4(x_4) \end{bmatrix}.$$

On solving, we obtain

$$H = \begin{bmatrix} 1 & i & -1 & -i \\ 1 & -1 & -1 & i \\ 1 & -i & -1 & i \\ 2i & 0 & 0 & 0 \end{bmatrix}.$$

Similarly, the other Haar-Vilenkin matrices are

$$P_1 = \begin{bmatrix} P_{11}(x_1) & P_{11}(x_2) & P_{11}(x_3) & P_{11}(x_4) \\ P_{12}(x_1) & P_{12}(x_2) & P_{12}(x_3) & P_{12}(x_4) \\ P_{13}(x_1) & P_{13}(x_2) & P_{13}(x_3) & P_{13}(x_4) \\ P_{14}(x_1) & P_{14}(x_2) & P_{14}(x_3) & P_{14}(x_4) \end{bmatrix}, \quad P_2 = \begin{bmatrix} P_{21}(x_1) & P_{21}(x_2) & P_{21}(x_3) & P_{21}(x_4) \\ P_{22}(x_1) & P_{22}(x_2) & P_{22}(x_3) & P_{22}(x_4) \\ P_{23}(x_1) & P_{23}(x_2) & P_{23}(x_3) & P_{23}(x_4) \\ P_{24}(x_1) & P_{24}(x_2) & P_{24}(x_3) & P_{24}(x_4) \end{bmatrix} \ldots .$$

Using Eqs. (13.7) and (13.8), we obtain

$$P_1 = \begin{bmatrix} \frac{1}{8} & \frac{3}{8} & \frac{5}{8} & \frac{7}{8} \\ \frac{1}{2}\left(\frac{1}{8}\right)^2 & \frac{1}{2}\left(\frac{3}{8}\right)^2 & \frac{1}{2}\left(\frac{5}{8}\right)^2 & \frac{1}{2}\left(\frac{7}{8}\right)^2 \\ \frac{1}{6}\left(\frac{1}{8}\right)^3 & \frac{1}{6}\left(\frac{3}{8}\right)^3 & \frac{1}{6}\left(\frac{5}{8}\right)^3 & \frac{1}{6}\left(\frac{7}{8}\right)^3 \\ \frac{1}{24}\left(\frac{1}{8}\right)^4 & \frac{1}{24}\left(\frac{3}{8}\right)^4 & \frac{1}{24}\left(\frac{5}{8}\right)^4 & \frac{1}{24}\left(\frac{7}{8}\right)^4 \end{bmatrix}, \quad P_2 = \begin{bmatrix} \frac{1}{2}\left(\frac{1}{8}\right)^2 & \frac{1}{2}\left(\frac{3}{8}\right)^2 & \frac{1}{2}\left(\frac{5}{8}\right)^2 & \frac{1}{2}\left(\frac{7}{8}\right)^2 \\ \frac{1}{2}\left(\frac{1}{8}\right)^2 & -\frac{1}{2}\left(\frac{3}{8}\right)^2 & \frac{1}{2}\left(\frac{5}{8}\right)^2 & -\frac{1}{2}\left(\frac{7}{8}\right)^2 \\ \frac{1}{3}\left(\frac{1}{8}\right)^2 & -\frac{1}{3}\left(\frac{3}{8}\right)^2 & \frac{1}{3}\left(\frac{5}{8}\right)^2 & -\frac{1}{3}\left(\frac{7}{8}\right)^2 \\ \frac{1}{4}\left(\frac{1}{8}\right)^2 & -\frac{1}{4}\left(\frac{3}{8}\right)^2 & \frac{1}{4}\left(\frac{5}{8}\right)^2 & -\frac{1}{4}\left(\frac{7}{8}\right)^2 \end{bmatrix}.$$

13.4.2 *Expanding a Function into Haar-Vilenkin Wavelet Series*

Let $f \in L^2[A, B]$. It can be expanded into Haar-Vilenkin Wavelet series as

$$f(x) = \sum_{i=1}^{m_0} a_i h_i(x), \tag{13.11}$$

where a_i denotes the Haar-Vilenkin wavelet coefficients. The discrete form of Eq. (13.11) is

$$\hat{f}(x_l) = \sum_{i=1}^{m_0} a_i h_i(x_l). \tag{13.12}$$

The matrix form of Eq. (13.12) is

$$f = aH, \tag{13.13}$$

where H is the Haar-Vilenkin matrix.
$a = (a_i)$, $f = (f_l)$ both are m_0 dimensional row vectors.
On solving the matrix (13.13), we get

$$a = fH^{-1}. \tag{13.14}$$

On replacing the value of a, in Eq. (13.11), we obtain the wavelet approximation of f. We can also check the degree of exactness of the approximation.

There are different ways to estimate the error function Δ of wavelet approximations. We have defined the error function as

$$\Delta = \int_A^B [f(x) - \hat{f}(x)]^2 \, dx, \tag{13.15}$$

where $\hat{f}(x)$ denotes the approximation of $f(x)$. The discrete form of Eq. (13.15) is

$$\Delta_{m_0} = \delta x \sum_{i=1}^{m_0} [f(x_i) - \hat{f}(x_i)]^2. \tag{13.16}$$

Example 13.3 Let $f(x) = \sqrt{x}$ for $x \in (0, 1)$.
Then Haar-Vilenkin matrix is formed as shown in Examples (13.1) and (13.2). The Haar-Vilenkin wavelet coefficients are calculated as in Eq. (13.14).

The error estimates Δ_{m_0} for the wavelet approximation are calculated as

m_0	Δ_{m_0}
4	1.270
5	0.312
6	0.0715
7	0.0183
8	0.0046

13.5 Solution of Ordinary Differential Equations

In this section we have expanded the highest order derivatives by Haar-Vilenkin wavelet series. Other lower order derivatives and the function are obtained through the integration and all the ingredients are substituted in the differential equation.

Consider the n-th order linear differential equation

$$\sum_{p=0}^{n} A_p(x)y^{(p)}(x) = f(x), \quad x \in [a,b], \tag{13.17}$$

with the initial conditions

$$y^{(p)}(a) = y_0^{(p)}, \quad p = 0, 1, 2, \ldots, n-1. \tag{13.18}$$

Here $A_p(x)$ are $f(x)$ are prescribed functions and $y_0^{(p)}$ are given constants.
Expand $y^{(n)}(x)$ by Haar-Vilenkin wavelet series as

$$y^{(n)}(x) = \sum_{i=1}^{m_0} a_i h_i(x), \tag{13.19}$$

where a_i are Haar-Vilenkin wavelet coefficients. On integration Eq. (13.19), $n-p$ times, we get

$$y^{(p)}(x) = \sum_{i=1}^{m_0} a_i P_{n-p,i}(x) + \Xi_p(x), \tag{13.20}$$

where

$$\Xi_p(x) = \sum_{m=0}^{n-p-1} \frac{1}{m!}(x-a)^m y_0^{(p+m)}. \tag{13.21}$$

We will replace x by x_l in Eqs. (13.17), (13.19), (13.20), and (13.21). We will substitute the values of (13.19), (13.20), and (13.21) in Eq. (13.17) and get a system of linear equations for calculating the Haar-Vilenkin coefficients a_i. After solving this system of equations the desired solution is calculated from (13.20). We can also check the exactness of this solution.

Example 13.4

$$\frac{dy}{dx} = x^2 - y, \quad y(0) = 0. \tag{13.22}$$

For $m_0 = 2$, we have calculated the values of H and P_1 in Example (13.1).

$$H = \begin{bmatrix} 1 & -1 \\ -\sqrt{2} & 0 \end{bmatrix}. \quad P_1 = \begin{bmatrix} \frac{1}{4} & \frac{3}{4} \\ \frac{1}{32} & \frac{9}{32} \end{bmatrix}.$$

By solving as above procedure, we have

$$y^{(1)}(x_l) = \sum_{i=1}^{2} a_i h_i(x_l) = a_1 h_1(x_l) + a_2 h_2(x_l). \tag{13.23}$$

$$\begin{aligned} y(x_l) &= \sum_{i=1}^{2} a_i P_{1i}(x_l) + \Xi(x_l) \\ &= \sum_{i=1}^{2} a_i P_{1i}(x_l) + y_0 \\ &= \sum_{i=1}^{2} a_i P_{1i}(x_l). \end{aligned} \tag{13.24}$$

On using the value of Eqs. (13.23) and (13.24) in (13.23), we obtain

$$\sum_{i=1}^{2} a_i h_i(x_l) + \sum_{i=1}^{2} a_i P_{1i}(x_l) = x_l^2.$$

We will obtain a system of equations

$$a(H + P_1) = F.$$

On solving, we will obtain the value of Haar-Vilenkin wavelet coefficients. On substituting these values in (13.22) we will get the required value and by using Matlab we have checked that these are very near to the exact value.

We will get the more exactness on taking the larger values of m_0.

References

1. Ciesielski, Z.: Haar orthogonal Functions in Analysis and Probability, Colloquia Societatis James Bolyai, 49, Alfred Haar Memorial Conference, Budapest, pp. 25–27 (1985)
2. Daubechies, I.: Ten Lectures on Wavelets, CBMS 61, SIAM, Philadelphia (1992)
3. Dubeau, F., Emejdani, S., Ksantini, R.: Non-uniform Haar wavelet. Appl. Math. Comput. **159**, 675–693 (2004)
4. Grozdanov, V., Stoilova, S.: Price and Haar type functions and uniform distribution of sequences. J. Inequal. Pure Appl. Math. **5**(2), 1–17 (2004)
5. Haar, A.: Zur Theorie der orthogonalen funktionen systeme. Math. An. **69**, 331–371 (1910)
6. Lepik, U., Hein, H.: Haar Wavelets with Applications. Springer International Publishing, Switzerland (2014)
7. Malozemov, V.N., Masharkii, S.M.: Generalized wavelet bases related to the discrete Vilenkin-Chresmathbben transform. St. Peterb. Math. J. **13**, 75–106 (2002)
8. Manchanda, P., Meenakshi, Siddiqi, A.H.: Haar-Vilenkin wavelet. Aligarh Bul. Math. **27**(1), 59–73 (2008)
9. Manchanda, P., Meenakshi: New classes of wavelets. In: Proceedings of the AIP Conference on Modelling of Engineering and Technological Problems, vol. 1146, pp. 253–271
10. Manchanda, P., Meenakshi, Siddiqi, A.H.: A Special Type of Multiresolution Analysis, accepted for publication
11. Schauder, M.J.: Einine eingenschaft der Haarschen orthogonalsystems. Math. Z. **28**, 317–320 (1928)

12. Schipp, F., Wade, W.R., Simon, P.: (with assistance by Pal, J.), Walsh Series: An Introduction to Dyadic Harmonic Analysis. Adam Hilger Ltd., Bristol and New York (1990)
13. Smailov, E.S.: On Paley-type theorem for multidimensional Fourier series on generalized Haar-type systems. Fundam. Prinkl. **7**(2), 533–563 (2000)
14. Uljanov, P.L.: Haar series and related questions, Colloqui Mathematica Societatis Janos Bolyai 49, Alfred Haar Memorial Conference, Budapest, pp. 57–96 (1985)
15. Vilenkin, N.Y.: On the theory of lacunary orthogonal system with gaps. Izo. Akad. Nauk. SSSR Ser. Math. **13**, 242–252 (1949)
16. Walnut, D.: An Introduction to Wavelet Analysis. Birkhäuser, Boston (2001)
17. Walsh, J.L.: A closed set of normal orthogonal function. Am. J. Math. **45**, 5–24 (1923)
18. Wojtaszczyk, P.: A Mathematical Introduction to Wavelets. London Mathematical Society Student Texts 37. Cambridge University Press, Cambridge (1997)

Chapter 14
Footprint-Based Personal Recognition Using Dactyloscopy Technique

Rohit Khokher and Ram Chandra Singh

Abstract The uniqueness of human footprint has drawn attention of academia and industry in recent years and is emerging as a latest biometric trait for biometric authentication. A robust technique to be used for identification and recognition of an individual using footprint as a biometric trait has been proposed in this work. Most of the footprint recognition methods require segmentation or connected component analysis. The determinant values that produce the features of the human footprint are generally utilized in the recognition processes. Static footprint images of 94 individuals (57 males and 37 females) of different regions of North India between age group 15–25 years have been acquired using Dactyloscopy technique. Biometric performance parameters such as false accept rate, false reject rate, genuine accept rate, half total error rate, and accuracy have been computed. The experimental results show that the performance parameters computed for Dactyloscopy technique could be used for biometric authentication. This study could be of potential use for forensic and non-forensic purposes and researchers working in foot biometrics.

Keywords Biometrics · Morphological operation · Dactyloscopy technique · FAR FRR · Accuracy

14.1 Introduction

An increased number of computer frauds like identity theft and computer hacking have been reported in past few years and therefore today's e-security are facing challenges against these threats and are in search of secure, accurate, and most effective

R. Khokher (✉)
Department of Computer Science and Engineering, Vidya College of Engineering, Vidya Knowledge Park, Baghpat Road, Meerut 250002, India
e-mail: khokherrohit@gmail.com

R.C. Singh
Department of Physics, School of Basic Sciences and Research, Sharda University, 32, 34, Knowledge Park-3, Greater Noida 201306, India
e-mail: rcsingh_physics@yahoo.com

P. Manchanda et al. (eds.), *Industrial Mathematics and Complex Systems*, Industrial and Applied Mathematics, DOI 10.1007/978-981-10-3758-0_14

alternatives to personal identification numbers and passwords. These fundamental problems can be resolved by biometric solutions because the biometric data of an individual is unique and inalienable. The physiological and behavioral characteristics of an individual are the two important capabilities to distinguish between an impostor and an authentic user. Biometric authentication has an advantage that the user has to be physically present during the process of identification. Therefore, it is inherently more reliable and capable than traditional knowledge based and token based techniques. Biometric personal identification has received growing interests, in recent years, from both the academia and the industry [1]. There are two types of biometric features: physiological (e.g., fingerprint, iris, and face) and behavioral (e.g., voice and handwriting). Each biometric feature has its own strengths and limitations and accordingly each biometric feature is used in authentication or identification applications. It is quite difficult to steal a biometric feature, create a copy, and use the fake one to attack the biometric systems. Recognition, identification, and verification of physiological biometric traits from video data and still images have been widely used in security access, multimedia, video indexing of large databases, and other commercial applications.

Footprint recognition is emerging as an important biometric trait and has drawn significant attention of the researchers working in the field of biometrics in past few years. Most studies are based on extracting some recognizable features since they are more robust than the features of the time domain. Biometric footprint recognizes an individual based on texture, foot shape, minutiae points, singular points, foot length, etc. For forensic applications, Kennedy [2] used, for the first time, barefoot inked images to extract geometrical features of foot impressions and since then foot biometrics has seen a considerable growth in this field. Kumar in his study [3] captured footprint images of left leg of 100 people from different angles; positioned and cropped these images according to the key points. A sequential modified Haar transform has been used to resize the footprint images to obtain modified Haar energy feature from the resized images. The Euclidean distance has been used to compare modified Haar energy feature with the feature vectors stored in the database. Krishnan [4] carried out his study on Gujjars of North India and studied the characteristics of their footprints. The static footprint features (e.g., foot shape and friction edge) have been used by King and Xiaopeng [5] to study the personal identification of an individual. Wang et al. [6] deliberated alternative system grounded on gait investigation. The dissemination of footprint substantial pressure surface reproduces the performance characteristics and the physiological characteristics of the humanoid figure. Recently, Khokher et al. [7] used principal component analysis (PCA) and independent component analysis (ICA) linear projection methods to extract texture- and shape-based features for personal identification of an individual. Uhl and Wild [8] explored an approach to study foot biometric characteristics, image enhancement, and feature extraction highlighting the characteristics of the foot geometry, their durability, and uniqueness properties. The gait recognition has also emerged as a remarkable signal processing tool for the biometric proof of identity [9]. Such as there is no minutia-based pattern matching system, study on gait-based identification by reflection of a person's walking style provides indication that such a system is accurate, possible to be advanced and used in the forthcoming days. Kuragano et al. [10] proposed a method to measure foot print similarity for gait analysis. Wild and Uhl [11] provided an overview of footprint and single-sensor based multimodal biometric recognition, and has developed a system for contemporary humanity, and as it is assumed that no complete biometric modality suitable for all the applications has been established. Khokher and Singh [12] in their study showed that the performance parameters computed for human footprint images using scanning technique show a better agreement with experimental results and could be used for biometric authentication. A correlation analysis has also been performed and a strong correlation is observed between actual height and toes, actual height and foot length, height and weight.

A growing interest, uniqueness of human footprint and limited study in this field motivated us to carryout this study. It is very difficult to achieve a high recognition rate by verifying raw footprint directly because people generally stand in various positions and postures with distances and angles between the two feet. To match the input pair of footprints with the centralized database, the input pair of footprints should be enhanced by using enhancement operations of preprocessing, normalization, orientation, and filtration. Such normalization may remove useful information for recognition, so geometric information of the footprint prior to normalization into an evaluation function for personal recognition decision is included. In this paper, we propose a footprint-based personal recognition using dactyloscopy technique to test its reliability. In the subsequent section, this paper deliberates the steps involved in acquiring footprint database in Sect. 14.2 and footprint enhancement process in Sect. 14.3. Section 14.4 includes feature extraction algorithm. Section 14.5 explains the simulation results and discussion of the study. The conclusion and future scope of this work is discussed in Sect. 14.6.

14.2 Data Acquisition

Dactyloscopy technique has been used to capture footprint images of 94 individuals (57 males and 37 females) of different genders, age groups, and of different regions of North India and create a database of these images (Table 14.1).

Table 14.1 DATABASE

S. No.	Gender	Age group (in years)	No. of individual
1.	Male	15–20	32
2.		20–25	25
3.	Female	15–20	17
4.		20–25	20

Fig. 14.1 Images captured using Dactyloscopy technique

In this technique, ink is applied on the lower part of the foot of an individual and its impression is taken on A4 size paper. After that, the images have been scanned at 300 dpi resolution as shown (Fig. 14.1).

To increase the accuracy and minimize the error in image capturing, the procedure of taking the footprint impressions and then scanning the same has been repeated five times for every individual. MATLAB functions, represented in italics, have been used for acquiring data and analysis of the footprint images.

14.3 Footprint Image Enhancement

The performance of the feature extraction algorithm depends on the quality of footprint images. In order to ensure that the terminal and bijunction feature extraction algorithm is robust to the quality of the input footprint images, an enhancement algorithm is necessary for the images acquired using Dactyloscopy technique to improve the clarity of the images. The proposed footprint image processing framework consists of the following stages: (i) preprocessing of image, (ii) normalization, (iii) orientation correction, (iv) frequency estimation, and (v) filtering. Preprocessing of image is important for reliable foot recognition. In the preprocessing stage, an improvement of the input image data is done that suppresses unwanted distortions or enhances image quality for further processing.

In stage (ii), normalization is done to reduce the variation in gray-level values along ridges and valleys by pixel-wise operation. A gray-level footprint image, $I_d(a, b)$ obtained by Dactyloscopy technique is defined as a $N \times N$ matrix which represents the intensity of the pixel at the xth row and y^{th} column. The mean and variance of a gray-level footprint image $I_d(a, b)$ are defined as,

$$M(I_d) = \frac{1}{N^2} \sum_{a=0}^{N-1} \sum_{b=0}^{N-1} I_d(a, b) \tag{14.1}$$

and

$$VAR(I_d) = \frac{1}{N^2} \sum_{a=0}^{N-1} \sum_{b=0}^{N-1} (I_d(a, b) - M(I_d))^2 \tag{14.2}$$

respectively. $I_d(a, b)$ is the gray-level value at pixel (a, b), M and VAR are the estimated mean and variance of I_d, respectively. The normalized image is defined as,

$$G_d(a, b) = \begin{cases} M_0 + \sqrt{\frac{VAR_0(I_d(a,b)-M)^2}{VAR}}, if\, I_d(a, b) > M \\ M_0 - \sqrt{\frac{VAR_0(I_d(a,b)-M)^2}{VAR}}, \text{ otherwise} \end{cases} \tag{14.3}$$

where $G_d(a, b)$ is the normalized gray-level value at pixel (a, b), M_0 and VAR_0 are the desired mean and variance values, respectively. Normalization is a pixel-wise operation which does not changes the clarity of the ridge and the valley structures. Figure 14.2 shows normalized image $G_d(a, b)$. A similar methodology is used by Hong et al. [13] for fingerprint image enhancement. Using normalization in direction and position, Nakajima et al. [14] improved Euclidean distance based footprint recognition method from roughly 30 to 85% on raw images.

In stage (iii), orientation has been done on the normalized image $G_d(a, b)$. The orientation is the intrinsic property of the footprint image which defines the invariant coordinates for ridges and valleys in local neighborhood. By studying the pixel-wise orientation of the image, the direction of each pixel along x-axis and y-axis is estimated to recognize the pattern of ridges in the footprints. The main steps involved in the orientation process are (a) divide $G_d(a, b)$ into blocks of size 5×5, (b) compute the gradients $\partial_x(a, b)$ and $\partial_y(a, b)$ at each pixel (a, b). Gaussian low-pass filter is used to compute gradients, (c) to smooth the gradients, Gaussian low-pass filter is used again to compute gradients, (d) estimate the local ridge orientation $V_x(a, b)$ and $V_y(a, b)$ along x-axis and y-axis, respectively, of each block centered at pixel (a, b) using the following equations [13]:

$$V_x(a, b) = \sum_{u=a-\frac{w}{2}}^{a+\frac{w}{2}} \sum_{v=b-\frac{w}{2}}^{b+\frac{w}{2}} 2\partial_x(u, v)\partial_y(u, v) \tag{14.4}$$

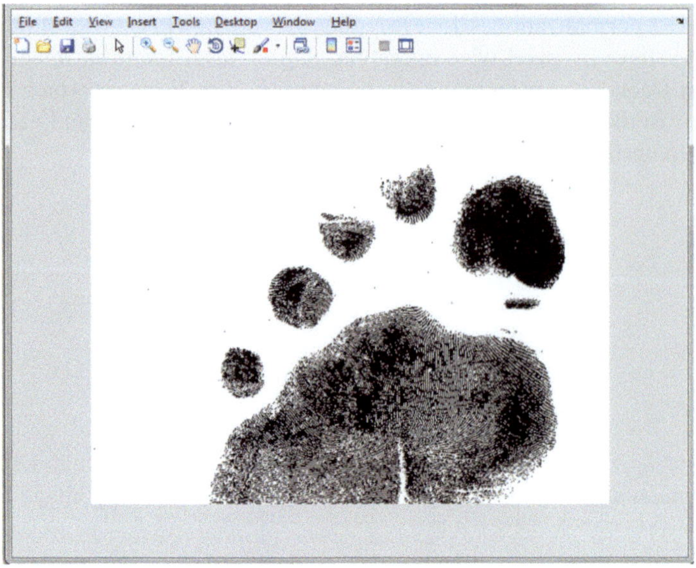

Fig. 14.2 Normalized image, $G_d(a, b)$

$$Vy(a,b) = \sum_{u=a-\frac{w}{2}}^{a+\frac{w}{2}} \sum_{v=b-\frac{w}{2}}^{b+\frac{w}{2}} (\partial_x^2(u,v) - \partial_y^2(u,v)) \tag{14.5}$$

$$\theta(a,b) = \frac{1}{2} tan^{-1} \left(\frac{V_y(a,b)}{V_x(a,b)} \right) \tag{14.6}$$

where $\theta(a, b)$ is least square estimate of the local ridge orientation at the block centered at pixel (a, b). The computed gradients from the Gaussian low-pass filter is passed to the *filter2*() function along with $G_d(a, b)$ to generate the local ridge orientation image $O_d(a, b)$ of same size as $G_d(a, b)$. The local ridge orientation image $O_d(a, b)$ is shown in Fig. 14.3.

In stage (iv), the local ridge frequency is estimated. A frequency image $F_d(a, b)$ is a $N \times N$ image where $F_d(a, b)$ represents the local ridge frequency and is defined as the frequency of the ridge and valley structures in a local neighborhood along a direction normal to the local ridges orientation. The main steps involved in frequency estimation are (a) the normalized image $G_d(a, b)$ and the orientation image $O_d(a, b)$ are taken as input of *freqest*() function to interpolate frequency of data intensity, (b) the function *freqest*() returns frequency estimated image $F_d(a, b)$ (see Fig. 14.4) of same size as $G_d(a, b)$.

In filtering process applied in stage (v), to remove the undesired noise efficiently and preserve the true ridge and valley structure, a band-pass filter is tuned to the corresponding frequency and orientation. The sinusoidal shaped waves of ridged and valleys vary slowly in a local constant orientation. Therefore, a band-pass filter

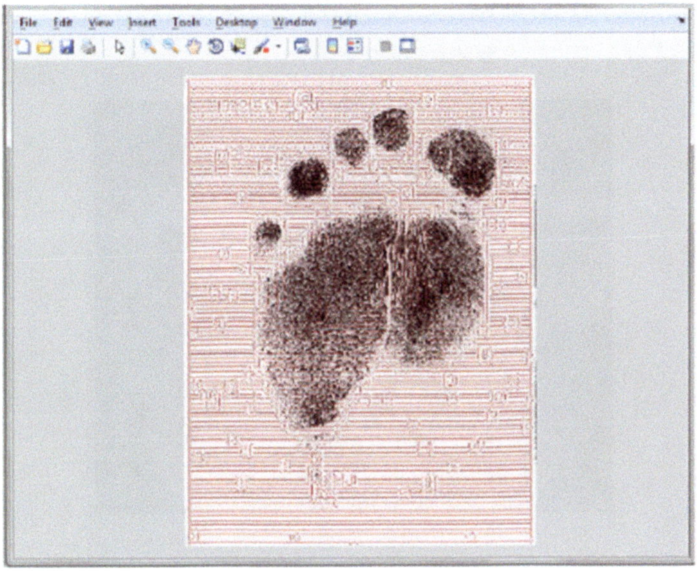

Fig. 14.3 Orientation image, $O_d(a, b)$

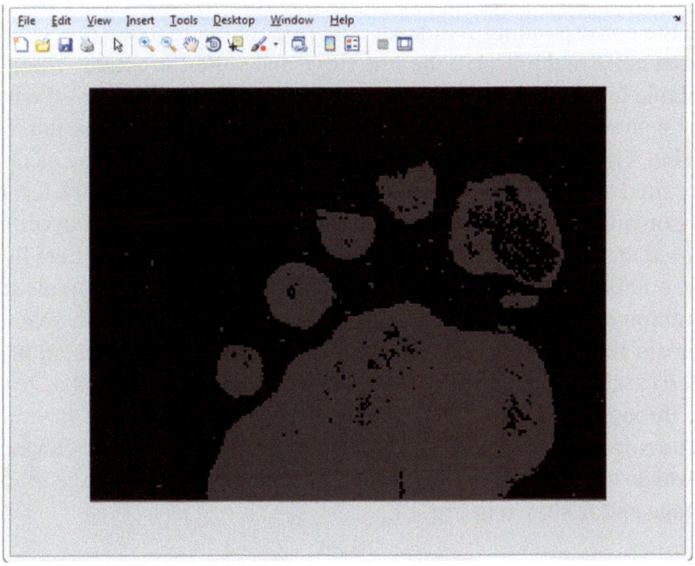

Fig. 14.4 Frequency estimated image $F_d(a, b)$

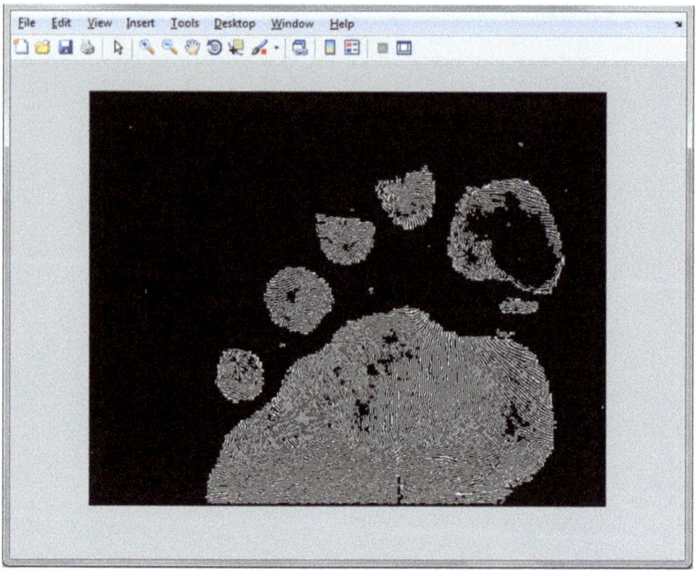

Fig. 14.5 Enhanced image, $E_d(a, b)$

is tuned to the corresponding frequency and orientation that can efficiently remove the undesired noise and preserve the true ridge and valley structure. There are many filters available in MATLAB which are used to enhance and process the image. Ten such filters namely, median, order-statistic, Wiener, average, Gaussian, Laplacian, Prewitt, Sobel, Gabor and circular-mean filters have been applied by us on footprint images acquired by Dactyloscopy technique to compute number of terminals and bijunctions of ridges in footprint images using feature extraction algorithm. It has been observed that the Gabor filter is good in comparison to other filters listed above because the number of variations between computed number of terminals and bijunctions are minimum. Gabor filter has optimal joint resolution in both spatial and frequency domain and has properties like frequency-selective and orientation-selective. The $O_d(a, b)$ and $F_d(a, b)$ images computed in Sects. 14.3 and 14.4, respectively, are passed through the Gabor filter to remove the undesired noise from image and preserve true ridge/valley structure. The output of Gabor filter is an enhanced image $E_d(a, b)$ which is stored in the database. An enhanced filter image $E_d(a, b)$ is a $N \times N$ image and is shown in Fig. 14.5.

14.4 Algorithm for Feature Extraction

The enhanced image $E_d(a, b)$ is passed through the feature extraction algorithm. Morphological thinning operation is applied on the enhanced image $E_d(a, b)$ to remove the foreground pixels and return the single pixel data of the footprint image

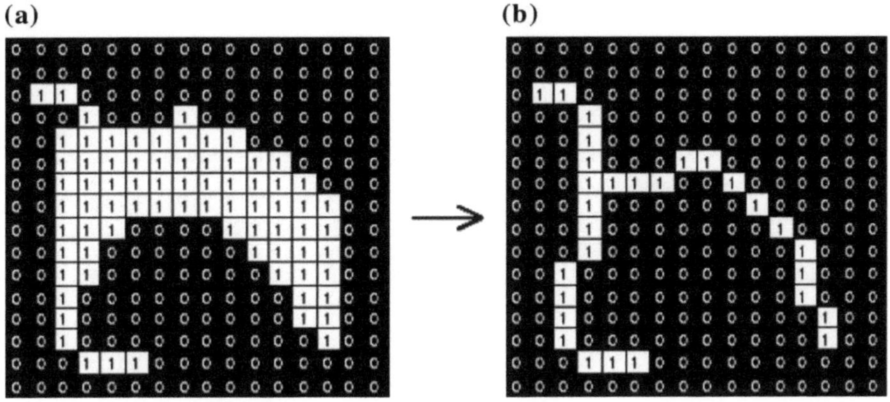

Fig. 14.6 Morphological thinning on enhanced image **a** $E_d(a, b)$ to get skeletal image, **b** $S_d(a, b)$

as skeleton image $S_d(a, b)$, as shown in Fig. 14.6. This operation is done using *bwmorph()* function which applies thin morphological operation to the binary image $E_d(a, b)$. The following steps of the algorithm are used to extract number of terminals and bijunctions in footprints obtained by Dactyloscopy technique:

Step 1: Skeleton image $S_d(a, b)$ is operated using *nlfilter()* function to perform general sliding-neighborhood operations.

Step 2: Binary image $L_d(a, b)$ is obtained using *nlfilter(*$S_d(a, b)$, [3 3], FUN) where *nlfilter()* applies the function FUN to each 3×3 sliding block of the $S_d(a, b)$.

Step 3: Counting of features are performed as

(a) if $L_d(a, b)$==1 then number of terminal++
(b) else if $L_d(a, b)$==3 then number of bijunctions++

Step 4: End

The computed features, i.e., number of terminals and bijunctions are stored in the database as templates for matching and identification of an individual.

14.5 Results and Discussion

To study the performance measurements and correlation between the weights, height, foot and body mass index, a database of footprint impressions of 94 individuals (57 males and 37 females) of age groups between 15–20 years and 20–25 years of different regions of North India has been created. The architecture of the footprint identification system used in this study is shown in Fig. 14.7.

If a user is using biometric application for the first time, the user needs to enroll his footprint images in a centralized database as a template linked internally to user's identity document (ID). During the time of authentification of the user, the biometric

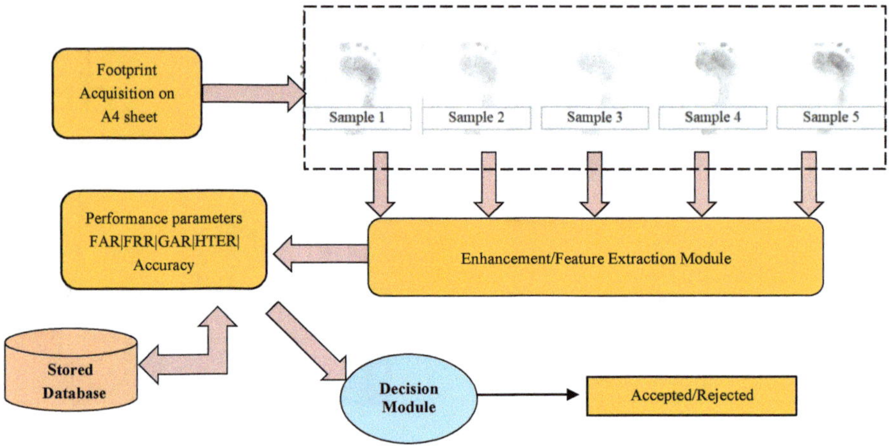

Fig. 14.7 The architecture of the footprint identification system using Dactyloscopy technique

input (footprint) of the user is compared to the data stored in the centralized database as templates by a matching algorithm that responds to false acceptance rate (FAR) and false reject rate (FRR), the metrics used to measure the performance of a biometric system.

The FAR is a measure of the probability that a biometric system incorrectly authorizes a non-authorized user. The acceptance (match) occurs due to incorrectly matching of the template and the biometric input. Similarly, the FRR is a measure of the probability that the system incorrectly rejects access to an authorized user. The rejection (no match) occurs due to failing to match the template with the biometric input. The number of error must be low as much as possible for the user's convenience. For age group 15–20 years, the FAR and FRR are 32 and 18.1%, respectively, for males; 30 and 15.2% for females. For age groups 20–25 years, the FAR and FRR for males are 31 and 11.3%, respectively and for females 30 and 13.6%. The FAR and FRR for the age group of 20–25 years for males are 31 and 11.3%, respectively, and for female 30 and 13.6%. The accuracy of a biometric system depends on these metrics. Practically, a biometric system with low FAR and high FRR must not allow access of any unauthorized user. Therefore, an authorized user have to put his feet on the device several times before access. It would also mean that the authorized users will have to put their feet on the device several times before they are allowed access. The development of a robust artifact removal algorithm and selection of an appropriate imaging sensor will help in achieving the accuracy of a biometric system.

The genuine accept rate (GAR), half total error rate (HTER), and accuracy are few other biometric metrics that determine the performance of a biometric system. The GAR is defined as a percentage of genuine users accepted by the system and is given by relation: GAR = 1-FRR. In a biometric authentication system, a reference threshold defined as a value that can decide whether a user is authorized or non-authorized by using biometric authentication is set at a particular value of FAR and

Table 14.2 Performance parameters (in percentage)

Gender	Age group (in years)	FAR	FRR	GAR	HTER	Accuracy
Male	15–20	32	18.1	81.9	25.0	74.9
	20–25	31	11.3	88.7	21.1	78.8
Female	15–20	30	15.2	84.8	22.6	77.4
	20–25	30	13.6	86.4	21.8	78.2

the GAR can be measured accordingly for this particular value of FAR. In this study, the threshold value is set as 20 pixels. When compared with other systems, the system with the highest GAR rate is considered to be most accurate. In this study, the GAR is found to be more than 80% for both the genders in the age groups of 15–20 and 20–25 years. Another possible way to measure the performance of a biometric system is to use the half total error rate (HTER) which combines the FRR and FAR and is defined by the formula: HTER = (FAR + FRR)/2. The FRR and FAR are strongly correlated to each other as they depend on the threshold value; when FAR increases, the FRR decreases and vice-versa. The average value of HTER is found to be 23.05% for males and 22.2% for females.

The accuracy of a biometric system is based on several verifying criteria including the identification rate, FAR, FRR, HTER, and additional biometric system standards and is normally expressed in percentage. The ability to identify an individual by an accuracy in a biometric system is in terms of percentage efficiency of a system. The accuracy of the biometric system depends on the threshold range, i.e., less the threshold range, more is the accuracy and vice-versa. The accuracy is calculated by the formula: Accuracy = 100 − (FAR + FRR)/2.

The results of Table 14.2 show that the average accuracy of for males is 76.8 and 77.8% for females. For the user's identification dactyloscopy technique can be used as one of the parameter. When some information is lost during the acquisition of image, the accuracy of dactyloscopy technique lowers down and can be increased by the advancement of feature extraction algorithm. The performance of other biometric traits such as fingerprint, face, iris, voice, key stroke, and hand geometry which are used for individual verification and identification by the commercialized security system are shown in Table 14.3.

Biometric systems are generally affected by demographic, performance, and environmental factors. The performance factors include capturing images of good quality, composition of target user size, time interval between enrollment and verification phases and robustness of recognition algorithm. Illumination conditions around the system, humidity, and temperatures are the environmental factors that affect the performance of a biometric system.

Table 14.3 Evaluation of biometric features

Biometric	Performance parameters (in percentage)		Samples
	FAR	FRR	
Face	1	10	37,437
Fingerprint	2	2	25,000
Hand geometry	2	2	129
Iris	0.94	0.99	1224
Keystrokes	7	0.1	15
Voice	2	10	30

14.6 Conclusion

This study focuses on a footprint-based recognition system of an individual using dactyloscopy technique. A database of footprint images of 94 individual of northern India of different gender and age group is acquired using dactyloscopy technique. The simulated results show a moderate accuracy to recognize an individual, an improvement is required before it can be implemented for commercial use. The result shows that an accuracy of 78.8% is achieved as the highest recognition rate in this study which could be used for forensic investigations. An accuracy of over 90% and almost 100% have been reported for face and fingerprint recognitions, respectively [15–18]. A limitation of this research is that all the images of footprint have been acquired in one standing position only. The results may deviate when the postures of acquiring footprint images varies.

The promising areas where commercialized biometric system using footprint rates can be used are public bathe, water parks, thermal bathe, swimming pools, spas and holy places like temples, mosque, and gurudwaras where person enters barefoot. A footprint of an individual visiting these areas can be acquired by simple installation of the sensor and scanner on the main entrance of the place. This study can be helpful for getting the evidences at the criminal sites and the footprint feature can be used to identify criminals. Footprint biometric trait has a promising feature in both types of biometric security systems, i.e., single-trait biometric system and multi-modal biometric system where footprint can be combined with other biometric traits and can give better results.

Acknowledgements The authors would like to thank all the subjects who consented to participate in this study. One of the authors (RK) thanks the Management of Vidya College of Engineering for extending all the necessary facilities required for this work.

References

1. Marcel, S.: Trusted biometrics under spoofing attacks. TABULA RASA. http://www.tabularasa-euproject.org/
2. Kennedy, B.R.: Uniqueness of bare feet and its use as possible as a possible means of identification. Forensic Sci. Int. **82**(1), 81–87 (1996)
3. Ambeth Kumar, V.D., Ramakrishnan, M.: Footprint recognition using modified sequential haar energy transform (MSHET). Int. J. Comput. Sci. **7**(3), 47–51 (2010)
4. Krishan, K.: Estimation of stature from footprint and foot outline dimensions in Gujjars of North India. Forensic Sci. Int. **175**(2–3), 93–101 (2008)
5. King, R.R., Xiaopeng, W.: Study of biometric identification method based on naked footprint. Int. J. Sci. Eng. **5**(2), 18–24 (2013)
6. Wang, W., Ping, X., Ding, Y.: Footprint heavy pressure surface pick-up and description. In: IEEE Computer Society-Third International Conference on Image and Graphics (ICIG'04) (2004)
7. Khokher, R., Singh, R.C., Kumar, R.: Footprint recognition with principal component analysis and independent component analysis. Macromol. Symp. **347**, 16–26 (2015)
8. Uhl, A., Wild, P.: Footprint-based biometric verification. J. Electron. Imaging-Soc. Photo-Opt. Instrum. Eng. **17**(1), 11–16 (2008)
9. Boulgouris, N.V., Hatzinakos, D., Plataniotis, K.N.: Gait recognition: a challenging signal processing technology for biometric identification. IEEE Signal Process. Mag. **22**(6), 78–90 (2005)
10. Kuragano, T., Yamaguchi, A., Furukawa, S.: A method to measure foot print similarity for gait analysis. In: IEEE Computer Society-2005. International Conference on Computational Intelligence for Modelling, Control and Automation, and International Conference Intelligent Agents, Web Technologies and Internet Commerce (CIMCAIAWIC'05) (2005)
11. Wild, P., Uhl, A.: Single-sensor hand and footprint-based multimodal biometric recognition: a Ph.D. thesis, Naturwissenschaftlichen Fakultät der Universität, Salzburg (2008)
12. Khokher, R., Singh, R.C.: Footprint-based personal recognition using scanning technique. Indian J. Sci. Technol. (In press) (2016)
13. Hong, L., Wan, Y., Jain, A.: Fingerprint image enhancement algorithm and performance evaluation. IEEE Trans. Pattern Anal. Mach. Intell. **20**(8), 777–789 (1998)
14. Nakajima, K., Mizukami, Y., Tanaka, K., Tamura, T.: Footprint-based personal recognition. IEEE Trans. Biomed. Eng. **47**(11), 1534–1537 (2000)
15. Jain, C.L., Halici, U., Hayashi, I., Lee, B.S.: Intelligent Biometric Techniques in Fingerprint and Face Recognition. CRC Press, Boca Raton (1999)
16. Marcel, S., Bengio, S.: Improving face verification using skin color information. IEEE Proc. Pattern Recognit. **2**, 378–381 (2002)
17. Rodriguez, Y., Marcel, S.: Face authentication using adapted local binary pattern histograms. In: 9th European Conference on Computer Vision, pp. 321–332 (2006)
18. Heusch, G., Rodriguez, Y., Marcel, S.: Local binary patterns as an image preprocessing for face authentication. In: Automatic Face and Gesture Recognition, pp. 6–14 (2006)

Chapter 15
An Iterative Algorithm for a Common Solution of a Split Variational Inclusion Problem and Fixed Point Problem for Non-expansive Semigroup Mappings

M. Dilshad, A.H. Siddiqi, Rais Ahmad and Faizan A. Khan

Abstract In this paper, we consider a split variational inclusion problem and a fixed point problem for non-expansive semigroup mappings in real Hilbert spaces. An iterative algorithm is introduced to approximate the common solution of split variational inclusion problem and a fixed point for a non-expansive semigroup mappings. Further, under some suitable conditions, it is proved that the sequences generated by the proposed algorithm converge strongly to a common solution of split variational inclusion problem and fixed point problem for a non-expansive semigroup mappings.

Keywords Split variational inclusion problem · Fixed point problem
Monotone operator · Non-expansive semigroup · Iterative algorithm
Strong convergence

15.1 Introduction

Throughout this paper, we assume that H_1 and H_2 are two real Hilbert spaces equipped with the norm $\|\cdot\|$ and inner product $\langle\cdot,\cdot\rangle$. A mapping $T : H_1 \to H_1$ is called *contraction*, if there exists a constant $k \in (0, 1)$ such that $\|Tx - Ty\| \leq k\|x - y\|$, for

---

M. Dilshad (✉) · F.A. Khan
Department of Mathematics, University of Tabuk,
Tabuk 71491, Kingdom of Saudi Arabia
e-mail: mdilshaad@gmail.com

F.A. Khan
e-mail: faizan911math@gmail.com

A.H. Siddiqi
School of Basic Sciences and Research, Sharda University,
Greater Noida 201306, India
e-mail: siddiqi.abulhasan@gmail.com

R. Ahmad
Department of Mathematics, Aligarh Muslim University,
Aligarh 202002, India
e-mail: raisain_123@rediffmail.com

P. Manchanda et al. (eds.), *Industrial Mathematics and Complex Systems*,
Industrial and Applied Mathematics, DOI 10.1007/978-981-10-3758-0_15

all $x, y \in H_1$. If $k = 1$, then T is called *non-expansive*. The mapping T is said to have a fixed point $x \in H_1$, if $Tx = x$.

Let C be a non-empty closed convex subset of the Hilbert space H_1. A family $S = \{T(s) : 0 \leq s < \infty\}$ of mappings from C into itself is called a *non-expansive semigroup* on C if it satisfies the following conditions:

(i) $T(0)x = x, \ \forall x \in C$.
(ii) $T(s + t) = T(s)T(t), \ \forall s, t \geq 0$.
(iii) $\|T(s)x - T(s)y\| \leq \|x - y\|, \forall x, y \in C$ and $s \geq 0$.
(iv) For all $x \in C$, $s \mapsto T(s)x$ is continuous.

The set of all common fixed point of a family S is denoted by $\text{Fix}(S)$, i.e.

$$\begin{aligned}\text{Fix}(S) &= \{x \in C : T(s)x = x, 0 \leq s < \infty\} \\ &= \bigcap_{0 \leq s < \infty} \text{Fix}(T(s)),\end{aligned}$$

where $\text{Fix}(T(s))$ is the set of fixed points of $T(s)$. It is trivial to show that $\text{Fix}(S)$ is closed and convex.

We consider the following fixed point problem for non-expansive semigroup S (in short FPP):

$$\text{Find } x \in H_1 \text{ such that } x \in \text{Fix}(S). \tag{15.1}$$

Shimizu and Takahashi [18] introduced and studied the following iterative method to prove a strong convergence theorem for FPP (15.1) in a real Hilbert spaces:

$$x_{n+1} = \alpha_n u + (1 - \alpha_n)\frac{1}{s_n}\int_0^{s_n} T(s)x_n ds, \quad \forall\, n \in N,$$

where $\{\alpha_n\}$ is a sequence in $(0, 1)$ and $\{s_n\}$ is a sequence of positive real numbers which diverges to $+\infty$. Recently, Chen and Song [8] introduced and studied the following iterative scheme for FPP (15.1) in real Hilbert spaces:

$$x_{n+1} = \alpha_n f x_n + (1 - \alpha_n)\frac{1}{s_n}\int_0^{s_n} T(s)x_n ds, \quad \forall\, n \in N,$$

where f is a contraction mapping.

Definition 15.1 A mapping $T : H_1 \to H_1$ is said to be

(i) monotone, if

$$\langle Tx - Ty, x - y\rangle \geq 0, \ \forall x, y \in H_1;$$

(ii) α-strongly monotone, if there exists a constant $\alpha > 0$ such that

$$\langle Tx - Ty, x - y\rangle \geq \alpha\|x - y\|^2, \ \forall x, y \in H_1;$$

(iii) β-inverse strongly monotone, if there exists a constant $\beta > 0$ such that

$$\langle Tx - Ty, x - y\rangle \geq \beta\|Tx - Ty\|^2, \ \forall x, y \in H_1;$$

(iv) firmly non-expansive, if

$$\langle Tx - Ty, x - y\rangle \geq \|Tx - Ty\|^2, \ \forall x, y \in H_1.$$

Definition 15.2 A multi-valued mapping $B : H_1 \to 2^{H_1}$ is said to be maximal monotone, if its $graph(B) = \{(x, y) \in H_1 \times H_1 : y \in Bx\}$ is not properly contained in the graph of any other monotone mapping.

It is well known that a monotone mapping B is *maximal* if and only if for $(x, u) \in H_1 \times H_1$, $\langle x - y, u - v\rangle \geq 0$, for every $(y, v) \in graph(B)$ implies that $u \in Bx$. If B is *maximal monotone*, then for each $x \in H_1$ and $\lambda > 0$ there is a unique $z \in H_1$ such that $x \in (I + \lambda B)^{-1}z$. In this case, the operator $J_\lambda^B = (I + \lambda B)^{-1}$, called the *resolvent* of B of parameter λ, is a non-expansive mapping and define everywhere.

The concept of split variational inequality problem is given by Censor et al. [5], which is to find a solution of variational inequality such that its image under a given bounded linear operator solves another variational inequality, i.e. to find $x^* \in C$ such that

$$\langle fx^*, x - x^*\rangle \geq 0, \ \forall\, x \in C, \tag{15.2}$$

such that

$$y^* = Ax^* \in Q \ \text{ solves } \ \langle gy^*, y - y^*\rangle \geq 0, \ \forall y \in Q, \tag{15.3}$$

where C is closed, convex subset of the Hilbert space H_1; Q is closed, convex subset of the Hilbert space H_2, $A : H_1 \to H_2$ is a bounded linear operator, $f : H_1 \to H_1$ and $g : H_2 \to H_2$ are two operators.

Recently, Moudafi [16] introduced the following split monotone variational inclusion (in short, SMVIP): Find $x^* \in H_1$ such that

$$0 \in fx^* + B_1x^*, \tag{15.4}$$

such that

$$y^* = Ax^* \in H_2 \ \text{ solves } \ 0 \in gy^* + B_2y^*, \tag{15.5}$$

where $B_1 : H_1 \to 2^{H_1}$ and $B_2 : H_2 \to 2^{H_2}$ are multi-valued monotone mappings.

Moudafi [16] introduced an iterative method for solving SMVIP (15.4) and (15.5) which is an important generalization of iterative algorithm given by Censor et al. [5]. Moudafi [16] emphasized that SMVIP (15.4) and (15.5) includes many special cases such as the split variational inequality problem, split common fixed point problem, split zero problem and split feasibility problem, see [2, 4–6, 15, 16], which have already been used in practice as a model in the intensity-modulation radiation therapy treatment planning, see [6, 7]. This formalism is also at the core of the modelling of many inverse problems arising for phase retrieval and other real-world problems; for instance, in sensor networks in computerized tomography and data compression, see [4, 9] and references therein.

Byrne et al. [2] studied the following split variational inclusion problem (SVIP): Find $x^* \in H_1$ such that

$$0 \in B_1 x^* \tag{15.6}$$

such that

$$y^* = Ax^* \in H_2 \text{ solves } 0 \in B_2 y^*. \tag{15.7}$$

The solution set of SVIP (15.6) and (15.7) is denoted by $\Xi = \{x^* \in H_1 : 0 \in B_1 x^*$ such that $0 \in B_2(Ax^*)$.

Byrne et al. [2] proposed the following iterative scheme to prove the strong and weak convergence theorem for SVIP (15.6) and (15.7). For given $x_0 \in H_1$ and $\lambda > 0$ compute iterative sequence $\{x_n\}$ generated by the following scheme:

$$x_{n+1} = J_\lambda^{B_1}\left[x_n + \gamma A^*\left(J_\lambda^{B_1} - I\right)Ax_n\right].$$

Motivated by the work of Moudafi [16], Byrne et al. [2] and by the ongoing research in this direction, we propose an iterative algorithm for approximating a common solution to fixed point problem FPP (15.1) and split variational inclusion problem SVIP (15.6) and (15.7). Furthermore, we prove that the sequences generated by the iterative algorithm converge strongly to a common solution of FPP and SVIP.

15.2 Preliminaries

The *metric projection* onto the set C $(C \subset H_1)$, denoted by P_C, is defined by, for all $x \in H_1$, $P_C x \in C$ and $\|x - P_C x\| = \inf_{y \in C}\|x - y\|$. P_C is also characterized by the fact $P_C x \in C$, $\langle x - P_C x, y - P_C x\rangle \le 0$ and $\|x - y\|^2 \ge \|x - P_C x\|^2 + \|y - P_C x\|^2$, for all $x \in H_1$, $y \in C$. It is well known that P_C is non-expansive and satisfies $\langle x - y, P_C x - P_C y\rangle \ge \|P_C x - P_C y\|^2$, for all $x, y \in H_1$.

Let $T : H_1 \to H_1$ be a non-expansive operator. Then, the inequality

$$\langle (x - Tx) - (y - Ty), Ty - Tx\rangle \le \frac{1}{2}\|(Tx - x) - (Ty - y)\|^2, \tag{15.8}$$

holds for all $(x, y) \in H_1 \times H_2$. Thus, we get for all $(x, y) \in H_1 \times \text{Fix}(T)$,

$$\langle x - Tx, y - Tx\rangle \le \frac{1}{2}\|Tx - x\|^2. \tag{15.9}$$

See ([11], Theorem 3.1) and ([10], Theorem 2.1) for the above inequalities.

It is also known that H_1 satisfies Opial's condition [17], i.e. for any sequence $\{x_n\}$ with $x_n \rightharpoonup x$, the inequality

$$\liminf_{n\to\infty} \|x - y\| < \liminf_{n\to\infty} \|x_n - y\| \tag{15.10}$$

holds for every $y \in H_1$ with $y \neq x$.

A mapping $T : H_1 \to H_1$ is said to be *averaged* if and only if it can be written as the average of identity mapping and a non-expansive mapping, namely $T = (1-\alpha)I + \alpha S$, where $\alpha \in (0, 1)$ and $S : H_1 \to H_1$ is non-expansive mapping. Thus firmly non-expansive mapping (in particular, projection on nonempty closed and convex subsets and resolvent operators of maximal monotone operators) are averaged. It can also be note that average mappings are non-expansive.

The following are some important properties of averaged operators, see e.g., [1, 3, 14]

Proposition 15.1 *(i) If $T = (1-\alpha)S + \alpha V$, where $S : H_1 \to H_1$ is averaged, $V : H_1 \to H_1$ is non-expansive and $\alpha \in (0, 1)$, then T is averaged.*

(ii) The composite $\{T_1, \ldots, T_N\}$ are averaged and have a nonempty common fixed point, then

$$\bigcap_{i=1}^{N} \mathrm{Fix}(T_i) = \mathrm{Fix}(T_1, T_2, \ldots, T_N).$$

(iii) If T is τ-inverse strongly monotone, then for $\gamma > 0$, γT is $\frac{\tau}{\gamma}$-inverse strongly monotone.

(iv) T is averaged if and only if, its complement $I - T$ is τ-inverse strongly monotone for some $\tau > \frac{1}{2}$.

Lemma 15.1 ([18]) *Let C be a non-empty bounded closed convex subset of a Hilbert space H_1 and let $S = \{T(s) : 0 \leq s < \infty\}$ be a non-expansive semigroup on C. Then for any $0 \leq h < \infty$ and $t > 0$,*

$$\limsup_{t\to\infty x\in C} \left\| \frac{1}{t}\int_0^t T(s)x ds - T(h)\left(\frac{1}{t}\int_0^t T(s)x ds\right) \right\| = 0.$$

Lemma 15.2 ([19]) *If $\{a_n\}$ be a sequence of nonnegative real numbers such that*

$$a_{n+1} \leq (1-\alpha_n)a_n + \delta_n, \quad n \geq 0,$$

where $\{\alpha_n\}$ is a sequence in $(0, 1)$ and $\{\delta_n\}$ is a sequence in R such that

(i) $\sum_{n=1}^{\infty} \alpha_n = \infty$;

(ii) $\limsup_{n\to\infty} \frac{\delta_n}{\alpha_n} \leq 0$ *or* $\sum_{n=1}^{\infty} |\delta_n| < \infty$

Then, $\lim_{n\to\infty} a_n = 0$.

Lemma 15.3 *The following inequality holds in real Hilbert space* H_1*:*

$$\|x + y\|^2 \leq \|x\|^2 + 2\langle y, x + y\rangle, \quad \forall x, y \in H_1.$$

Lemma 15.4 ([12]) *Assume that* T *is non-expansive self-mapping of a closed convex subset* C *of a Hilbert space* H_1*. If* T *has a fixed point, then* $I - T$ *is demiclosed, that is, whenever* $\{x_n\}$ *is a sequence in* C *converging weakly to some* $x \in C$ *and the sequence* $\{(I - T)x_n\}$ *converges strongly to some* y*, it follows that* $(I - T)x = y$*, where* I *is the identity mapping on* H_1*.*

Following lemma can be proved easily using the definition of resolvent operator.

Lemma 15.5 $x^* \in H_1$ *and* $y^* = Ax^*$ *are the solution of SVIP (15.6), (15.7) if and only if*

$$x^* = J_\lambda^{B_1} x^* \text{ and } y^* = Ax^* = J_\lambda^{B_2} y^*, \ for\ some\ \lambda > 0.$$

15.3 Main Result

In this section, we prove a strong convergence of sequences based on the proposed iterative algorithm for computing the common approximate solution of FPP (15.1) and SVIP (15.6), (15.7).

Theorem 15.1 *Let* H_1 *and* H_2 *be two real Hilbert spaces. Assume that* $B_1 : H_1 \to 2^{H_1}$ *and* $B_2 : H_2 \to 2^{H_2}$ *are maximal monotone operators and* $A : H_1 \to H_2$ *is a bounded linear operator. Let* $S = \{T(s) : 0 \leq s < \infty\}$ *be a non-expansive semigroup on* H_1 *such that* $\{\text{Fix}(S)\} \cap \{\Xi\} \neq \emptyset$*. Let* $f : H_1 \to H_1$ *be a contraction mapping with constant* $\alpha \in (0, 1)$*. Let* $\{s_n\}$ *be a positive real sequence which diverges to* ∞*. For a given arbitrary* $x_0 \in H_1$*,* $\{u_n\}, \{x_n\}$ *are the iterative sequences generated by the iterative algorithm:*

$$\begin{aligned} u_n &= J_\lambda^{B_1}\left(x_n + \gamma A^*\left(J_\lambda^{B_2} - I\right)Ax_n\right); \\ x_{n+1} &= \alpha_n f x_n + (1 - \alpha_n)\frac{1}{s_n}\int_0^{s_n} T(s)u_n ds, \end{aligned} \tag{15.11}$$

where $\gamma \in (0, \frac{1}{L})$*,* L *is the spectral radius of the operator* $A^\star A$ *and* $A^\star$ *is the adjoint of* A *and* $\{\alpha_n\}$ *is a sequence in* $(0, 1)$ *satisfying*

(i) $\lim_{n\to\infty} \alpha_n = 0, \sum_{i=1}^{\infty} \alpha_n = \infty$ *and* $\sum_{i=1}^{\infty} |\alpha_n - \alpha_{n-1}| < \infty$*;*

(ii) $\sum_{i=1}^{\infty} \frac{|s_{n+1} - s_n|}{s_{n+1}} < \infty$*.*

Then, the sequence $\{x_n\}$ *converges to* $z \in \{\text{Fix}(S)\} \cap \{\Xi\}$*, where* $z = P_{\{\text{Fix}(S)\} \cap \{\Xi\}} f z$*.*

Proof Let $x^* \in \{\text{Fix}(S)\} \cap \{\Xi\} \Rightarrow x^* \in \Xi \Rightarrow x^* = J_\lambda^{B_1} x^*,\ Ax^* = J_\lambda^{B_2} Ax^*$. Then by (15.11), we have

$$\begin{aligned}
\|u_n - x^*\|^2 &= \left\| J_\lambda^{B_1}[x_n + \gamma A^\star \left(J_\lambda^{B_2} - I\right) Ax_n] - x^* \right\|^2 \\
&= \left\| J_\lambda^{B_1}[x_n + \gamma A^\star \left(J_\lambda^{B_2} - I\right) Ax_n] - J_\lambda^{B_1} x^* \right\|^2 \\
&\le \left\| x_n + \gamma A^\star \left(J_\lambda^{B_2} - I\right) Ax_n - x^* \right\|^2 \\
&\le \|x_n - x^*\|^2 + \gamma^2 \left\| A^\star \left(J_\lambda^{B_2} - I\right) Ax_n \right\|^2 \\
&\quad + 2\gamma \left\langle x_n - x^*, A^\star \left(J_\lambda^{B_2} - I\right) Ax_n \right\rangle \\
&\le \|x_n - x^*\|^2 + 2\gamma \left\langle x_n - x^*, A^\star \left(J_\lambda^{B_2} - I\right) Ax_n \right\rangle \\
&\quad + \gamma^2 \left\langle \left(J_\lambda^{B_2} - I\right) Ax_n, AA^\star \left(J_\lambda^{B_2} - I\right) Ax_n \right\rangle. \qquad (15.12)
\end{aligned}$$

By using (15.9), we have

$$\begin{aligned}
&2\gamma \left\langle x_n - x^*, A^\star \left(J_\lambda^{B_2} - I\right) Ax_n \right\rangle \\
&= 2\gamma \left\langle A(x_n - x^*), \left(J_\lambda^{B_2} - I\right) Ax_n \right\rangle \\
&= 2\gamma \left\langle A(x_n - x^*) + \left(J_\lambda^{B_2} - I\right) Ax_n - \left(J_\lambda^{B_2} - I\right) Ax_n, \left(J_\lambda^{B_2} - I\right) Ax_n \right\rangle \\
&= 2\gamma \left\{ \left\langle J_\lambda^{B_2} Ax_n - Ax^*, \left(J_\lambda^{B_2} - I\right) Ax_n \right\rangle - \left\| \left(J_\lambda^{B_2} - I\right) Ax_n \right\|^2 \right\} \\
&\le 2\gamma \left\{ \frac{1}{2} \left\| \left(J_\lambda^{B_2} - I\right) Ax_n \right\|^2 - \left\| \left(J_\lambda^{B_2} - I\right) Ax_n \right\|^2 \right\} \\
&\le -\gamma \left\| \left(J_\lambda^{B_2} - I\right) Ax_n \right\|^2, \qquad (15.13)
\end{aligned}$$

and

$$\begin{aligned}
\gamma^2 \left\langle \left(J_\lambda^{B_2} - I\right) Ax_n, AA^\star \left(J_\lambda^{B_2} - I\right) Ax_n \right\rangle &\le L\gamma^2 \left\langle \left(J_\lambda^{B_2} - I\right) Ax_n, \left(J_\lambda^{B_2} - I\right) Ax_n \right\rangle \\
&= L\gamma^2 \left\| \left(J_\lambda^{B_2} - I\right) Ax_n \right\|^2. \qquad (15.14)
\end{aligned}$$

Using (15.13) and (15.14), (15.12) becomes

$$\|u_n - x^*\|^2 \le \|x_n - x^*\|^2 + \gamma(L\gamma - 1) \left\| \left(J_\lambda^{B_2} - I\right) Ax_n \right\|^2. \qquad (15.15)$$

Since $\gamma \in \left(0, \frac{1}{L}\right)$, this implies that

$$\|u_n - x^*\|^2 \leq \|x_n - x^*\|^2. \tag{15.16}$$

Let $t_n = \frac{1}{s_n}\int_0^{s_n} T(s)u_n ds$. Then using $T(s)x^* = x^*$ and (15.16), we have

$$\begin{aligned}\|t_n - x^*\| &= \left\|\frac{1}{s_n}\int_0^{s_n} T(s)u_n ds - x^*\right\| \\ &\leq \frac{1}{s_n}\int_0^{s_n} \|T(s)u_n - T(s)x^* ds\| \\ &\leq \|u_n - x^*\| \\ &\leq \|x_n - x^*\|.\end{aligned} \tag{15.17}$$

Now, we show that $\{x_n\}$ is bounded.

$$\begin{aligned}\|x_{n+1} - x^*\| &= \|\alpha_n f x_n + (1-\alpha_n)t_n - x^*\| \\ &\leq \alpha_n\|f x_n - x^*\| + (1-\alpha_n)\|t_n - x^*\| \\ &\leq \alpha_n\|f x_n - f x^*\| + \alpha_n\|f x^* - x^*\| + (1-\alpha_n)\|t_n - x^*\| \\ &\leq \alpha_n\alpha\|x_n - x^*\| + \alpha_n\|f x^* - x^*\| + (1-\alpha_n)\|x_n - x^*\| \\ &= [1 - \alpha_n(1-\alpha)]\,\|x_n - x^*\| + \alpha_n\|f x^* - x^*\| \\ &\leq \max\left\{\|x_n - x^*\|\,,\ \frac{\|f x^* - x^*\|}{1-\alpha}\right\} \\ &\ \ \vdots \\ &\leq \left\{\|x_0 - x^*\|\,,\ \frac{\|f x^* - x^*\|}{1-\alpha}\right\}.\end{aligned}$$

Hence, $\{x_n\}$ is bounded, and consequently $\{u_n\}$, $\{f x_n\}$, $\{t_n\}$ are bounded.

Now, we estimate

$$\begin{aligned}&\|t_{n+1} - t_n\| \\ &= \left\|\frac{1}{s_{n+1}}\int_0^{s_{n+1}} T(s)u_{n+1}ds - \frac{1}{s_n}\int_0^{s_n} T(s)u_n ds\right\| \\ &= \left\|\frac{1}{s_{n+1}}\int_0^{s_{n+1}} [T(s)u_{n+1} - T(s)u_n]ds + \left(\frac{1}{s_{n+1}} - \frac{1}{s}\right)\int_0^{s_{n+1}} T(s)u_n ds\right. \\ &\quad \left. + \frac{1}{s_{n+1}}\int_{s_n}^{s_{n+1}} T(s)u_n ds\ \right\| \\ &= \left\|\frac{1}{s_{n+1}}\int_0^{s_{n+1}} [T(s)u_{n+1} - T(s)u_n]ds + \left(\frac{1}{s_{n+1}} - \frac{1}{s}\right)\int_0^{s_n} [T(s)u_n - T(s)x^*]ds\right. \\ &\quad \left. + \frac{1}{s_{n+1}}\int_{s_n}^{s_{n+1}} [T(s)u_n - T(s)x^*]ds\right\|\end{aligned}$$

$$\leq \|u_{n+1} - u_n\| + \frac{|s_{n+1} - s_n| s_n}{s_{n+1} s_n} \|u_n - x^*\| + \left(\frac{|s_{n+1} - s_n|}{s_{n+1}} \right) \|u_n - x^*\|$$
$$\leq \|u_{n+1} - u_n\| + \frac{2|s_{n+1} - s_n|}{s_{n+1}} \|u_n - x^*\|. \tag{15.18}$$

Since $\gamma \in \left(0, \frac{1}{L}\right)$, the mapping $J_\lambda^{B_2} \left(I + \gamma + A^* \left(J_\lambda^{B_1} - I \right) A \right)$ is average and hence non-expansive, then we have

$$\begin{aligned} &\|u_{n+1} - u_n\| \\ &= \left\| J_\lambda^{B_1} \left[x_{n+1} + \gamma A^\star \left(J_\lambda^{B_2} - I \right) A x_n \right] - J_\lambda^{B_1} \left[x_n + \gamma A^\star \left(J_\lambda^{B_2} - I \right) A x_n \right] \right\| \\ &= \left\| J_\lambda^{B_1} \left[\left(I + \gamma A^\star \left(J_\lambda^{B_2} - I \right) \right) A \right] x_{n+1} - J_\lambda^{B_1} \left[\left(I + \gamma A^\star \left(J_\lambda^{B_2} - I \right) \right) A \right] x_n \right\| \\ &\leq \|x_{n+1} - x_n\|. \end{aligned} \tag{15.19}$$

Further using (15.18), (15.19) and boundedness of $\{u_n\}$, $\{f x_n\}$ and $\{t_n\}$, we have

$$\begin{aligned} &\|x_{n+1} - x_n\| \\ &= \|\alpha_n f x_n + (1 - \alpha_n) t_n - \alpha_{n-1} f x_{n-1} - (1 - \alpha_{n-1}) t_{n-1}\| \\ &= \alpha_n \|f x_n - f x_{n-1}\| + (\alpha_n - \alpha_{n-1}) \|f x_{n-1}\| + (1 - \alpha_n) \|t_n - t_{n-1}\| \\ &\quad + (\alpha_n - \alpha_{n-1}) \|t_{n-1}\| \\ &= \alpha_n \alpha \|x_n - x_{n-1}\| + (1 - \alpha_n) \|t_n - t_{n-1}\| + |\alpha_n - \alpha_{n-1}| \{\|f x_{n-1}\| + \|t_{n-1}\|\} \\ &\leq \alpha_n \alpha \|x_n - x_{n-1}\| + (1 - \alpha_n) \|u_n - u_{n-1}\| + 2 \frac{|s_n - s_{n-1}|}{s_n} \|u_{n-1} - p\| \\ &\quad + 2|\alpha_n - \alpha_{n-1}| K \\ &\leq [1 - \alpha_n (1 - \alpha)] \|x_n - x_{n-1}\| + 2 \frac{|s_n - s_{n-1}|}{s_n} \|u_n - x^*\| + 2|\alpha_n - \alpha_{n-1}| K \\ &= [1 - \alpha_n (1 - \alpha)] \|x_n - x_{n-1}\| + 2 \left\{ M \frac{|s_n - s_{n-1}|}{s_n} + K |\alpha_n - \alpha_{n-1}| \right\}. \end{aligned}$$

Let $\beta_n = [1 - \alpha_n (1 - \alpha)]$, $\delta_n = 2 \left\{ M \frac{|s_n - s_{n-1}|}{s_n} + K |\alpha_n - \alpha_{n-1}| \right\}$, then using Lemma 15.2, we get

$$\|x_{n+1} - x_n\| \to 0, \quad as \quad n \to \infty. \tag{15.20}$$

Now, since

$$\begin{aligned} x_{n+1} - x_n &= \alpha_n f x_n + (1 - \alpha_n) t_n - x_n \\ &= \alpha_n (f x_n - x_n) + (1 - \alpha_n)(t_n - x_n), \end{aligned}$$

then, we have

$$(1-\alpha_n)\|t_n - x_n\| \le \alpha_n \|fx_n - x_n\| + \|x_{n+1} - x_n\|.$$

Since $\alpha_n \to 0$ and $\|x_{n+1} - x_n\| \to 0$ as $n \to \infty$, we have

$$\|t_n - x_n\| \to 0, \text{ as } n \to \infty. \tag{15.21}$$

Next, we estimate

$$\begin{aligned}
\|T(s)x_n - x_n\| &= \Bigg\| T(s)x_n - T(s)\frac{1}{s_n}\int_0^{s_n} T(s)u_n ds + T(s)\frac{1}{s_n}\int_0^{s_n} T(s)u_n ds \\
&\quad - \frac{1}{s_n}\int_0^{s_n} T(s)u_n ds + \frac{1}{s_n}\int_0^{s_n} T(s)u_n ds - x_n \Bigg\| \\
&\le \left\| T(s)x_n - T(s)\frac{1}{s_n}\int_0^{s_n} T(s)u_n ds \right\| \\
&\quad + \left\| T(s)\frac{1}{s_n}\int_0^{s_n} T(s)u_n ds - \frac{1}{s_n}\int_0^{s_n} T(s)u_n ds \right\| \\
&\quad + \left\| \frac{1}{s_n}\int_0^{s_n} T(s)u_n ds - x_n \right\| \\
&\le \left\| x_n - \frac{1}{s_n}\int_0^{s_n} T(s)u_n ds \right\| \\
&\quad + \left\| T(s)\frac{1}{s_n}\int_0^{s_n} T(s)u_n ds - \frac{1}{s_n}\int_0^{s_n} T(s)u_n ds \right\| \\
&\quad + \left\| \frac{1}{s_n}\int_0^{s_n} T(s)u_n ds - x_n \right\| \\
&\le 2\left\| x_n - \frac{1}{s_n}\int_0^{s_n} T(s)u_n ds \right\| \\
&\quad + \left\| T(s)\frac{1}{s_n}\int_0^{s_n} T(s)u_n ds - \frac{1}{s_n}\int_0^{s_n} T(s)u_n ds \right\|.
\end{aligned} \tag{15.22}$$

Let $U = \{w \in H_1 : \|w - x^*\| \le m\}$, where $m = \min\left\{\|x_0 - x^*\|, \frac{\|fx^* - x^*\|}{1-\alpha}\right\}$. Since $\{x_n\}$, $\{fx_n\}$ are bounded, then U is a nonempty bounded closed convex subset of H_1 which is $T(s)$-invariant, for each $0 \le s < \infty$ and contains $\{x_n\}$. So without loss of generality, we may assume that $S = \{T(s) : 0 \le s < \infty\}$ is a non-expansive semigroup on U. By Lemma 15.1, we have

$$\lim_{n\to\infty} \left\| T(s)\frac{1}{s_n}\int_0^{s_n} T(s)u_n ds - \frac{1}{s_n}\int_0^{s_n} T(s)u_n ds \right\| = 0. \tag{15.23}$$

By using (15.21), (15.22) and (15.23), we have

$$\lim_{n\to\infty} \|T(s)x_n - x_n\| = 0.$$

It follows from (15.15) and Lemma 15.4 that

$$\begin{aligned}
&\|x_{n+1} - x^*\|^2 \\
&= \|\alpha_n f x_n + (1-\alpha_n)t_n - x^*\|^2 \\
&= \alpha_n \|f x_n - x^*\|^2 + (1-\alpha_n)\|t_n - x^*\|^2 \\
&\le \alpha_n \|f x_n - x^*\|^2 + (1-\alpha_n)\|u_n - x^*\|^2 \\
&\le \alpha_n \|f x_n - x^*\|^2 + (1-\alpha_n)\left[\|x_n - x^*\|^2 + \gamma(L\gamma - 1)\left\|\left(J_\lambda^{B_2} - I\right)Ax_n\right\|^2\right] \\
&\le \alpha_n \|f x_n - x^*\|^2 + \|x_n - x^*\|^2 + \gamma(L\gamma - 1)\left\|\left(J_\lambda^{B_2} - I\right)Ax_n\right\|^2. \qquad (15.24)
\end{aligned}$$

This implies that

$$\begin{aligned}
&\gamma(-L\gamma + 1)\left\|\left(J_\lambda^{B_2} - I\right)Ax_n\right\|^2 \\
&\le \alpha_n \|f(x_n) - x^*\|^2 + \|x_n - x^*\|^2 - \|x_{n+1} - x^*\|^2 \\
&= \alpha_n \|f(x_n) - x^*\|^2 + \|x_{n+1} - x_n\|(\|x_n - x^*\| + \|x_{n+1} - x^*\|).
\end{aligned}$$

Since $(1 - L\gamma) > 0,\ \alpha_n \to 0,\ \|x_{n+1} - x_n\| \to 0$, this implies that

$$\left\|\left(J_\lambda^{B_2} - I\right)Ax_n\right\| \to 0, \quad as\ n \to \infty. \qquad (15.25)$$

Also, we have

$$\begin{aligned}
\|u_n - x^*\| &= \left\|J_\lambda^{B_1}\left(x_n + \gamma + A^*\left(J_\lambda^{B_2} - I\right)Ax_n\right) - x^*\right\| \\
&= \left\|J_\lambda^{B_1}\left(x_n + \gamma A^*\left(J_\lambda^{B_2} - I\right)Ax_n\right) - J_\lambda^{B_1}x^*\right\| \\
&\le \left\langle u_n - x^*,\ x_n + \gamma A^*\left(J_\lambda^{B_2} - I\right)Ax_n - x^*\right\rangle \\
&= \frac{1}{2}\Big\{\|u_n - x^*\|^2 + \left\|x_n + \gamma A^*\left(J_\lambda^{B_1} - I\right)Ax_n - x^*\right\|^2 \\
&\quad - \left\|(u_n - x^*) - \left[x_n + \gamma + A^*\left(J_\lambda^{B_1} - I\right)Ax_n\right]\right\|\Big\} \\
&\le \frac{1}{2}\Big\{\|u_n - x^*\| + \|x_n - x^*\| + \gamma(l\gamma - 1)\left\|\left(J_\lambda^{B_2} - I\right)Ax_n\right\|^2 \\
&\quad - \left\|u_n - x_n - \gamma A^*\left(J_\lambda^{B_2} - I\right)Ax_n\right\|\Big\}
\end{aligned}$$

$$\leq \frac{1}{2}\Big\{\|u_n - x^*\| + \|x_n - p\| + \gamma(l\gamma - 1)\left\|\left(J_\lambda^{B_2} - I\right)Ax_n\right\|^2 - \Big[\|u_n - x_n\|^2 + \|x_n - x^*\|^2 - \|u_n - x_n\|^2 + 2\gamma\|A(u_n - x_n)\|\left\|\left(J_\lambda^{B_2} - I\right)Ax_n\right\|\Big]\Big\}. \tag{15.26}$$

Therefore, we have

$$\|u_n - x^*\|^2 \leq \|x_n - x^*\|^2 - \|u_n - x_n\|^2 + 2\gamma\|A(u_n - x_n)\|\left\|\left(J_\lambda^{B_2} - I\right)Ax_n\right\|. \tag{15.27}$$

From (15.24) and (15.27), we have

$$\begin{aligned}\|x_{n+1} - x^*\|^2 &= \alpha_n\|fx_n - x^*\| + (1 - \alpha_n)\Big[\|x_n - x^*\| - \|u_n - x_n\|^2 \\ &\quad + 2\gamma\|A(u_n - x_n)\|\left\|\left(J_\lambda^{B_2} - I\right)Ax_n\right\|\Big] \\ &\leq \alpha_n\|fx_n - x^*\| + \|x_n - x^*\|^2 - \|u_n - x_n\|^2 \\ &\quad + 2\gamma\|A(u_n - x_n)\|\left\|\left(J_\lambda^{B_2} - I\right)Ax_n\right\|.\end{aligned}$$

This implies that

$$\begin{aligned}\|u_n - x_n\| &\leq \alpha_n\|fx_n - x^*\| + \|x_n - x^*\|^2 - \|x_{n+1} - x^*\|^2 \\ &\quad + 2\gamma\|A(u_n - x_n)\|\left\|\left(J_\lambda^{B_2} - I\right)Ax_n\right\| \\ &\leq \alpha_n\|fx_n - x^*\| + \|x_{n+1} - x_n\|\left(\|x_n - x^*\|^2 + \|x_{n+1} - x^*\|^2\right) \\ &\quad + 2\gamma\|A(u_n - x_n)\|\left\|\left(J_\lambda^{B_2} - I\right)Ax_n\right\|.\end{aligned}$$

By (15.20) and (15.25), we have

$$\|u_n - x_n\| \to 0, \quad \text{as } n \to \infty. \tag{15.28}$$

Now, we can write

$$\begin{aligned}\|T(s)t_n - x_n\| &\leq \|T(s)t_n - T(s)x_n\| + \|T(s)x_n - x_n\| \\ &\leq \|t_n - x_n\| + \|T(s)x_n - x_n\| \to 0 \text{ as } n \to \infty,\end{aligned}$$

and

$$\begin{aligned}\|T(s)t_n - t_n\| &\leq \|T(s)t_n - T(s)x_n\| + \|T(s)x_n - x_n\| + \|x_n - t_n\| \\ &\leq \|t_n - x_n\| + \|T(s)x_n - x_n\| + \|x_n - t_n\| \to 0 \text{ as } n \to \infty.\end{aligned}$$

Since $\{t_n\}$ is bounded, there exists a subsequence $\{t_{n_i}\} \subseteq \{t_n\}$ such that $t_{n_i} \rightharpoonup w \in C$.

Now, we prove that $w \in \{\text{Fix}(S)\} \cap \{\Xi\}$. First, we show that $w \in \text{Fix}(S)$. Assume that $w \neq \text{Fix}(S)$. Since $t_{n_i} \rightharpoonup w$ and $T(s)w \neq w$. From opial condition, we have

$$\begin{aligned}
\liminf_{i\to\infty} \|t_{n_i} - w\| &\le \liminf_{i\to\infty} \left\|t_{n_i} - T(s)w\right\| \\
&\le \liminf_{i\to\infty} \left\{\left\|t_{n_i} - T(s)t_{n_i}\right\| + \left\|T(s)t_{n_i} - T(s)w\right\|\right\} \\
&\le \liminf_{i\to\infty} \|t_{n_i} - w\|,
\end{aligned}$$

which is a contradiction. Hence $w \in \text{Fix}(S)$.

Also, we have

$$\begin{aligned}
&\|T(s)u_n - u_n\| \\
&\le \|T(s)u_n - T(s)x_n\| + \|T(s)x_n - x_n\| + \|x_n - u_n\| \\
&= \|x_n - u_n\| + \|T(s)x_n - x_n\| + \|x_n - u_n\| \to 0 \ \ as \ \ n \to \infty.
\end{aligned} \tag{15.29}$$

Since $\{u_n\}$ is bounded, we consider a weak cluster point w of $\{u_n\}$. Hence, there exists a subsequence $\{u_{n_k}\}$ of $\{u_n\}$, which converges weakly to w.

Now $T(s)$ being non-expansive, by (15.29) and Lemma 15.4, we obtain $w \in \text{Fix}(S)$. On the other hand

$$u_{n_k} = J_\lambda^{B_1}\left(x_{n_k} + \gamma\left(J_\lambda^{B_1} - I\right)Ax_n\right),$$

$$\frac{(x_{n_k} - u_{n_k}) + A^\star(J_\lambda^{B_2} - I)Ax_{n_k}}{\lambda} \in B_1 u_{n_k}.$$

By taking $k \to 0$ and by (15.25), (15.28) and using that graph of maximal monotone operator is weakly strong closed, we obtain $0 \in B_1 w$. Since $\{x_n\}$ and $\{u_n\}$ have same asymptotical behaviour, so $\{Ax_{n_k}\}$ weakly converges to Aw.

Next, using (15.25) and the fact that the resolvent operator is non-expansive and Lemma 15.1, we obtain $Aw \in B_2(Aw)$. Thus $w \in \{\text{Fix}(S)\} \cap \{\Xi\}$.

Now, we claim that $\lim_{n\to\infty}\sup\langle fz - z,\ x_n - z\rangle \le 0$, where $z = P_{\{\text{Fix}(S)\}\cap\{\Xi\}}\, fz$. Indeed, we have

$$\begin{aligned}
\limsup_{n\to\infty}\langle fz - z, x_n - z\rangle &= \limsup_{n\to\infty}\langle fz - z,\ t_n - z\rangle \\
&\le \limsup_{n\to\infty}\langle fz - z,\ t_{n_i} - z\rangle \\
&= \limsup_{n\to\infty}\langle fz - z, w - z\rangle \\
&\le 0,
\end{aligned} \tag{15.30}$$

because $z \in P_{\{\text{Fix}(S)\}\cap\{\Xi\}}\, fz$.

Finally, we have

$$
\begin{aligned}
\|x_{n+1} - z\|^2 &= \langle \alpha_n f x_n + (1-\alpha_n)t_n - z,\ x_{n+1} - z\rangle \\
&= \alpha_n \langle f x_n - z,\ x_{n+1} - z\rangle + (1-\alpha_n)\langle t_n - z,\ x_{n+1} - z\rangle \\
&= \alpha_n \langle f x_n - f z,\ x_{n+1} - z\rangle \\
&\quad + \alpha_n \langle f z - z,\ x_{n+1} - z\rangle + (1-\alpha_n)\langle t_n - z,\ x_{n+1} - z\rangle \\
&\leq \frac{\alpha_n}{2}\left\{\|f x_n - f z\|^2 + \|x_{n+1} - z\|^2\right\} \\
&\quad + \alpha_n \langle f z - z,\ x_{n+1} - z\rangle + \frac{(1-\alpha_n)}{2}\left\{\|t_n - z\|^2 + \|x_{n+1} - z\|^2\right\} \\
&\leq \frac{\alpha_n}{2}\left\{\alpha^2 \|x_n - z\|^2 + \|x_{n+1} - z\|^2\right\} \\
&\quad + \frac{(1-\alpha_n)}{2}\left\{\|t_n - z\|^2 + \|x_{n+1} - z\|^2\right\} + \alpha_n \langle f z - z,\ x_{n+1} - z\rangle \\
&\leq \frac{\alpha^2 \alpha_n}{2}\|x_n - z\|^2 + \frac{\alpha_n}{2}\|x_{n+1} - z\|^2 + \frac{(1-\alpha_n)}{2}\|x_n - z\|^2 \\
&\quad + \frac{(1-\alpha_n)}{2}\|x_{n+1} - z\|^2 + \alpha_n \langle f z - z,\ x_{n+1} - z\rangle \\
&= \frac{1}{2}[1 - \alpha_n(1-\alpha^2)]\ \|x_n - z\|^2 + \frac{(1-\alpha_n)}{2}\|x_{n+1} - z\|^2 \\
&\quad + \frac{\alpha_n}{2}\|x_{n+1} - z\|^2 + \alpha_n \langle f z - z,\ x_{n+1} - z\rangle.
\end{aligned}
$$

This implies that

$$
\|x_{n+1} - z\|^2 \leq [1 - \alpha_n(1-\alpha^2)]\ \|x_n - z\|^2 + 2\alpha_n \langle f z - z,\ x_{n+1} - z\rangle. \quad (15.31)
$$

By using (15.28) and Lemma 15.3, we see that $x_n \to z$. Further, from $\|u_n - x_n\| \to 0, u_n \rightharpoonup w \in \{\mathrm{Fix}(S)\} \cap \{\Xi\}, x_n \to z$ as $n \to \infty$ that $z = w$. This completes the proof.

Remark 15.1 The results presented in this paper can be seen as the extension and generalization of the previous known results in this field, see e.g. [2, 13].

References

1. Bauschke, H.H., Combettes, P.L.: Convex Analysis and Monotone Operator Theory in Hilbert Spaces. Springer, New York (2011)
2. Byrne, C.: Iterative oblique projection onto convex sets and the split feasibility problem. Inverse Probl. **18**, 441–453 (2002)
3. Byrne, C.: A unified treatment for some iterative algorithms in signal processing and image reconstruction. Inverse Probl. **20**, 103–120 (2004)
4. Byrne, C., Censor, Y., Gibali, A., Reich, S.: Weak and strong convergence of algorithms for split common null point problem. J. Nonlinear Convex Anal. **13**, 759–775 (2012)

5. Censor, Y., Gibali, A., Reich, S.: Algorithms for the split variational inequality problem. Numer. Algorithms **59**, 301–323 (2012)
6. Censor, Y., Elfving, T.: A multi-projection algorithm using Bregman projection in product space. Numer. Algorithms **8**, 221–239 (1994)
7. Censor, Y., Bortfeld, T., Martin, B., Trofimov, A.: A unified approach for inversion problems in intensity modulated radiation therapy. Phys. Med. Biol. **51**, 2353–2365 (2006)
8. Chen, R., Song, Y.: Convergence to common fixed point of nonexpansive semigroups. J. Comput. Appl. Math. **200**, 566–575 (2007)
9. Combettes, P.L.: The convex feasibility problem in image recovery. Adv. Imaging Electron Phys. **95**, 155–453 (1996)
10. Crombez, G.: A geometrical look at iterative methods for operators with fixed points. Numer. Funct. Anal. Optim. **26**, 157–175 (2005)
11. Crombez, G.: A hierarchical presentation of operators with fixed points on Hilbert spaces. Numer. Func. Anal. Optim. **27**, 259–277 (2006)
12. Goebel, K., Kirk, W.A.: Topics on Fixed Point Theory. Cambridge University Press, Cambridge (1990)
13. Kazmi, K.R., Rizvi, S.H.: An iterative method for split variational inclusion problem and fixed point problem for a nonexpansive mapping. Optim. Lett. doi:10.1007/s115900-013-0629-2
14. Lopez, G., Martin-Marquez, V., Xu, H.K.: Iterative algorithm for the multi-sets feasibility problems. In: Censor, Y., Jiang, M., Wang, G. (eds.) Biomedical Mathematics, Promising Direction in Imaging Therapy Planning and Inverse Problems, pp. 243–279. Medical Physics Publishing, Madison (2010)
15. Moudafi, A.: The split common fixed point problem for demicontractive mappings. Inverse Probl. **26**, 1–6 (2010)
16. Moudafi, A.: Split monotone variational inclusions. J. Optim. Theory Appl. **150**, 275–283 (2011)
17. Opial, Z.: Weak convergence of the sequence of successive approximation for nonexpansive mappings. Bull. Amer. Math. Soc. **73**(4), 595–597 (1967)
18. Shimizu, T., Takahashi, W.: Strong convergence to common fixed points of families of nonexpansive mappings. J. Math. Anal. Appl. **211**, 71–83 (1997)
19. Xu, H.K.: Viscosity approximation methods for nonexpansive mappings. J. Math. Anal. Appl. **298**, 279–291 (2004)

Chapter 16
The Impact of Vaccination to Control Human Papillomavirus Dynamics

Sudip Chakraborty, Joydeep Pal, Sonia Chowdhury and Priti Kumar Roy

Abstract Human papillomavirus (HPV) has been a prevalent virus for a long time. The number of cancer cases from benign tumours is constantly increasing and HPV is playing an important role, more aggressively is evolving as cervical cancer in young females. More than 170 types of HPV virus exists and among them some are "High-risk" variety, some are "Low-risk" variety. The control of this virus outbreak remains a challenge till date. The aim of this study is to investigate the role of vaccination as a control strategy in decreasing of the spread of the disease. In this research article, we formulated a model considering high-risk-type HPV, low-risk-type HPV, and low–high risk (infected by both low-risk and high-risk)-type HPV. We derive the basic reproduction ratio and also show that there exists a disease-free equilibrium, which is locally asymptotically stable. Furthermore, an analysis is then performed on crucial parameters in order to determine their importance and potential impact on HPV dynamics. Our analytical and numerical analysis reveals that HPV infection can be reduced by using vaccination as a control strategy.

Keywords Human papillomavirus (HPV) · Cervical cancer · Low–high risk type Vaccination · Basic reproduction ratio · Next-generation method

16.1 Introduction

Human papillomavirus (HPV) is a family of virus that includes more than 170 different types of virus and among them 40 types of virus are sexually transmitted [1]. Genital HPVs, which are transmitted sexually, are the primary factors in cervical cancer worldwide [2]. Cervical cancer is now one of the most common forms of cancer

S. Chakraborty · J. Pal · P.K. Roy (✉)
Centre for Mathematical Biology and Ecology, Department of Mathematics, Jadavpur University, Kolkata 700032, India
e-mail: pritiju@gmail.com

S. Chowdhury
Department of Mathematics, University of Kalyani, Kalyani 741235, West Bengal, India

P. Manchanda et al. (eds.), *Industrial Mathematics and Complex Systems*, Industrial and Applied Mathematics, DOI 10.1007/978-981-10-3758-0_16

worldwide among young women [3]. It is also the cause of non-cervical cancers in young men with occurrences of anal cancers, oropharyngeal cancers, and penile and prostate cancer [4]. It is revealed clinically, that Human papillomavirus (HPV) type 16 and 18 are responsible for about half of the cervical cancer cases in the United states and Europe [5]. Sexually transmitted HPV types fall into two categories: some types of HPV which can cause warts, are treated as low-risk genotype of HPV, while other types which lead to different kinds of cancer in females including cervical cancer, are treated as high-risk genotype of HPV [1, 6]. The virus types 2, 3, 4, 7, 8, 11, 22 are some low-risk HPV types and the virus type 16, 18, 31, 33, 45 are some high-risk HPV types [5]. For instance, HPV infections are recovered automatically by an individual immune system before they have the chance to develop into a productive infection. Around one-half of these infections are with a high-risk HPV type [7]. Most high-risk HPV infections occur without any symptoms and within 1–2 years it cures without any interventions and do not cause cancer. Some HPV infections, however, can exist for many years. Persistent infections with high-risk papillomavirus types can lead to cellular changes which if untreated, may progress to cancer. Generally, all types of cervical cancer are caused by HPV and it is also diagnosed to be clinically related factor for 95% anal cancer, 70% oropharyngeal cancers, and other rarer cancer cases in sexually active men. Cervical cancer is more prevalent in young women from 20–24 years [8].

Vaccination programs are recognized as being among the world's most successful public health programs. Its impact on lowering rates of infection (and sequential reduction in complications and health burden) makes vaccination one among the most cost effective and economically attractive of all health interventions strategies for various infectious diseases [9]. For HPV, vaccination seems to be the best approach to prevent cervical cancer. Merck and GlaxoSmithKline have developed two vaccines, Gardasil and Carvarix, respectively for treating HPV-induced cancers. These vaccines mainly targets on type 16 and 18 HPV ailments. Additionally, Gardasil also protects against type 6 and 11 [6]. Successful vaccination increases virus-neutralizing antibodies in serum [10]. These vaccines are nearly 100% effective in women at preventing diseases caused by virus-specific strains, including precancerous lesions of the cervix, vagina as well as genital warts. However, the HPV vaccine was confronted with some social myths and side effects like pain and syncope's which led some controversy in some Western countries recently [11, 12].

Studies on presence of HPV DNA in cervical samples show that 10% or more of all clinical lesions contain at least two different HPV types. McLaughlin et al. [13] have investigated if multiple HPV types can coexist in the same cell and interact with one another. His studies provide valuable insights into the interactions that may occur between different HPV types. Robert J. Smith [14] and Mori [15] has also give an idea about that coinfection of multiple genotypes of HPV is commonly observed among women with abnormal cervical cytology. A broad body of biological and clinical literature exist globally for HPV but mathematical modeling on the subject, generated quite recent interest among epidemiologists. In India, this seems to be a unexplored area, and predicting the possibilities of applying vaccination programs to fight against HPV.

Different models have been developed to analyze the transmission dynamics of sexually transmitted disease (STD) as well as the effectiveness of some intervention strategies against the spread of these diseases [16–19]. Mathematical study on the disease is quite limited and mostly focused on the biological and epidemiological backgrounds including some preventive strategies through vaccination. Previously, works involving mathematical studies associated to cervical cancer has been concentrated generally on epidemiology, with importance on the transmission among individuals and the efficacy of HPV vaccines [20–22].

In this research article, we have introduced a mathematical model to describe the interpretation between low-risk type and high-risk type HPV and their dynamics including a new class low–high risk HPV population and how vaccination becomes important in the effort of reducing the HPV infection in the population which is a novelty in itself.

16.2 Formulation of the General Mathematical Model

We consider a HPV population model taking susceptible, vaccinated, infected, and recovered classes. Let the total population at any time t be denoted by $N(t)$. The total population is divided into six compartments, that are Susceptible(S), Vaccinated(V), Infected by Low-risk HPV(I_L), Infected by High-risk HPV(I_H), Infected by both Low–High risk HPV(I_{LH}), and Recovered(R). It is assumed that susceptible individuals are recruited into the population at a rate Π. The susceptible individuals are vaccinated at a rate ω and the vaccinated individuals return to the susceptible class after losing their immunity at a rate σ. The susceptible individuals are infected by the classes I_L, I_H and I_{LH} at the rate λ respectively. Furthermore, it is assumed that vaccinated individuals may also be infected at a low rate $(1-\phi)$, where $\phi \in (0, 1)$ measures the efficacy of the vaccine. Now, the infected individuals by low-risk HPV may also be infected High-risk HPV at a rate β and move to the class I_{LH} and vice versa. The infected individuals by low-risk, High-risk, and both low–High risk HPV can recover at the rates r_1, r_2 and r_3 respectively. However, some recovered individuals revert to the susceptible class at a rate α after a wane their immunity. All individuals naturally die at a rate μ and sick individuals die of cancer at a rate ξ.

Thus, we have the following mathematical model:

$$\frac{dS}{dt} = \Pi - \lambda S(I_L + I_H + I_{LH}) - \omega S + \alpha R + \sigma V - \mu S,$$
$$\frac{dV}{dt} = \omega S - (1-\phi)V(I_L + I_H + I_{LH}) - (\sigma + \mu)V,$$
$$\frac{dI_L}{dt} = \lambda S I_L + (1-\phi)V I_L - \beta I_L I_H - (r_1 + \mu)I_L,$$
$$\frac{dI_H}{dt} = \lambda S I_H + (1-\phi)V I_H - \beta I_L I_H - (r_2 + \mu + \xi)I_H,$$

$$\frac{dI_{LH}}{dt} = \lambda S I_{LH} + (1-\phi)V I_{LH} + \beta I_L I_H - (r_3 + \mu + \xi) I_{LH},$$
$$\frac{dR}{dt} = r_1 I_L + r_2 I_H + r_3 I_{LH} - (\alpha + \mu) R. \quad (16.1)$$

where $N = S + V + I_L + I_H + I_{LH} + R$ and $S(0) > 0, V(0) \geq 0, I_L(0) \geq 0,$ $I_H(0) \geq 0, I_{LH}(0) \geq 0, V(0) \geq 0.$

16.2.1 Theoretical Analysis of the System

We begin by showing that all feasible solutions are uniformly bounded in a proper subset of $\Lambda \in \mathbb{R}_6^+$. The feasible region Λ with $\Lambda = \left\{(S, V, I_L, I_H, I_{LH}, R) \in \mathbb{R}_6^+ : N \leq \frac{\Pi}{\mu}\right\}$ is considered. Now, taking the sum all the equations of the model we have

$$\frac{dN(t)}{dt} = \Pi - \mu N(t) - \beta I_L I_H - \xi(I_H + I_{LH}),$$
$$\leq \Pi - \mu N(t). \quad (16.2)$$

Applying integration on (16.2), we obtain

$$N(t) \leq N(0)e^{-\mu t} + \frac{\Pi}{\mu}(1 - e^{-\mu t}), \quad (16.3)$$

where N(0) represents the initial value of the respective variables. Then $0 \leq N \leq \frac{\Pi}{\mu}$ as $t \to \infty$. Therefore, $\frac{\Pi}{\mu}$ is an upper bound of $N(t)$ provided $N(0) \leq \frac{\Pi}{\mu}$. Hence, the feasible solution of the above model system enters the region Λ is positively invariant set. Thus, the system is biologically meaningful and mathematically well posed in the domain of Λ. In this domain, it is sufficient to consider the dynamics of the flow generated by the model system (16.2).

16.2.1.1 Stability of the Disease-Free Equilibrium

In this system, the disease-free equilibrium is given by

$$E_0 = (S_0, V_0, 0, 0, 0, 0) = \left(\frac{\Pi(\sigma+\mu)}{\mu(\sigma+\omega+\mu)}, \frac{\Pi\omega}{\mu(\sigma+\omega+\mu)}, 0, 0, 0, 0\right).$$

The stability of this equilibrium will be investigated using the next-generation method [23, 25]. Using the common notations of next-generation method for the above model, the associated matrices F and V for the new infection terms and the remaining transition terms are respectively given by

$$F = \begin{pmatrix} \lambda S_0 + (1-\phi)V_0 & 0 & 0 \\ 0 & \lambda S_0 + (1-\phi)V_0 & 0 \\ 0 & 0 & \lambda S_0 + (1-\phi)V_0 \end{pmatrix}$$

$$= \begin{pmatrix} \frac{\lambda\Pi(\sigma+\mu)+(1-\phi)\Pi\omega}{\mu(\lambda+\omega+\mu)} & 0 & 0 \\ 0 & \frac{\lambda\Pi(\sigma+\mu)+(1-\phi)\Pi\omega}{\mu(\lambda+\omega+\mu)} & 0 \\ 0 & 0 & \frac{\lambda\Pi(\sigma+\mu)+(1-\phi)\Pi\omega}{\mu(\lambda+\omega+\mu)} \end{pmatrix}$$

and $V = \begin{pmatrix} r_1+\mu & 0 & 0 \\ 0 & r_2+\mu+\xi & 0 \\ 0 & 0 & r_3+\mu+\xi \end{pmatrix}$.

It follows that, $FV^{-1} = \begin{pmatrix} \frac{\lambda\Pi(\sigma+\mu)+(1-\phi)\Pi\omega}{\mu(\lambda+\omega+\mu)(r_1+\mu)} & 0 & 0 \\ 0 & \frac{\lambda\Pi(\sigma+\mu)+(1-\phi)\Pi\omega}{\mu(\lambda+\omega+\mu)(r_2+\mu+\xi)} & 0 \\ 0 & 0 & \frac{\lambda\Pi(\sigma+\mu)+(1-\phi)\Pi\omega}{\mu(\lambda+\omega+\mu)(r_3+\mu+\xi)} \end{pmatrix}$.

Thus, the reproduction number

$$R_0 = \rho(FV^{-1}) = \frac{\lambda\Pi(\sigma+\mu)+(1-\phi)\Pi\omega}{\mu(\lambda+\omega+\mu)(r_2+\mu+\xi)} \tag{16.4}$$

as $r_2 < r_3 < r_1$ and ξ is very low and where $\rho(FV^{-1})$ is the spectral radius of the matrix FV^{-1}.

Now, we have the following theorem.

Theorem 16.2.1 *If $R_0 < 1$, infection-free equilibrium is stable, while if $R_0 > 1$, the infection-free equilibrium is unstable and the infected state equilibrium exists.*

16.2.1.2 Existence of the Endemic Equilibrium

In this system, endemic equilibrium point E is the steady-state solution where the disease persists in the population. For the existence of endemic equilibrium $E = (S^*;\ V^*;\ I_L^*;\ I_H^*;\ I_{LH}^*;\ R^*)$, its coordinates should satisfy the conditions; $E = (S^*;\ V^*;\ I_L^*;\ I_H^*;\ I_{LH}^*;\ R^*) \neq 0$, where $S^* > 0$; $V^* > 0$; $I_L^* > 0$; $I_H^* > 0$; $I_{LH}^* > 0$; $R^* > 0$. The endemic equilibrium point is obtained by setting equation of the system (16.1) to zero. We then solve for state variables in terms of S and obtain the following;

$$\begin{aligned} V^* &= \frac{\omega S^*}{(1-\phi)Z + C}; \\ I_L^* &= \frac{XS^* - E}{\beta}; \\ I_H^* &= \frac{XS^* - D}{\beta}; \\ I_{LH}^* &= \frac{(XD - S^*)(XS^* - E)}{\beta(F - XS^*)}; \\ R^* &= \frac{(r_1+r_2)\beta Z - (XS^* - D)r_1 - (XS^* - E)r_2}{\beta G} \end{aligned} \tag{16.5}$$

where,
$A = \omega + \mu, C = \sigma + \mu, D = r_1 + \mu, E = r_2 + \mu + \xi, F = r_3 + \mu + \xi, G = \alpha + \mu, X = \lambda + \frac{(1-\phi)\omega}{(1-\phi)Z+C}$ and $Z = I_L^* + I_H^* + I_{LH}^*$.
Now substituting the values we have

$$S^* = \frac{(a_1 + a_2 Z)(a_3 Z + C)^2}{(a_4 + a_5 Z)(a_3 Z + C) + a_6}. \tag{16.6}$$

where, $a_1 = Dr_1 + Er_2,\ a_2 = \Pi\beta G + \alpha\beta(r_1 + r_2)Z,\ a_3 = 1 - \phi,\ a_4 = \beta AG - \sigma\omega, a_5 = \beta\lambda G, a_6 = (r_1 + r_2)[\lambda + (1 - \phi)\omega]$.

Hence putting the value of S^* in $Z = I_L^* + I_H^* + I_{LH}^*$ we have an eighth degree polynomial of Z as

$$h_0 + h_1 Z + h_2 Z^2 + h_3 Z^3 + h_4 Z^4 + h_5 Z^5 + h_6 Z^6 + h_7 Z^7 + h_8 Z^8 = 0. \tag{16.7}$$

The polynomial (16.7) has at least one positive root if $h_0 < 0$ and $h_0 + h_1 + h_3 + h_5 + h_7 > 0$. This implies that there exists at least one endemic equilibrium point. Hence, the existence of the endemic equilibrium point will be governed by the following theorem. The model system (16.1) has at least one endemic equilibrium point, if $h_0 < 0$ and $h_0 + h_1 + h_3 + h_5 + h_7 > 0$. Then, the theorem below gives a condition for the existence of the endemic equilibrium point for HPV model with vaccination. The endemic equilibrium point of HPV model with vaccination exists if and only if $R_0 > 1$.

16.3 Numerical Simulation

Numerical simulations of the model system are carried out using a set of parameter values given in Table 16.1. Some parameter values were obtained from different literature and others were estimated. We simulate the model system by using ODE45 solver coded in MATLAB program language by using the parameter values shown in Table 16.1 and the following initial conditions; $S(0) = 2000$; $V(0) = 100$; $I_L(0) = 200$; $I_H(0) = 50$; $I_{LH}(0) = 20$ and $R(0) = 200$.

16.3.1 Numerical Results in the Presence of Vaccination

Here, we show the trend of the state variables of the modified HPV model. The decrease in the number of susceptible individuals implies that many of the susceptible individuals are vaccinated as in Fig. 16.1a. As initially susceptible proportion remains larger, so the vaccinated individuals increases. With continuous vaccination to individuals the no. of vaccinated people decreases over time. It is reflected in the Fig. 16.1b. Figure 16.1c shows that, the number of infected by low-risk HPV

Table 16.1 List of parameters used in the model (16.1)

Parameter	Definition	$Value\ (month)^{-1}$	Reference
Π	Rate of recruitment into the susceptible population	40	*Estimated*
λ	Rate of infection	0.001	*Estimated*
ω	Rate of vaccination	0.3	[24]
α	Rate of reversion from recovered to susceptible population	0.2	[24]
σ	Rate of returning from vaccinated to susceptible population	0.02	[24]
ϕ	Rate of non-infection of vaccinated population	0.99	[24]
β	Contact rate between I_L and I_H classes to make I_{LH} class	0.01	[14]
r_1	Recovery rate of I_L population	4	*Estimated*
r_2	Recovery rate of I_H population	3.6	*Estimated*
r_3	Recovery rate of I_{LH} population	3.65	*Estimated*
μ	Natural death rate	0.2	[24]
ξ	Disease death rate	0.01	[24]

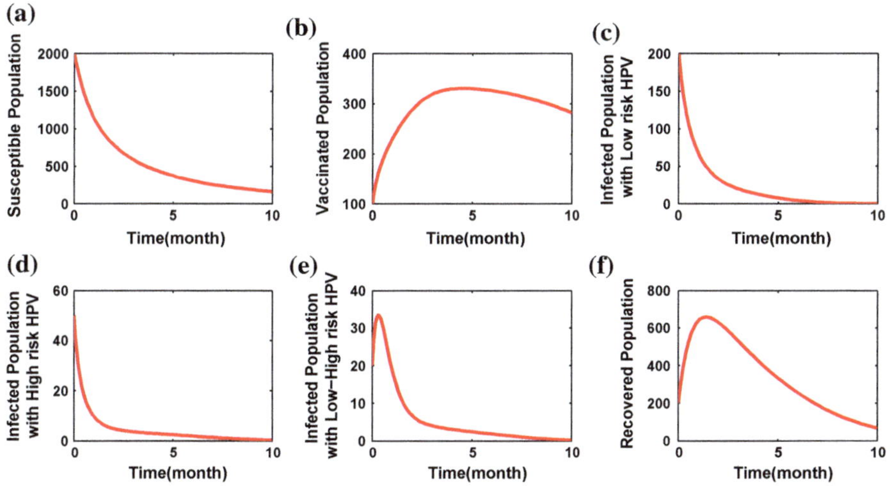

Fig. 16.1 Simulation results showing the trends of the state variables of the HPV model for **a** Susceptible population, **b** Vaccinated population, **c** Infected with low-risk HPV, **d** Infected with high-risk HPV, **e** Infected with low–high risk HPV and **f** Recovered population of the model system (16.1) where $R_0 = 0.3374$

individuals decreases. Figure 16.1d shows that the number of infected by high-risk HPV individuals decreases very fast due to vaccination, disease death etc. Figure 16.1e indicates that, the number of individuals with both low- and high-risk HPV individuals increases initially due to interaction of I_L and I_H classes then decreases for vaccination. Furthermore, the number of recovered individuals increases because of the high rate of recovery for I_L class and then it decreases as many recovered individuals goes to susceptible class again, as shown in Fig. 16.1f.

16.3.2 Variation of Population Under Different Vaccination Rates

In Fig. 16.2, we show the role of vaccination in reducing HPV infection in the population. Figure 16.2a shows that, when the vaccination rate ω increases the susceptible population decreases rapidly. This implies that most of the susceptible individuals become vaccinated. When the value of ω increases, the vaccination population increases as shown in Fig. 16.2b. It is to be noted that vaccine can protect HPV-infected people with certain High-risk seropositive cases. However, the vaccine cannot protect the spectrum of low-risk HPV. So obviously people can get infected with low-risk virus from vaccinated class also and this is reflected with variation of parameter in Fig. 16.2c. Figure 16.2d shows that, the rate of decreasing of number of infected by high-risk HPV population goes higher as the rate of ω increases and vaccination plays an very important role in this population. Similarly, Fig. 16.2e indicates that, the number of individuals of I_{LH} class decreases in higher rate as ω increases. As vaccination rate increases the number of recovered individuals also increases, as shown in the Fig. 16.2f.

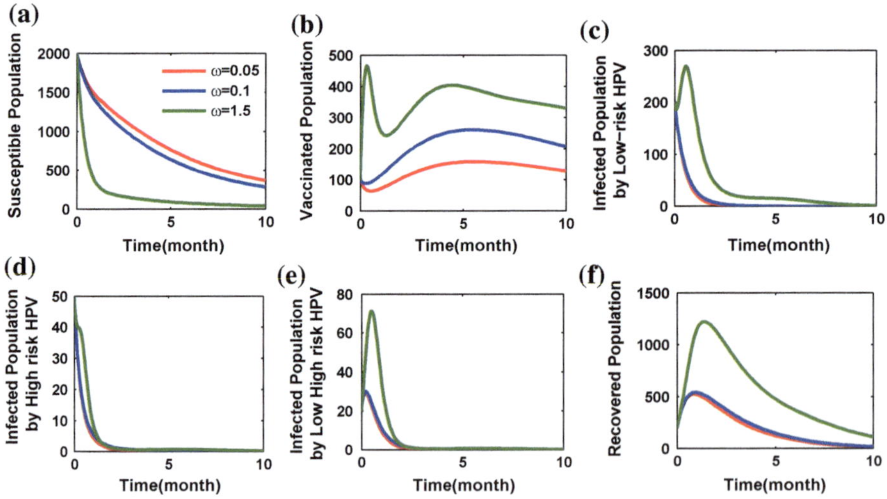

Fig. 16.2 Simulation results showing the effect of varying vaccination rate (ω) on **a** Susceptible population, **b** Vaccinated population **c** Infected with low-risk HPV, **d** Infected with high-risk HPV, **e** Infected with low–high risk HPV and **f** Recovered population

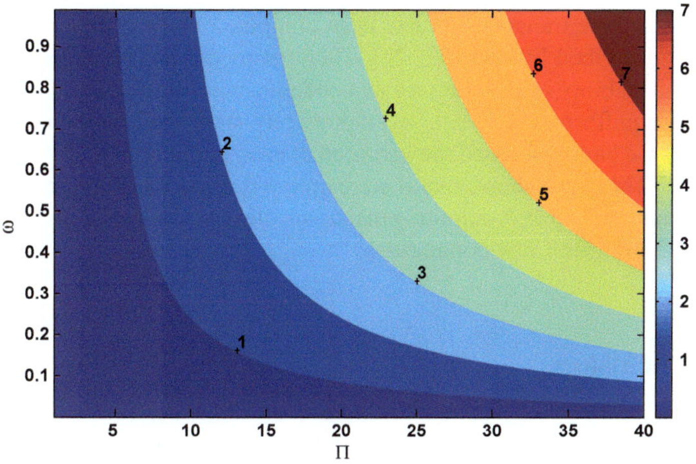

Fig. 16.3 Contour plots of R_0 as a function of Π and ω

Color legend is for differentiating the different region of R_0. From our analytic study if $R_0 < 1$ then the system will be stable. So above the dark blue region the system will be unstable. The contour plot exhibits the R_0 as a function of Π and ω (Fig. 16.3), it shows that when the value of Π lies within 10 the system becomes stable ($R_0 < 1$) with increasing value of the vaccination programme. However, if we increase the value of Π beyond this value the system loses its stability although the vaccination rate may be higher.

16.4 Discussion

In this research article, we have formulated a mathematical model of HPV dynamics considering six population classes which are interrelated to the transmission of the disease. Here, we have studied the impact of vaccination as a control strategy against the transmission of the HPV infection. We have taken vaccinated population as an individual class. We have introduced a new class considering those individuals who are co-infected by both low-risk and high-risk HPV. Our analytical results indicated taking the reproduction number $R_0 < 1$, all the population exhibits stable equilibrium. It is observed that the parameter ω is an important parameter in respect of the infection control. We have derived numerically the stable region for ω and Π. Moreover, it is observed that the 100% vaccination rate can maintain its stability for values of Π below 10. It is also observed that the infection readily decreases after an initial 1-month period when the vaccination imparts its biological efficacy on the coinfection cases (i.e. infected by the low–high risk HPV) to reduce the prevalence of infection. The model analysis showed that there exists a domain where the model is epidemiologically and mathematically well posed. The threshold parameter that operates the disease transmission has been computed by next generation operator approach as described by Van [23, 25]. Then the model has been analyzed for the existence and stability of disease-free and endemic equilibrium. It is proved that the disease-free equilibrium is locally asymptotically stable under certain conditions. Numerically we have discussed the variation of each population under different vaccination rate. The result shows that if we increase the vaccination rate ω in a certain range i.e. from 0.05 to 1.5, it definitely makes an affect on the susceptible, vaccinated, high-risk and low–high risk populations.

16.5 Conclusion

In this research work, we have investigated the impact of vaccination on the prevention of human papillomavirus (HPV) infection dynamics in humans. We have worked out the feasibility of the vaccination strategy where we have shown analytically and numerically how vaccination can ensure a predictable preventive policy against the disease transmission among sexually active population who are at a greater risk to develop cervical cancer from the high-risk and low–high risk strains of the papillomavirus. Here, we developed a mathematical model system of HPV which can capture some relevant properties of the disease with its clinical manifestation and biological background. Although the total eradication of this HPV disease remains a challenge in some countries but from the result of our study, we request that our policymakers should initiate some programs on implementing vaccination for this disease. Vaccination is indeed needed to reduce the transmission of HPV infection in HPV prevalent area as it has significant impact on the reduction of infection transmission, which we have showed in the research article. It is also revealed from our numerical and analytical study that coinfection plays a crucial role in propagating the complexities of the disease that can welcome more extreme cases. Through vaccination we can also address the problem of high risk HPV in a better way.

Acknowledgements The research is supported by Government of India, DST PURSE and UGC DRS Programme, Department of Mathematics, Jadavpur University. Sonia Chowdhury is supported by UGC - Dr.D.S. Kothari Post Doctoral Fellowship (No. F.4-2/2006(BSR)/MA/14-15/0038, 5 May 2015).

References

1. Bergot, A.S., Kassianos, A., Frazer, I.A., Mittal, D.: New Approaches to immunotherapy for HPV associated cancers. Open Access, Cancers **3**, 346–3495 (2011). doi:10.3390/cancers3033461
2. Crow, J.M.: HPV: the global burden. Nature **488**, S2–S3 (2012). doi:10.1038/488S2a
3. Bosch, F.X., Lorincz, A., Munoz, N., Meijer, C.J.L.M., Shah, K.V.: The causal relation between human papillomavirus and cervical cancer. J. Clin. Pathol. **55**(4), 244–265 (2002)

4. Chaturvedi, A.K.: Beyond cervical cancer: burden of other HPV-related cancers among men and women. J. Adolesc. Health **46**(4), S20–S26 (2010)
5. Insinga, R.P., Dasbach, E.J., Elbasha, E.H.: Epidemiologic natural history and clinical management of Human Papillomavirus (HPV) Disease: a critical and systematic review of the literature in the development of an HPV dynamic transmission model. BMC Infect. Dis. **9**, article 119 (2009)
6. Lowy, D.R., Schiller, J.T.: Prophylactic human papillomavirus vaccines. J. Clin. Investigat. **116**, 1167–1173 (2006)
7. Hariri, S., Unger, E.R., Sternberg, M., et al.: Prevalence of genital human papillomavirus among females in the United States, the National Health and Nutrition Examination Survey, 2003–2006. J. Infect. Dis. **204**(4), 566–573 (2011)
8. Dunne, E.F., Unger, E.R., Sternberg, M., McQuillan, G., Swan, D.C., Patel, S.S., Markowitz, L.E.: Prevalence of HPV infection among females in the United States. JAMA **297**(8), 813–819 (2007)
9. SchlipkÖter, U.: Communicable diseases: achievements and challenges for public health. Public Health Rev. **32**(1), 90 (2010)
10. Burd, E.M.: Human Papillomavirus and Cervical Cancer. Clin. Microbiol. Rev. **16**, 1–17 (2003)
11. Reiter, P.L., Brewer, N.T., Gottlieb, S.L., McRee, A.L., Smith, J.S.: How much will it hurt? HPV vaccine side effects and influence on completion of the three-dose regimen. Vaccine **27**(49), 6840–6844 (2009)
12. Brinth, L., Theibel, A.C., Pors, K., Mehlsen, J.: Suspected side effects to the quadrivalent human papilloma vaccine. Dan. Med. J. **62**(4), A5064 (2015)
13. McLaughlin-Drubin, M.E., Meyers, C.: Evidence for the coexistence of two genital HPV types within the same host cell in vitro. Virology **321**, 173–180 (2004)
14. Smith, R.J., Li, J., Mao, J., Saha, B.: Can. Appl. Math. Q. **21**(2) (2015)
15. Mori, Seiichiro, Kusumoto-Matsuo, Rika, Ishii, Yoshiyuki, Takeuchi, Takamasa, Kukimoto, Iwao: Replication interference between human papillomavirus types 16 and 18 mediated by heterologous E1 helicases. Virol. J. **11**, 11 (2014)
16. Mathematical Models for Therapeutic Approaches to Control HIV Disease Transmission published by Springer in 2016. ISBN: 978-981-287-851-9 (Print) 978-981-287-852-6 (Online)
17. Roy, P.K., Chatterjee, A.N.: Effect of HAART on CTL mediated immune cells: an optimal control theoretic approach. Electr. Eng. Appl. Comp. Springer **90**, 595–607 (2011)
18. Chatterjee, A.N., Roy, P.K.: Anti-viral drug treatment along with immune activator IL-2: A control based mathematical approach for HIV infection. Int. J. Control **85**(2), 220–237 (2012)
19. Chowdhury, S., Roy, P.K., Smith, R.J.: Mathematical modelling of enfuvirtide and protease inhibitors as combination therapy for HIV. Int. J. Nonlinear Sci. Numer. Simul. **17**(6), 259–275 (2016)
20. Barnabas, R.V., Laukkanen, P., Koskela, P., Kontula, O., Lehtinen, M., Garnett, G.P.: Epidemiology of HPV 16 and cervical cancer in Finland and the potential impact of vaccination: mathematical modelling analyses. PLOS Med **3**(0624), 0632 (2006)
21. Baussano, I., Ronco, G., Segnan, N., French, K., Vineis, P., Garnett, G.P.: HPV-16 infection and cervical cancer: modeling the influence of duration of infection and precancerous lesions. Epidemics **2**, 21–28 (2010)
22. Brown, V., White, K.A.: The HPV vaccination strategy: could male vaccination have an impact? Comput. Math. Methods Med. **11**, 223–237 (2010)
23. Tchuenche, J.M., Dube, N., Bhunu, C.P., Smith, R.J., Bauch, C.T.: The impact of media coverage on the transmission dynamics of human influenza. BMC Public Health **11**(1), 1 (2011)
24. Shaban, N., Mofi, H.: Int. J. Math. Anal. **8**(9), 441–454 (2014)
25. Van den Driessche, P., Watmough, J.: Reproduction numbers and Sub-threshold endemic equilibria for compartmental models of disease transmission. Math. Bio-sci. **180**, 29–48 (2002)

Chapter 17
Novel Solution of Nonlinear Equations Using Genetic Algorithm

Chhavi Mangla, Harsh Bhasin, Musheer Ahmad and Moin Uddin

Abstract Nonlinear equations represent highly complex systems and their solutions by conventional methods have high computational complexity. Methods like Bisection, Regula Falsi, Newton–Raphson, Secant, Muller, etc., are used to solve such problems. This work find gaps in the existing methods and justifies the applicability of Genetic Algorithm to the problem. A Genetic Algorithm-based method has been proposed, which is more efficient and produces better results as compared to the existing methods.

Keywords Nonlinear equations · Soft computing techniques · Genetic Algorithm

17.1 Introduction

Linear and nonlinear system of equations are used in numerous engineering applications. Finding a robust and efficient solution to such system, is a tedious task and, at times, too complex to be handled by the conventional methods like Newton's Method, Bisection method, Regula Falsi, etc. The existing conventional approaches dealing the above problem can be categorized as follows:

(a) Calculus-based methods, which include conventional Newton's method, Secant method, Bisection method, etc.

C. Mangla (✉) · M. Ahmad
Department of Applied Sciences and Humanities, Jamia Millia Islamia, New Delhi 110025, India
e-mail: cmangla89@gmail.com

M. Ahmad
e-mail: mushr_mt@yahoo.co.in

H. Bhasin · M. Uddin
Department of Computer Science, Jamia Hamdard, New Delhi, India
e-mail: i_harsh_bhasin@yahoo.com

M. Uddin
e-mail: prof_moin@yahoo.com

P. Manchanda et al. (eds.), *Industrial Mathematics and Complex Systems*, Industrial and Applied Mathematics, DOI 10.1007/978-981-10-3758-0_17

(b) Heuristic methods, which include evolutionary computational techniques like Particle swarm optimization, Genetic Algorithm etc.

The high computational complexity, both in terms of time and space, and the constraints involved in the conventional methods make their application to the said problems difficult. This opens the window of Genetic Algorithm (GA) to the fascinating world of mathematics. Note that GA is heuristic search process based on the Darwin's theory of evolution. It has been found that the application of GA produces a robust, efficient, and effective result in finite time. Also, it can deal with the problem of local maxima. The problem of finding out the solution of a given equation or that of a given set of equations is also a search problem. Moreover, one of the factors that justifies the applicability of GA to the problem is the vastness of the search space.

This work proposes a soft computing technique for finding efficient solutions to solve the given system of equations. This approach uses principle of evolutionary computation and has been effectively applied to find the approximate solutions of such problems. The empirical analysis has been carried out to reach the solution to the problem. The crossover and mutation rates in the experiments carried out, were kept constant but, in the ongoing work, the variations of these, are also being analyzed. The process has been successfully implemented and results are encouraging. The advantage of the current approach is that no additional constraint involving differentiability of the equation is required. Therefore, this approach can be used for noncontinuous equations also.

The organization of the paper is as follows: In Sect. 17.2, a brief literature review providing the state of the art of available work is presented. Section 17.3 provides an overview of GA. In Sect. 17.4, the proposed algorithm is presented. Section 17.5 includes the experiment and results. The last section presents the conclusion, applications and future scope.

17.2 State of Art

A systematic review provides a great source to understand, evaluate and interpret all available research work related to the research area. It also helps to access the state of the art and to find problems, if any, in the existing works. Moreover, it also helps to justify to the proposed approach.

In this view, an extensive literature review has been carried out. The results of the review have been presented in the following table (Table 17.1). Though, many other papers were studied and analyzed, the methodology are more or less similar to those adopted in the papers given in the following table. The important points and the issues pertaining to the methodologies were considered while designing the experiment. However, review of these papers has not been mentioned in the table [1, 4, 5, 7, 9, 11, 12, 14, 20].

Table 17.1 State of art

S.No.	Year	Work proposed	Verification/validation
1	2001	The work uses a gradient descent method in order to solve nonlinear system of equations [2]	The proposed work has been verified on the following test problems:
			• Extended rosenbrock function
			• Broyden tridiagonal function
2	2005	The author presents a methodology for sorting equations from system of nonlinear equations, which can be solved using fixed point method. The work involves combining machine learning on the basis of Genetic and Genetic Algorithm, which help in managing a population of possible solution processes [19]	The work has been verified on the Combined cycle gas turbine simulation problem
3	2006	In this paper, real-coded multi-crossover Genetic Algorithm has been developed for estimating the various parameters of nonlinear system [3]	The proposed work has been verified on the following test problems:
			• First-order plus dead-time system
			• Nonlinear and unstable plant problem
4	2006	In the paper, the nonlinear system of equations, at each step, is first transformed into a constrained nonlinear programming problem, and then with a line search strategy, it is solved using sequential quadratic programming (SQP) algorithm [15]	For validation of the current approach, few examples on system of nonlinear equations in two variables with quadratic degree are taken into consideration
5	2008	A new perspective has been proposed in the work, by considering nonlinear system of equations as multi-objective optimization problem [6]	The proposed work has been verified on the following test problems:
			• Interval arithmetic benchmark
			• Neuropsychology application
			• Kinematic application
			• Combustion application
			• Chemical equilibrium application
6	2011	In the proposed work, the author combines two heuristic optimization tools, Genetic Algorithm, and Particle Swarm Optimization for solving complex nonlinear equation system [1]	Methodology has been verified with a suit of 17 unconstrained test problems
7	2013	The paper describes a specialized application of Genetic Algorithm for approximating solution of optimum problems by introducing pairs of harmonious and symmetric individuals [18]	Few examples involving 2 variables have been used for verification of the proposed technique

(continued)

Table 17.1 (continued)

S.No.	Year	Work proposed	Verification/validation
8	2013	In this paper, the author first converted single and simple set of nonlinear system of equations into unconstrained optimization problem, and complex set of systems into constrained optimization problem. Afterward, Genetic Algorithm tool is applied to solve the system [16]	The proposed work has been verified on the following test problems:
			• Fluid mechanics application
			• Arithmetic applications
			• Combustion applications
			• Neurophysiology applications
9	2014	The article presents the estimation of root of nonlinear equations using Genetic Algorithm via. population size, crossover rate, degree of mutation, and coefficient size [10]	The proposed work has been verified on the following test problems:
			• Traveling Salesman Problem
			• Neurophysiology Application
			• Tank Reactor System
10	2015	The author developed a new approach in which optimum solution of nonlinear system of equations is obtained by a method based on variants of Genetic Algorithm using evolutionary computational technique [17]	The work has been verified on a set of different 20 nonlinear equations in single variable

17.3 Genetic Algorithm

GA is a heuristic-based search process based on the theory of natural selection and survival of the fittest [5, 13]. These were developed by John Holland [8] in 1960 at University of Michigan, USA. Many computational problems require searching the optimal solution from a large search space. For such problems, GA has proved to be robust and efficient way to evolve optimal solution. Thus, GA has been successfully applied to numerous areas like artificial intelligence, financial time series analysis, image processing, multi-modal optimization, robotics, portfolio management, etc.

17.3.1 Mechanism of Genetic Algorithm [8]

In GA, a population of candidate solutions for a specified optimization problem is randomly created. Each individual in GA, is then represented as a chromosome, which is a candidate of the solution. The selection of chromosome is done in a competitive manner based on their fitness. This is followed by the application of the genetic search operators namely, selection, crossover, and mutation, are applied over such chosen chromosomes to create a new generation of chromosomes in which the expected quality in terms of their fitness value is better than the previous generation. This process is repeated until the termination criterion is met and the best class of chromosome, which is actually a double type, is reported as solution to the concerned problem.

17.4 Proposed Work

Nonlinear system of equations:
A system of nonlinear equation may be defined as follows.

$$\begin{aligned} f_1(x) &= 0 \\ f_2(x) &= 0 \\ &\vdots \\ f_n(x) &= 0 \end{aligned}$$

where each function f_i is nonlinear function, which act as mapping a vector $x = (x_1, x_2, ..., x_n)^t$ of the n-dimensional space $\boldsymbol{R^n}$ to real line. Some of the functions may be linear and others nonlinear. The solution for nonlinear system involves finding solution in such a way that each of the above function $f_i(x)$ is equal to zero.

The proposed work presents a solution to such nonlinear system using soft computing technique called Genetic Algorithm. The algorithm is as follows:
Step 1: Transform each function $f_i(x)$ into $z_i(x)$ as

$$z_i = abs f_i(x) \text{ for each } i = 1, 2, \ldots, n.$$

Thus, the above system becomes a multi-objective optimization problem:

$$min \quad z_1, min \quad z_2, \ldots, min \quad z_n$$

Step 2: An initial generation of population is created which act as a chromosome for the GA.
Step 3: On the basis of fitness value, some chromosomes are chosen from the initial population. Using crossover and mutation rate (which are predefined), a new offspring generation is created until it does not exceed the predefined number of generations.
Step 4: The fitness value for each individual from offspring population is evaluated on the basis of multi- objective problem and the feasibility of the solution is checked.

Step 5: The termination criterion is to minimize z which is the sum of the individual fitness functions:

$$z = z_1 + z_2 + \cdots + z_n$$

If the termination criterion is satisfied, then go to *step 6*, else go to *step 3*.
Step 6: Report the solution.
It may be noted that the initial population is crafted as per the given problem and the plausibility of finding the solution, in given time, by the selected initial population. The initial population can neither have too many chromosomes or too less chromosomes. An appropriate number of chromosomes has been found by the initial empirical analysis and hence, is used in the experiment. The values of various parameters and the experimental setup has been reported in the next section.

17.5 Experimentation and Results

To validate the performance of the proposed algorithm, an extensive empirical analysis has been carried out. The experimentation is performed by setting the various parameters for GA as depicted in Table 17.2. In order to validate and verify the proposed approach, the following equations have been selected such that some of its roots are nonintegral and the range of the roots variate considerably.

$$4x^3 - 7x^2 + 0.578 = 0 \tag{17.1}$$
$$8x^3 - 6x^2 - 3x - 54 = 0 \tag{17.2}$$
$$x^4 - 5x^3 + 5x^2 + x - 20 = 0 \tag{17.3}$$
$$x^4 - 19x^3 - 21x^2 + 400 = 0 \tag{17.4}$$
$$x^3 - 977x^2 + 975x + 976 = 0 \tag{17.5}$$
$$x^2 + 105x + 500 = 0 \tag{17.6}$$
$$3x^2 + 3001x + 1000 = 0 \tag{17.7}$$
$$x^2 - 9999x - 10000 = 0 \tag{17.8}$$

Table 17.2 Parameters of GA

Parameters	Settings
Population size	200
Scaling function	Rank
Selection function	Tournament/Roulette wheel
Mutation function	Gaussian
Crossover function	Single point
Generations	50
Ratios/Fractions	Default

Run	Equation 1		Equation 2		Equation 3		Equation 4		Equation 5		Equation 6		Equation 7		Equation 8	
	$4x^3 - 7x^2 + 0.578 = 0$		$8x^3 - 6x^2 - 3x - 54 = 0$		$x^4 - 5x^3 + 5x^2 + x - 20 = 0$		$x^4 - 19x^3 - 21x^2 + 400 = 0$		$x^3 - 977x^2 + 975x + 976 = 0$		$x^2 + 105x + 500 = 0$		$3x^2 + 3001x + 1000 = 0$		$x^2 - 9999x + 10000 = 0$	
	Roulette wheel	Tournament	Roulette wheel	Tournament	Roulette wheel	Tournament	Roulette wheel	Tournament	Roulette wheel	Tournament	Roulette wheel	Tournament	Roulette wheel	Tournament	Roulette wheel	Tournament
1	5.98E-04	5.34E-04	0.033326931	0.003166334	3.65E-04	0.001905272	0.037028397	0.011440981	0.04855961	0.130875777	0.011866363	0.0118153	0.042014086	0.072020418	0.549022137	5.450676723
2	0.001015286	2.15E-04	0.004687464	0.004417561	0.003073118	0.018139144	0.041608376	0.092908315	0.093473069	0.009727351	0.01094501	0.014184123	0.257774228	0.020217312	4.599665488	2.557047716
3	1.28E-04	2.78E-04	0.017708135	0.013762712	0.003254299	0.003397448	0.02914816	0.111254358	0.871039964	0.18387613	0.010563544	0.001873713	0.226261441	0.972527167	0.066157621	1.219463758
4	3.61E-04	2.08E-04	0.055042024	0.029175972	0.026103102	5.07E-04	0.015455424	0.009994285	0.542712299	0.486578202	0.03140099	0.005542405	1.437793368	0.636488139	0.500740742	0.826209334
5	3.71E-05	7.48E-04	0.009955038	0.011536961	0.001603386	0.007028449	0.023340539	0.003606382	0.757575394	0.750443034	0.0148965	0.043525391	0.733901795	0.014793639	1.170495869	0.946480707
6	1.81E-04	0.001277686	0.005630487	0.03701429	4.26E-04	5.47E-05	0.015870281	0.061378784	0.350492243	0.144914077	0.02113509	0.020257538	0.638210124	0.415655031	0.566498846	1.634308317
7	3.70E-05	2.19E-04	0.009439954	0.031667505	5.76E-04	8.83E-04	0.036726074	0.021607272	0.42410352	0.02579882	0.005480728	0.005318297	0.337192834	0.24231522	1.180480789	2.176940559
8	9.42E-04	7.07E-04	0.006658034	0.008060728	0.010934656	9.48E-04	0.007442816	0.113600262	0.337729594	0.184551363	0.004933129	0.004669686	1.848425212	0.419926402	2.943666412	6.158315769
9	1.14E-04	1.32E-04	0.009610953	0.044735833	0.005749778	7.25E-05	0.075505906	0.086162715	0.14800182	0.129752995	0.022144714	0.005268549	1.145906535	0.934884853	0.447374986	0.298738315
10	6.19E-04	3.15E-04	0.011073762	0.024822936	0.0098523	0.004964706	0.119729099	0.027655139	0.003199781	0.28665307	0.001447914	0.008396412	1.385984646	0.253387529	0.180911883	2.039606171
11	3.32E-04	8.40E-04	6.96E-04	0.007864019	0.01129805	0.003281258	0.074081363	0.170732951	0.117081871	0.014206256	0.004604027	0.011599696	0.103567201	0.368983253	1.615699623	1.340085345
12	3.10E-04	7.31E-04	0.002115306	0.002873204	0.011012569	0.0109757	0.037230849	0.015887012	0.41175105	0.157685563	0.011867148	0.009811539	0.341275246	1.221902788	1.294091029	1.071591741
13	8.33E-04	6.32E-05	0.004444177	0.005785098	8.09E-04	0.001326393	0.049989538	0.04525294	0.467725345	0.329285663	0.02537635	0.003038464	0.055002703	0.477568475	0.438926258	0.396239232
14	0.002667048	0.001073115	0.011084448	0.002197831	0.002744721	0.021894721	0.006984952	0.03773901	0.540725452	0.752300553	0.006785156	0.003587474	0.311693172	0.423543027	0.733682386	3.084228883
15	0.001302695	2.95E-04	0.007916082	0.004298767	3.36E-04	0.004355178	0.102499715	0.022778922	0.121393403	0.408866193	0.011313984	0.020217975	0.304391378	0.3251893	0.538794698	2.042227161
16	6.62E-04	2.53E-04	0.00778388	0.024326752	0.002090556	7.60E-04	0.036664335	0.020681203	0.161713836	0.390867792	0.016496719	0.001131538	0.165614571	0.38925646	2.155948066	5.26393597
17	0.001872616	0.001196379	0.003735694	0.001524764	0.004940544	0.001078893	0.093395135	0.031111558	0.272437952	0.166066966	0.044817249	0.03009141	0.203232105	1.906353766	2.06289431	0.106995675
18	2.02E-04	4.02E-06	0.036042451	0.001403127	0.00783049	0.031655539	0.174649326	0.086394732	0.225891539	0.578557811	0.018962539	4.96E-04	0.670742473	2.474231218	2.775767745	2.289408428
19	0.001091792	4.39E-05	0.011107875	0.006508315	0.002295816	0.003460069	0.00176387	0.269211813	0.232318683	0.277182892	0.047854338	0.008843916	0.189548273	0.6416364	7.089191358	1.265365298
20	1.08E-04	0.001074263	0.009504255	0.004900331	0.007964626	0.005515344	0.008960238	0.018966729	0.417598804	0.614187947	0.052603983	0.019919692	0.195463277	0.703770112	0.440692505	4.330705093
21	8.05E-06	1.33E-04	0.001181209	0.001244931	0.005485354	0.002008177	0.056836553	0.005312429	0.363805641	0.689669235	0.021906736	0.002286821	0.911252482	0.756953586	0.654660959	0.183577869
22	0.001310297	1.53E-04	0.013868423	0.008295426	0.001939157	0.008029166	0.072599062	0.027347851	0.324268055	2.054142062	0.006648566	0.012158793	0.362732913	0.392632751	0.501014846	0.888961082
23	0.001865682	1.17E-04	0.003699953	7.66E-04	0.004490619	0.001729614	0.077047693	0.012928025	0.343167425	0.562801177	0.005814469	0.036077585	0.082891667	0.015585774	2.456839274	1.515244697
24	4.08E-04	0.001615001	0.001350475	0.004890399	0.003575038	3.36E-04	0.042489718	5.92E-04	0.05880009	0.004223747	0.002432153	0.004119513	0.179789436	0.048430585	1.707972407	0.406440834
25	7.14E-04	2.04E-04	0.017605263	0.003322564	0.001055152	0.007548391	0.078767219	0.092414075	0.093460523	0.10851801	0.007843749	0.018385105	1.455539072	0.382487187	0.457114782	0.012015753

Fig. 17.1 Experimentation and results

Table 17.3 Objective value

Equation number	Equation	Objective function value
1	$4x^3 - 7x^2 + 0.578 = 0$	$4.02E - 06$
2	$8x^3 - 6x^2 - 3x - 54 = 0$	$6.96E - 04$
3	$x^4 - 5x^3 + 5x^2 + x - 20 = 0$	$5.47E - 05$
4	$x^4 - 19x^3 - 21x^2 + 400 = 0$	$5.92E - 04$
5	$x^3 - 977x^2 + 975x + 976 = 0$	$3.20E - 03$
6	$x^2 + 105x + 500 = 0$	$4.96E - 04$
7	$3x^2 + 3001x + 1000 = 0$	$1.48E - 02$
8	$x^2 - 9999x - 10000 = 0$	$1.20E - 02$

The equation has at least one solution, however, the proposed approach does not find all the solution, but stops as soon as it finds even a single solution. The experimentation has been carried out for roulette wheel selection function and tournament selection function separately. 25 runs for each selection function has been carried out and the results has been noted in Fig. 17.1. Table 17.3 shows the objective function value for each equation, and the corresponding best value. These values are obtained by taking the minimum of corresponding best values, for each roulette wheel and tournament selection function, for each equation which has been shown in Fig. 17.1.

17.6 Conclusion

This paper presents a novel approach to get an approximate solution for system of nonlinear equations using soft computing technique called Genetic Algorithm. GA optimizes the time complexities for solving such nonlinear system. The solution strategy involves an initial assumption of various parameters of GA, which are population size, crossover, and mutation functions, along with number of generations. The parameter selection helps in improving the efficiency of the algorithm. The computational cost can be reduced by treating the problem as multi-objective optimization problem and hence, applying the heuristic search technique. In addition to that, it is proposed to use the variants of GA like Diploid Genetic Algorithm for solving these problems [21, 22]. In the work being continued, the crossover rate and mutation rate has been varied to see the effects of these rates on the quality of the solution obtained. Also, as a part of future scope, the work can be taken further for handling higher dimension and complex nonlinear systems.

References

1. Abd-El-Wahed, W.F., Mousa, A.A., El-Shorbagy, M.A.: Integrating particle swarm optimization with genetic algorithms for solving nonlinear optimization problems. J. Comput. Appl. Math. **235**, 1446–1453 (2011)
2. Bianchini, M., Fanelli, S.: Optimal algorithms for well-conditioned nonlinear systems of equations. IEEE Trans. Comput. **50**(7), 689–698 (2001)
3. Chang, W.D.: An improved real-coded genetic algorithm for parameters estimation of nonlinear systems. Mech. Syst. Signal Process. **20**, 236–246 (2006)
4. Effati, S., Nazemi, A.R.: A new method for solving a system of the nonlinear equations. Appl. Math. Comput. **168**, 877–894 (2005)
5. Goldberg, D.E.: Genetic Algorithms in Search, Optimization and Machine Learning. Addison-Wesley, Reading (1989)
6. Grosan, C., Abraham, A.: A new approach for solving nonlinear equations systems. IEEE Trans. Syst. Man Cybern.-Part A: Syst. Humans **38**(3), 698–714 (2008)
7. Guessan, A.N.: Analytical existence of solutions to a system of nonlinear equations with application. J. Comput. Appl. Math. **234**, 297–304 (2010)
8. Holland, J.: An Introductory Analysis with Applications to Biology, Control, and Artificial Intelligence. Adaptation in Natural and Artificial Systems. MIT Press, Cambridge (1975)
9. Ji, Z., Li, Z., Ji, Z.: Research on genetic algorithm and data information based on combined framework for nonlinear functions optimization. Proc. Eng. **23**, 155–160 (2011)
10. Joshi, G., Krishna, M.B.: Solving system of non-linear equations using genetic algorithm. In: International Conference on Advances in Computing, Communications and Informatics (ICACCI) (2014)
11. Konaka, A., Coitb, D.W., Smith, A.E.: Multi-objective optimization using genetic algorithms: a tutorial. Reliab. Eng. Syst. Saf. **91**, 992–1007 (2006)
12. Mastorakis, N.E.: Solving non-linear equations via genetic algorithms. In: Proceedings of the 6th WSEAS International Conference on Evolutionary Computing, pp. 24–28. Lisbon, Portugal (2005)
13. McCall, J.: Genetic algorithms for modelling and optimization. J. Comput. Appl. Math. **184**, 205–222 (2005)
14. Mousa, A.A., El-Desoky, I.M.: GENLS: Co-evolutionary algorithm for nonlinear system of equations. Appl. Math. Comput. **197**, 633–642 (2008)
15. Nie, P.: An SQP approach with line search for a system of nonlinear equations. Math. Comput. Modell. **43**, 368–373 (2006)
16. Pourrajabian, A., Ebrahimi, R., Mirzaei, M., Shams, M.: Applying genetic algorithms for solving nonlinear algebraic equations. Appl. Math. Comput. **219**, 11483–11494 (2013)
17. Raja, M.A.Z., Sabir, Z., Mehmood, N., Al-Aidarous, E.S., Khan, J.A.: Design of stochastic solvers based on genetic algorithms for solving nonlinear equations. Neural Comput. Appl. **26**, 1–23 (2015)
18. Ren, H., Wua, L., Bi, W., Argyros, I.K.: Solving nonlinear equations system via an efficient genetic algorithm with symmetric and harmonious individuals. Appl. Math. Comput. **219**, 10967–10973 (2013)
19. Rovira, A., Valdés, M., Casanova, J.: A new methodology to solve non-linear equation systems using genetic algorithms. Application to combined cycle gas turbine simulation. Int. J. Numer. Meth. Eng. **63**, 1424–1435 (2005)
20. Zhang, X., Wu, Z.: Study neighborhood field optimization algorithm on nonlinear sorptive barrier design problems. Neural Comput. Appl. (2015)
21. Bhasin, H., Mehta, S.: On the applicability of diploid genetic algorithms. AI Soc. **31**(2), 265–274 (2015)
22. Bhasin, H., Behal G., Aggarwal, N., Saini, R.K., Choudhary, S.: On the applicability of diploid genetic algorithms in dynamic environments. Soft Comput. **20**(9), 3403–3410 (2015)

Chapter 18
An M/M/c/N Feedback Queuing Model with Reverse Balking and Reneging

Rakesh Kumar and Bhupender Kumar Som

Abstract In this paper, a finite capacity Markovian multi-server feedback queuing system with reverse balking and reneging is studied. The steady-state solution of the model is derived recursively. Some important measures of the performance like expected system size, expected rate of reneging, and expected rate of reverse balking are derived. The sensitivity analysis of the model is performed. Some useful comparisons are performed. Finally special cases of the model are discussed.

Keywords Reverse balking · Reneging · Customers' impatience
Sensitivity analysis · Queuing models

18.1 Introduction

Business environment is highly challenging these days due to uncertainty. Uncertainty appears in all dimensions of an operating firm, for example, uncertain economic environment, uncertain natural calamities and uncertain customer behavior. Hence the margin of error is very low for business organizations. Every firm is looking for risk management and precise prediction of future. Customer behavior is one of the most uncertain characteristics of business environment. Customers have become more selective. Brand switching is more frequent. Due to higher level of expectations customers get more impatient with a particular firm. Thus customers' impatience has become a burning problem for corporate world. A customer is said to be impatient if he tends to join the queue only when a short wait is expected and tends to remain in the line if his wait has been sufficiently small.

R. Kumar (✉)
School of Mathematics, Shri Mata Vaishno Devi University, Katra 182320, Jammu and Kashmir, India
e-mail: rakesh_stat_kuk@yahoo.co.in

B.K. Som
JIMS, 3, Institutional Area, Sec-5 Rohini, New Delhi 110085, India
e-mail: bksoam@live.com

P. Manchanda et al. (eds.), *Industrial Mathematics and Complex Systems*, Industrial and Applied Mathematics, DOI 10.1007/978-981-10-3758-0_18

Impatience is of three forms. The first is balking, deciding not to join the queue at all up on arrival; the second is reneging, the reluctance to remain in the waiting line after joining and waiting, and the third is jockeying between lines when each of a number of parallel service channels has its own queue, Gross and Harris [1]. Wang et al. [5] have presented a nice review on queuing systems with impatient customers. They have surveyed queuing systems according to various dimensions like customer impatience behaviors, solution methods of queuing models with impatient customers, and associated optimization aspects.

In the case of balking as described in aforementioned paragraphs, the arriving customer balks with more probability if there is large number of customers in the queuing system and vice-versa. But when we talk about the businesses like investment, there are more chances for customers to invest with the firms having large number of customers associated with them. Thus the probability of joining of customers in such firms is high that is, the probability of balking will be low.

If we view the investment firm in terms of a queuing system, the probability of balking (not joining the firm) will be less when the system size (the number of customers with the firm) is large and vice-versa. This kind of balking is referred to as Reverse Balking.

Recently, Jain et al. [2] incorporate the concept of reverse balking in queuing theory. Queues with reverse balking find their applications in investment business, restaurants, hospitals, schools, business of quality products, etc. Kumar et al. [3] extend the work of Jain et al. [2] by studying a single server queue with reverse balking and reverse reneging. Kumar et al. [4] study a single server Markovian queue with reverse balking and feedback customers. They derive the steady-state solution of the model and present some important performance measures. They study this model with reference to its applications in insurance business.

In this paper we generalize the work of Kumar et al. [4] by considering the multi-server case along with reneging. We study an M/M/c/N feedback queuing system with reverse balking and reneging. We perform the steady-state analysis of the model.

Rest of the paper is structured as follows: in Sect. 18.2, the queuing model is described; in Sect. 18.3, the mathematical model of the queuing system is presented; the steady-state solution of the model is obtained in Sect. 18.4; Sect. 18.5 deals with measures of performance; the sensitivity analysis and comparisons are provided in Sect. 18.6; in Sect. 18.7, special cases of the model are discussed. Finally, the paper is concluded in Sect. 18.8.

18.2 Model Description

Consider a multi-server queuing system in which arrivals to the queuing system occur one by one in accordance to a Poisson process with mean rate λ. The inter-arrival times are independently, identically, and exponentially distributed with parameter λ. There is a finite number of servers, (say, c) and the service times at each server are independently, identically and exponentially distributed with parameter μ such that mean service rate $\mu_n = n\mu$ for $n < c$ and $\mu_n = c\mu$ for $n \geq c$. The capacity of the system is finite, say N. Customers are served in order of their arrival, that is, the queue discipline is First-Come, First-Served. When the system is empty customers balk with probability q' and may not balk with probability $p'(= 1 - q')$. When there is at least one customer in the system, the customers balk with a probability $1 - \frac{n}{N-1}$ and join the system with probability $\frac{n}{N-1}$. Such kind of balking is referred to as reverse balking. The customers may get impatient due to certain reasons and decide to leave the queue before receiving service, that is, the customers wait up to certain time (T, say) and may leave the system before getting service due to impatience (reneging). The reneging times, T are independently, identically, and exponentially distributed with parameter ξ. Further, a customer may not be satisfied with the service and may decide to rejoin the queue with a probability $q\ (= 1 - p)$ as a feedback customer.

18.3 Mathematical Model

In this section the mathematical model of the queuing system is presented. Let $P_n(t)$ be the probability that there are n customers in the system at time t. The differential-difference equations governing the model are

$$\frac{d}{dt}P_0(t) = -\lambda p' P_0(t) + \mu p P_1(t), \quad n = 0 \tag{18.1}$$

$$\frac{d}{dt}P_1(t) = \lambda p' P_0(t) - \left[\left(\frac{1}{N-1}\right)\lambda + \mu p\right] P_1(t) + (2\mu p)P_2(t), \quad n = 1 \tag{18.2}$$

$$\begin{aligned}\frac{d}{dt}P_n(t) = {} & \left(\frac{n-1}{N-1}\right)\lambda P_{n-1}(t) - \left[\left(\frac{n}{N-1}\right)\lambda + n\mu p\right] P_n(t) + \\ & \{(n+1)\mu p\}P_{n+1}(t), \quad 2 \leq n < c \leq N-1\end{aligned} \tag{18.3}$$

$$\begin{aligned}\frac{d}{dt}P_n(t) = {} & \left(\frac{n-1}{N-1}\right)\lambda P_{n-1}(t) - \left[\left(\frac{n}{N-1}\right)\lambda + c\mu p + (n-c)\xi\}\right] P_n(t) + \\ & [c\mu p + \{(n+1) - c\}\xi]P_{n+1}(t), \quad c \leq n \leq N-1\end{aligned} \tag{18.4}$$

$$\frac{d}{dt}P_N(t) = \lambda P_{N-1}(t) - [c\mu p + (N-c)\xi]P_N(t), \quad n = N \tag{18.5}$$

18.4 Steady-State Solution

In this section, we present the steady-state equations of the queuing model and solve them recursively in order to obtain the steady-state probabilities of system size.

In steady-state as $\lim_{t\to\infty} P_n(t) = P_n$. Therefore, $\lim_{t\to\infty} \frac{dP_n(t)}{dt} = 0$. Hence, the Eqs. (18.1)–(18.5) reduce to the following:

$$0 = -\lambda p' P_0 + \mu p P_1, \quad n = 0 \tag{18.6}$$

$$0 = \lambda p' P_0 - \left[\left(\frac{1}{N-1}\right)\lambda + \mu p\right] P_1 + (2\mu p) P_2, \quad n = 1 \tag{18.7}$$

$$0 = \left(\frac{n-1}{N-1}\right)\lambda P_{n-1} - \left[\left(\frac{n}{N-1}\right)\lambda + n\mu p\right] P_n + \{(n+1)\mu p\} P_{n+1}, \quad 2 \le n < c \le N-1 \tag{18.8}$$

$$0 = \left(\frac{n-1}{N-1}\right)\lambda P_{n-1} - \left[\left(\frac{n}{N-1}\right)\lambda + c\mu p + (n-c)\xi\}\right] P_n + [c\mu p + \{(n+1) - c\}\xi] P_{n+1}, \quad c \le n \le N-1 \tag{18.9}$$

$$0 = \lambda P_{N-1} - [c\mu p + (N-c)\xi] P_N, \quad n = N \tag{18.10}$$

Solving the Eqs. (18.6)–(18.10) recursively, we get

$$P_n = \begin{cases} \left[\frac{(n-1)!}{(N-1)^{n-1}} \prod_{r=1}^{n} \frac{\lambda}{r\mu p}\right] p' P_0, & n < c \le N-1 \\ \left[\frac{(n-1)!}{(N-1)^{n-1}} \prod_{s=c}^{n} \frac{\lambda}{c\mu p + (s-c)\xi} \prod_{r=1}^{c-1} \frac{\lambda}{r\mu p}\right] p' P_0, & c \le n \le N-1 \\ \left[\frac{(N-2)!}{(N-1)^{N-2}} \prod_{s=c}^{N} \frac{\lambda}{c\mu p + (s-c)\xi} \prod_{r=1}^{c-1} \frac{\lambda}{r\mu p}\right] p' P_0, & n = N \end{cases}$$

Using condition of normality $\sum_{n=0}^{N} P_n = 1$, we obtain

$$P_0 = \frac{1}{1 + Q_1 + Q_2 + P_N} \tag{18.11}$$

where

$$Q_1 = \left[\frac{(n-1)!}{(N-1)^{n-1}} \prod_{r=1}^{n} \frac{\lambda}{r\mu p}\right] p',$$

$$Q_2 = \left[\frac{(n-1)!}{(N-1)^{n-1}} \prod_{s=c}^{n} \frac{\lambda}{c\mu p + (s-c)\xi} \prod_{r=1}^{c-1} \frac{\lambda}{r\mu p} \right] p'$$

and

$$P_N = \left[\frac{(N-2)!}{(N-1)^{N-2}} \prod_{s=c}^{N} \frac{\lambda}{c\mu p + (s-c)\xi} \prod_{r=1}^{c-1} \frac{\lambda}{r\mu p} \right] p'.$$

Thus, we have obtained explicitly the steady-state system size probabilities.

18.5 Measures of Performance

In this section, we derive some important measures of performance. Once the steady-state probabilities are obtained it is easy to derive certain measures of performance as follows.

18.5.1 Expected System Size (L_s)

$$\begin{aligned} L_s &= \sum_{n=0}^{N} n P_n \\ &= \sum_{n=0}^{c-1} n P_n + \sum_{n=c}^{N-1} n P_n + N P_N \end{aligned}$$

or

$$\begin{aligned} L_s = &\sum_{n=0}^{c-1} \left[n \frac{(n-1)!}{(N-1)^{n-1}} \prod_{r=1}^{n} \frac{\lambda}{r\mu p} \right] p' P_0 \\ &+ \sum_{n=c}^{N-1} \left[n \frac{(n-1)!}{(N-1)^{n-1}} \prod_{s=c}^{n} \frac{\lambda}{c\mu p + (s-c)\xi} \prod_{r=1}^{c-1} \frac{\lambda}{r\mu p} \right] p' P_0 \\ &+ \left[N \frac{(N-2)!}{(N-1)^{N-2}} \prod_{s=c}^{N} \frac{\lambda}{c\mu p + (s-c)\xi} \prod_{r=1}^{c-1} \frac{\lambda}{r\mu p} \right] p' P_0 \end{aligned}$$

18.5.2 Expected Rate of Reneging (R_r)

$$R_r = \sum_{n=c}^{N} (n-c)\xi P_n$$

or

$$R_r = \sum_{n=c}^{N-1} (n-c)\xi \left[\frac{(n-1)!}{(N-1)^{n-1}} \prod_{s=c}^{n} \frac{\lambda}{c\mu p + (s-c)\xi} \right] p' P_0 +$$
$$(N-c)\xi \left[\frac{(N-2)!}{(N-1)^{N-2}} \prod_{s=c}^{N} \frac{\lambda}{c\mu p + (s-c)\xi} \prod_{r=1}^{c-1} \frac{\lambda}{r\mu p} \right] p' P_0$$

18.5.3 Expected Rate of Reverse Balking (R'_b)

$$R'_b = q'\lambda P_0 + \sum_{n=1}^{N-1} \left(1 - \frac{n}{N-1}\right) \lambda P_n$$

or

$$R'_b = q'\lambda P_0 + \sum_{n=1}^{c-1} \left(1 - \frac{n}{N-1}\right) \lambda \left[\frac{(n-1)!}{(N-1)^{n-1}} \prod_{r=1}^{n} \frac{\lambda}{r\mu p} \right] p' P_0$$
$$+ \sum_{n=c}^{N-1} \left(1 - \frac{n}{N-1}\right) \lambda \left[\frac{(n-1)!}{(N-1)^{n-1}} \prod_{s=c}^{n} \frac{\lambda}{c\mu p + (s-c)\xi} \prod_{r=1}^{c-1} \frac{\lambda}{r\mu p} \right] p' P_0$$

18.6 Sensitivity Analysis

In this section, we perform sensitivity analysis of the model. We study effect of various parameters on measures of performance. We also compare this model with the one studied by Kumar et al. [4]. From Table 18.1, we can observe that as the mean arrival rate increases the expected system size, the expected rate of reneging and expected rate of reverse balking increases.

The variation in performance measures with respect to parameter μ is presented in Table 18.2. All the performance measures show a decreasing trend as μ (service rate) increases.

Table 18.1 Variation in L_s, R_r and R'_b w.r.t. λ. For $\mu = 3$, $\xi = 0.4$, $N = 10$, $q' = 0.1$, $c = 3$ and $q = 0.2$

λ	L_s	R'_b	R_r
2	0.4626	0.8900	0.0001
2.5	0.5313	1.2254	0.0001
3	0.5914	1.5820	0.0003
3.5	0.6450	1.9542	0.0005
4	0.6941	2.3381	0.0008
4.5	0.7398	2.7308	0.0013
5	0.7833	3.1299	0.0020
5.5	0.8255	3.5335	0.0030
6	0.8673	3.9400	0.0044
6.5	0.9097	4.3479	0.0064
7	0.9535	4.7555	0.0091
7.5	1.0000	5.1613	0.0129
8	1.0506	5.5631	0.0180
8.5	1.1068	5.9586	0.0249
9	1.1705	6.3450	0.0342
9.5	1.2443	6.7185	0.0466
10	1.3308	7.0749	0.0630

Table 18.2 Variation in L_s, R_r and R'_b w.r.t. μ. For $\lambda = 3$, $\xi = 0.4$, $N = 10$, $q' = 0.1$, $c = 3$, and $q = 0.2$

μ	L_s	R'_b	R_r
2	0.7393	1.8206	0.0012
2.5	0.6551	1.6918	0.0005
3	0.5914	1.5820	0.0003
3.5	0.5404	1.4878	0.0001
4	0.4983	1.4062	0.0001
4.5	0.4626	1.3349	0.0001
5	0.4320	1.2723	0.0000
5.5	0.4054	1.2167	0.0000
6	0.3819	1.1671	0.0000
6.5	0.3611	1.1226	0.0000
7	0.3425	1.0824	0.0000
7.5	0.3257	1.0459	0.0000
8	0.3105	1.0127	0.0000
8.5	0.2967	0.9823	0.0000
9	0.2841	0.9544	0.0000
9.5	0.2726	0.9286	0.0000
10	0.2619	0.9049	0.0000

Table 18.3 Variation in L_s, R_r and R'_b w.r.t. ξ. For $\lambda = 4$, $\mu = 3$, $N = 10$, $q' = 0.1$, $c = 3$, and $q = 0.2$

ξ	L_s	R'_b	R_r
0.1	0.6946	2.3380	0.0002
0.2	0.6944	2.3381	0.0004
0.3	0.6942	2.3381	0.0006
0.4	0.6941	2.3381	0.0008
0.5	0.6939	2.3381	0.0010
0.6	0.6938	2.3382	0.0011
0.7	0.6937	2.3382	0.0013
0.8	0.6936	2.3382	0.0014
0.9	0.6935	2.3382	0.0015
1	0.6934	2.3382	0.0016

Table 18.4 Variation in L_s, R_r and R'_b w.r.t. q' for $\lambda = 4$, $\mu = 3$, $N = 10$, $\xi = 0.4$, $c = 3$, and $q = 0.2$

q'	L_s	R'_b	R_r
0.1	0.6941	2.3381	0.0008
0.2	0.6629	2.4127	0.0008
0.3	0.6268	2.4993	0.0007
0.4	0.5843	2.6011	0.0007
0.5	0.5336	2.7224	0.0006
0.6	0.4722	2.8694	0.0005
0.7	0.3962	3.0514	0.0004
0.8	0.2997	3.2824	0.0003
0.9	0.1732	3.5853	0.0002
1	0.0000	4.0000	0.0000

Table 18.3 shows the variation in performance measures with respect to reneging rate, ξ. We can see that there is increase in expected rate of reneging as well as expected rate of reverse balking but the expected system size decreases with increase in ξ.

The variation in performance measures with respect to the probability of balking, q' (when the system is empty) is presented in Table 18.4. As the balking probability increases the expected system size and the expected rate of reneging decreases but the expected rate of reverse balking increases rapidly with increase in q'. This justifies the functioning of the present model.

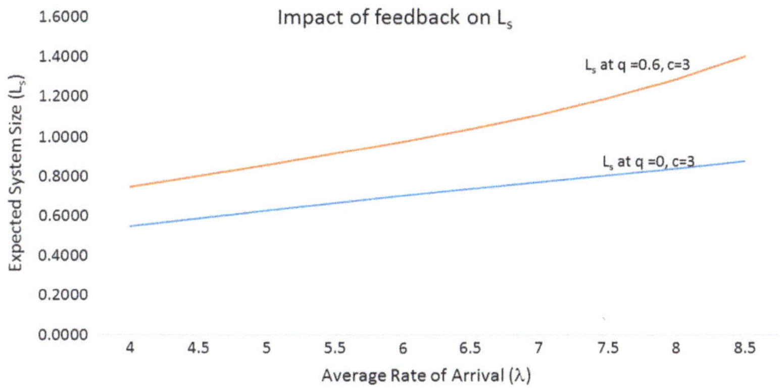

Fig. 18.1 Effect of feedback on expected system size

In Fig. 18.1 the effect of feedback on expected system size is shown. We take $\mu = 3, N = 10, \xi = 0.6$ and $q' = 0.3$. The feedback customers increase the expected system size. We can see from the figure that when the feedback probability is 0.6 the expected system size is relatively higher in comparison to the case when there is no feedback.

18.6.1 Comparison with the Model Studied by Kumar et al. [4]

In this subsection, we perform certain comparisons. When $c = 1$ and $\xi = 0$ the model studied in this paper reduces to the one studied by Kumar et al. [4].

Figure 18.2 shows the effect of reneging on expected system size in multiple server case ($c = 3$). It can be observed that the expected system size is always lower in case of reneging than the case of without reneging.

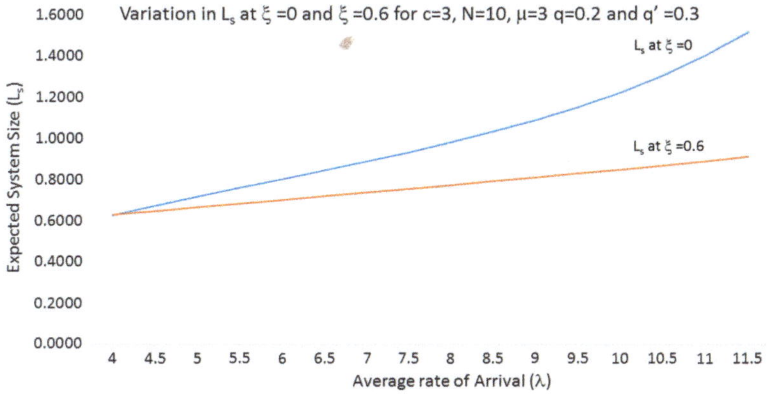

Fig. 18.2 Effect of reneging on expected system size

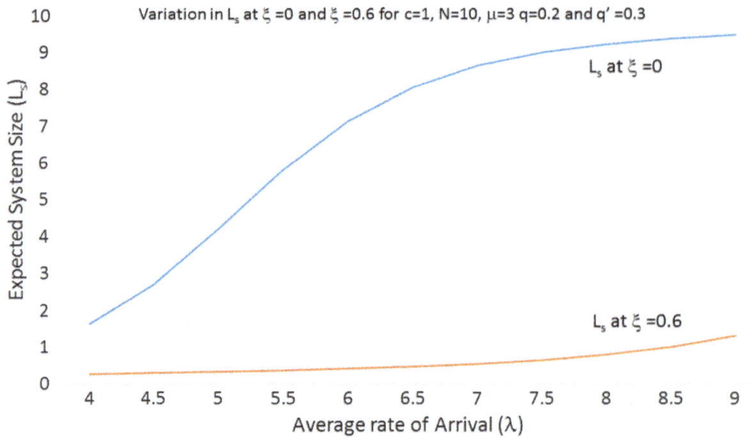

Fig. 18.3 Comparison of the model with the model at [4]

In Fig. 18.3 we consider the single server case with feedback and study the variation in expected system size with and without reneging. In case of reneging the system size is significantly lower than the case when there is no reneging. This comparative scenario is useful to the people dealing with implementation of various queuing systems.

18.7 Special Cases

In this section, the special cases of the model are discussed.

Case 1: When there is a single server and no reneging

The model reduces to a single server feedback queuing model with reverse balking as studied by Kumar et al. [4].

Case 2: When there is a single server, no reneging and no feedback

In this case, the queuing model resembles with the one studied by Jain et al. [2].

18.8 Conclusions and Future Work

An $M/M/c/N$ queuing system with reverse balking, reneging, and feedback is studied. The steady-state probabilities are obtained. Sensitivity analysis of the model is also performed. Analysis of numerical results establishes the role of reverse balking in this queuing system.

The transient analysis of the model can also be carried out. The same model can be studied in non-Markovian environment. The model can also be extended to study the effect of heterogeneous servers.

Acknowledgements The authors are very much thankful to the anonymous referees for their constructive comments which helped to bring this paper in the present form.

References

1. Gross, D., Harris, C.M.: Fundamentals of Queuing Theory. Wiley, New York (1985)
2. Jain, N.K., Kumar, R., Som, B.K.: An M/M/1/N queuing system with reverse balking. Am. J. Operat. Res. **4**(2), 17–20 (2014)
3. Kumar, R., Som, B.K.: An M/M/1/N queuing system with reverse balking and reverse reneging. AMO - Adv. Model. Optim. **16**(2), 339–353 (2014)
4. Kumar, R., Som, B.K., Jain, S.: An M/M/1/N feedback queuing system with reverse balking. J. Reliab. Stat. Stud. **8**(1), 31–38 (2015)
5. Wang, K., Li, N., Jiang, Z.: Queuing system with impatient customers: a review. In: Proceedings of 2010 IEEE International Conference on Service Operations and Logistics and Informatics, pp. 82–87 (2010)

Chapter 19
Solution of Fuzzy Heat Equation Under Fuzzified Thermal Diffusivity

U.M. Pirzada and D.C. Vakaskar

Abstract This paper presents a solution for a fuzzy partial differential equation with fuzzy boundary and initial conditions. The solution of fuzzy heat equation is proposed using Seikkala differentiability of a fuzzy-valued function. The effect of fuzzified thermal diffusivity is studied.

Keywords Fuzzy numbers · Heat equation · Thermal diffusivity

19.1 Introduction

Seldom it is observed that for many physical systems, involving incomplete and imprecise description which is reflected in their mathematical model. It well known that fuzzy theory is one of the most powerful tool to study and analysis problems involving uncertainty, impreciseness, ambiguity. This motivates us to study these systems as fuzzy systems. The uncertain dynamical systems often lead to uncertain (fuzzy) partial differential equations. Fuzzy partial differential equations (FPDEs) are the generalization of partial differential equations (PDEs) in fuzzy sense. Modeling of real situation in terms of partial differential equations involves uncertain variables and parameters (known partially or approximately). This impreciseness or uncertainties can be described mathematically using fuzzy numbers. For example, in case of heat equation, temperature variable can be treated as a fuzzy variable as it is defined by linguistic states like cool, cold, normal, hot, etc. The diffusivity coefficient can be regarded as fuzzy because it may not be precisely available.

U.M. Pirzada (✉)
School of Science and Engineering, Navrachana University of Vadodara, Vadodara 391410, India
e-mail: salmapirzada@yahoo.com

D.C. Vakaskar
Department of Applied Mathematics, Faculty of Technology and Engineering, M.S. University of Baroda, Vadodara 390001, India
e-mail: dcvakaskar@gmail.com

P. Manchanda et al. (eds.), *Industrial Mathematics and Complex Systems*, Industrial and Applied Mathematics, DOI 10.1007/978-981-10-3758-0_19

Buckley and Feuring [7] examined solutions of elementary fuzzy partial differential equations in [7]. They checked the Buckley–Feuring (BF) solution exist or not. If the BF-solution does not exist Seikkala solution is found. The solution is based on the Seikkala derivative defined in [15].

In [1], Allahviranloo (2002) proposed a difference method to solve FPDEs. This method was based on Seikkala derivative of fuzzy functions. The Adomian method was studied to find the approximate solution of fuzzy heat equation in [2] (2009) based on Hukuhara derivatives. While in [3], Allahviranloo and Afshar (2010) presented numerical methods for solving the fuzzy partial differential equations. These numerical methods were based on the derivative due to Bede and Gal [6]. Mahmoud and Iman [13] (2013) presented finite volume method that solves some FPDEs such as fuzzy hyperbolic equations, fuzzy parabolic equations, and fuzzy elliptic equations. They have obtained explicit, implicit, and Crank–Nicolson schemes for solving fuzzy heat equation. Study of heat, wave, and Poisson equations with uncertain parameters are given in [5] (2013). Recently, Allahviranloo et al. have studied fuzzy solutions for fuzzy heat equation with fuzzy initial value based on generalized Hukuhara differentiability in [4] (2014). Applications to FPDEs are presented with a new inference method in [8](2009). B.A. Faybishenko [12] (2012) presented a hydrogeologic system as a fuzzy system. He derived a fuzzy logic form of parabolic-type partial differential equation and solved the basic principles of fuzzy arithmetic. Pirzada and Vakaskar have studied solution of fuzzy partial differential equations using Adomian decomposition method in [14] (2015). They have used Seikkala differentiability to find the solution.

Thermal diffusivity computes how fast a body can change its temperature. For certain composite material, crisp estimate of thermal diffusivity is difficult to obtain. Some sort of perceptional uncertainty is there in the measures of thermal diffusivity. Hence, in such a situation it may be advisable to be used as a fuzzy diffusivity. The simultaneous effect of varying thermal diffusivity can be studied by defining it as a fuzzy number. With this motivation, we study the solution of fuzzy heat equation under the effect of fuzzy thermal diffusivity constant. To the best of our knowledge, this work is not explored by any previous researcher. The paper is organized as follows:

The concepts of fuzzy numbers and fuzzy-valued function are given in Sect. 19.2. Fuzzy modeling of heat equation and solution are explained in Sect. 19.3. Illustration is given in Sect. 19.4. The conclusions are presented in Sect. 19.5.

19.2 Preliminaries

In this section, we state some basic concepts regarding fuzzy numbers and fuzzy-valued functions. The following definition of fuzzy number is stated from [14].

Definition 19.1 ([14]) Let $\mathbb{R}$ be the set of real numbers and $\tilde{a} : \mathbb{R} \to [0, 1]$ be a fuzzy set. We say that $\tilde{a}$ is a fuzzy number if it satisfies the following properties:

(i) $\tilde{a}$ is normal, that is, there exists $x_0 \in \mathbb{R}$ such that $\tilde{a}(x_0) = 1$;
(ii) $\tilde{a}$ is fuzzy convex, that is, $\tilde{a}(tx + (1-t)y) \geq min\{\tilde{a}(x), \tilde{a}(y)\}$, whenever x, y $\in \mathbb{R}$ *and* $t \in [0, 1]$;
(iii) $\tilde{a}(x)$ is upper semi-continuous on $\mathbb{R}$, that is, $\{x/\tilde{a}(x) \geq \alpha\}$ is a closed subset of $\mathbb{R}$ for each $\alpha \in (0, 1]$;
(iv) $cl\{x \in \mathbb{R}/\tilde{a}(x) > 0\}$ forms a compact set,

where cl denotes closure of a set. The set of all fuzzy numbers on $\mathbb{R}$ is denoted by $F(\mathbb{R})$.

The α-level set for a fuzzy number is defined as follows:

Definition 19.2 For $\alpha \in (0, 1]$, α-level set $\tilde{a}_\alpha$ of any $\tilde{a} \in F(\mathbb{R})$ is defined as

$$\tilde{a}_\alpha = \{x \in \mathbb{R}/\tilde{a}(x) \geq \alpha\}.$$

The 0-level set $\tilde{a}_0$ is defined as the closure of the set $\{x \in \mathbb{R}/\tilde{a}(x) > 0\}$.

Remark 19.1 (i) We can easily see that, for any $\tilde{a} \in F(\mathbb{R})$ and for each $\alpha \in (0, 1]$, $\tilde{a}_\alpha$ is compact convex subset of $\mathbb{R}$, and $\tilde{a}_\alpha = [a_1(\alpha), a_2(\alpha)]$.
(ii) The fuzzy number $\tilde{a} \in F(\mathbb{R})$ can be generated from its α-level sets by a well-known decomposition theorem (Ref. [10]).

The following theorem of Goetschel and Voxman [11], shows the characterization of a fuzzy number in terms of its α-level sets.

Theorem 19.1 *For* $\tilde{a} \in F(\mathbb{R})$, *define two functions* $\tilde{a}_1(\alpha), \tilde{a}_2(\alpha) : [0, 1] \to \mathbb{R}$. *Then*

(i) $\tilde{a}_1(\alpha)$ *is bounded left continuous non-decreasing function on (0,1];*
(ii) $\tilde{a}_2(\alpha)$ *is bounded left continuous non-increasing function on (0,1];*
(iii) $\tilde{a}_1(\alpha)$ *and* $\tilde{a}_2(\alpha)$ *are right continuous at* $\alpha = 0$*;*
(iv) $\tilde{a}_1(\alpha) \leq \tilde{a}_2(\alpha)$.

Moreover, if the pair of functions $\tilde{a}_1(\alpha)$ *and* $\tilde{a}_2(\alpha)$ *satisfy the conditions (i)-(iv), for each* $\alpha \in [0, 1]$, *then there exists a unique* $\tilde{a} \in F(\mathbb{R})$ *such that* $\tilde{a}_\alpha = [\tilde{a}_1(\alpha), \tilde{a}_2(\alpha)]$, *for* $\alpha \in [0, 1]$.

Definition 19.3 The membership function of a triangular fuzzy number $\tilde{a}$ is defined as

$$\tilde{a}(r) = \begin{cases} \frac{(r-a^L)}{(a-a^L)} & if \ a^L \leq r \leq a \\ \frac{(a^U-r)}{(a^U-a)} & if \ a < r \leq a^U \\ 0 & otherwise \end{cases}$$

which is denoted by $\tilde{a} = (a^L, a, a^U)$. The α-level set of $\tilde{a}$ is then

$$\tilde{a}_\alpha = [(1-\alpha)a^L + \alpha a, (1-\alpha)a^U + \alpha a].$$

Definition 19.4 Using Zadeh's extension principle, addition, multiplication of two fuzzy numbers $\tilde{a}, \tilde{b}$ and scalar multiplication of fuzzy number $\tilde{a}$ with a scalar $\lambda \in \mathbb{R}$ by their α-level sets are defined as follows:

$$
\begin{aligned}
(\tilde{a} \oplus \tilde{b})_\alpha &= [\tilde{a}_1(\alpha) + \tilde{b}_1(\alpha), \tilde{a}_2(\alpha) + \tilde{b}_2(\alpha)] \\
(\tilde{a} \otimes \tilde{b})_\alpha &= [\min\left\{\tilde{a}_1(\alpha)\tilde{b}_1(\alpha), \tilde{a}_1(\alpha)\tilde{b}_2(\alpha), \tilde{a}_2(\alpha)\tilde{b}_1(\alpha), \tilde{a}_2(\alpha)\tilde{b}_2(\alpha)\right\}, \\
&\qquad \max\left\{\tilde{a}_1(\alpha)\tilde{b}_1(\alpha), \tilde{a}_1(\alpha)\tilde{b}_2(\alpha), \tilde{a}_2(\alpha)\tilde{b}_1(\alpha), \tilde{a}_2(\alpha)\tilde{b}_2(\alpha)\right\}] \\
(\lambda \odot \tilde{a})_\alpha &= [\lambda \cdot \tilde{a}_1(\alpha), \lambda \cdot \tilde{a}_2(\alpha)],\ if\ \lambda \geq 0 \\
&= [\lambda \cdot \tilde{a}_2(\alpha), \lambda \cdot \tilde{a}_1(\alpha)],\ if\ \lambda < 0,
\end{aligned}
$$

where α-level sets of $\tilde{a}$ and $\tilde{b}$ are $\tilde{a}_\alpha = [\tilde{a}_1(\alpha), \tilde{a}_2(\alpha)]$, $\tilde{b}_\alpha = [\tilde{b}_1(\alpha), \tilde{b}_2(\alpha)]$, for $\alpha \in [0, 1]$.

Definition 19.5 Let V be a real vector space and $F(\mathbb{R})$ be a set of fuzzy numbers. Then a fuzzy-valued function $\tilde{f} : V \to F(\mathbb{R})$ is defined on V. Corresponding to such a function $\tilde{f}$ and for each $\alpha \in [0, 1]$, we denote two real-valued functions $\tilde{f}_1(x, \alpha)$ and $\tilde{f}_2(x, \alpha)$ on V for all $x \in V$. These functions $\tilde{f}_1(x, \alpha)$ and $\tilde{f}_2(x, \alpha)$ are called α-level functions of the fuzzy-valued function $\tilde{f}$.

Seikkala differentiability of fuzzy-valued function $\tilde{U}(x, t)$ is defined as follows.

Definition 19.6 ([7]) Let I_1, I_2 be subsets of $\mathbb{R}$. Let $\tilde{U}$ be a fuzzy-valued function defined on $I_1 \times I_2$, I_1, I_2 are intervals. Let α-level sets $\tilde{U}_\alpha(x, t) = [u_1(x, t, \alpha), u_2(x, t, \alpha)]$ for all $\alpha \in [0, 1]$. We assume that $u_i(x, t, \alpha)$ have continuous partial derivatives, for all $(x, t) \in I_1 \times I_2$, for each α, $i = 1, 2$. Define

$$
\left[\frac{\partial \tilde{U}}{\partial t}(x, t)\right]_\alpha = \left[\frac{\partial u_1}{\partial t}(x, t, \alpha), \frac{\partial u_2}{\partial t}(x, t, \alpha)\right]
$$

for all $(x, t) \in I_1 \times I_2$, all α. If, for each fixed $(x, t) \in I_1 \times I_2$, $\left[\frac{\partial U}{\partial t}(x, t)\right]_\alpha$ defines the α-level set of a fuzzy number, then we say that $\tilde{U}(x, t)$ is partially differentiable with respect to t. Similarly, we can define partial differentiability of $\tilde{U}$ with respect to x. Also, we can define higher order partial derivatives in same manner.

The sufficient conditions for $[\frac{\partial U}{\partial t}(x, t)]_\alpha$ to define α-level sets of a fuzzy number are

(i) $\frac{\partial u_1}{\partial t}(x, t, \alpha)$ is an increasing function of α for each $(x, t) \in I_1 \times I_2$;
(ii) $\frac{\partial u_2}{\partial t}(x, t, \alpha)$ is a decreasing function of α for each $(x, t) \in I_1 \times I_2$; and
(iii) $\frac{\partial u_1}{\partial t}(x, t, 1) \leq \frac{\partial u_2}{\partial t}(x, t, 1)$ for all $(x, t) \in I_1 \times I_2$.

19.3 Fuzzy Heat Equation

19.3.1 Fuzzy Model

In a thin uniform metal rod with nonuniform temperature, heat (thermal energy) is flowed from regions of higher temperature to regions of lower temperature. This temperature distribution is mathematically modeled as an one-dimensional heat equation

$$\frac{\partial u}{\partial t} = P \frac{\partial^2 u}{\partial x^2}, \tag{19.1}$$

where P is the thermal diffusivity and $u(x, t)$ is the temperature. The term temperature is defined as a fuzzy variable with different linguistic states, like cold, normal, hot etc. To capture more realistic phenomena, we study the heat equation in fuzzy sense by treating the variable temperature as a fuzzy variable. In this case, we have a fuzzy heat equation

$$\frac{\partial \tilde{U}}{\partial t} = P \odot \frac{\partial^2 \tilde{U}}{\partial x^2}, \tag{19.2}$$

where $\tilde{U}(x, t)$ is the fuzzy temperature represented by fuzzy numbers, P is the thermal diffusivity and an operator $\odot$ defines multiplication of a fuzzy number with a real number. Here, $\frac{\partial \tilde{U}}{\partial t}$ is a partial derivative of fuzzy function $\tilde{U}$ with respect to variable t where as $\frac{\partial^2 \tilde{U}}{\partial x^2}$ is a second-order partial derivative of $\tilde{U}$ with respect to x.

To estimate how fast a body can change its temperature, thermal diffusivity is useful. Many times it is not reasonable to define it using one crisp number. By fuzzy theory, it is possible to study the simultaneous effect of varying thermal diffusivity by treating as a fuzzy number. Hence, we modify our fuzzy model of heat equation with fuzzy diffusivity as

$$\frac{\partial \tilde{U}}{\partial t} = \tilde{P} \otimes \frac{\partial^2 \tilde{U}}{\partial x^2}, \tag{19.3}$$

where $\tilde{U}(x, t)$ is the fuzzy temperature, $\tilde{P}$ is the fuzzy thermal diffusivity and $\otimes$ is a multiplication operator between two fuzzy numbers. Moreover, we see that boundary conditions and initial conditions are not often precise or known completely. We express this impreciseness in the boundary and initial conditions in terms of fuzzy numbers. That is, $\tilde{U}(0, t) = \tilde{T}_0$, $\tilde{U}(l, t) = \tilde{T}_l$ and $\tilde{U}(x, 0) = \tilde{f}(x)$, where $\tilde{T}_0$ and $\tilde{T}_l$ are fuzzy constants and $\tilde{f}(x)$ is a fuzzy function.

19.3.2 Solution Concept

We consider the fuzzy heat equation in the form

$$\frac{\partial \tilde{U}}{\partial t} = \tilde{P} \otimes \frac{\partial^2 \tilde{U}}{\partial x^2}, \tag{19.4}$$

where $\tilde{P}$ is a fuzzy diffusivity, $\tilde{U}(x,t)$ is fuzzy temperature at $(x,t) \in I_1 \times I_2$, subject to certain fuzzy boundary and initial conditions and $\otimes$ is fuzzy multiplication operator.

The Eq. (19.4) is well defined in fuzzy sense since we assume that Seikkala derivatives of $\tilde{U}$ with respect to variables t and x exist. We find the Seikkala solution (S-solution) of (19.4) subject to specific fuzzy boundary and initial conditions.

Let $\tilde{U}_\alpha(x,t) = [u_1(x,t,\alpha), u_2(x,t,\alpha)]$, $[\frac{\partial U}{\partial t}(x,t)]_\alpha = [\frac{\partial u_1}{\partial t}(x,t,\alpha), \frac{\partial u_2}{\partial t}(x,t,\alpha)]$, $[\frac{\partial^2 U}{\partial x^2}(x,t)]_\alpha = [\frac{\partial^2 u_1}{\partial x^2}(x,t,\alpha), \frac{\partial^2 u_2}{\partial x^2}(x,t,\alpha)]$ and $\tilde{P}_\alpha = [p_1(\alpha), p_2(\alpha)]$, for all $(x,t) \in I_1 \times I_2$ and all $\alpha \in [0,1]$. Using differentiability of $\tilde{U}$ and fuzzy arithmetic, the fuzzy equation (19.4) can be written as the system of parametric form of heat equations

$$\frac{\partial u_1}{\partial t} = \min\left\{ p_1(\alpha)\frac{\partial^2 u_1}{\partial x^2}, p_1(\alpha)\frac{\partial^2 u_2}{\partial x^2}, p_2(\alpha)\frac{\partial^2 u_1}{\partial x^2}, p_2(\alpha)\frac{\partial^2 u_2}{\partial x^2} \right\}, \tag{19.5}$$

$$\frac{\partial u_2}{\partial t} = \max\left\{ p_1(\alpha)\frac{\partial^2 u_1}{\partial x^2}, p_1(\alpha)\frac{\partial^2 u_2}{\partial x^2}, p_2(\alpha)\frac{\partial^2 u_1}{\partial x^2}, p_2(\alpha)\frac{\partial^2 u_2}{\partial x^2} \right\}, \tag{19.6}$$

for all $(x,t) \in I_1 \times I_2$ and all $\alpha \in [0,1]$. We assume that $p_1(\alpha), p_2(\alpha) > 0$.

Case (i): $\frac{\partial^2 u_1}{\partial x^2}, \frac{\partial^2 u_2}{\partial x^2} > 0$, we further simply the system as

$$\frac{\partial u_1}{\partial t} = p_1(\alpha)\frac{\partial^2 u_1}{\partial x^2}, \tag{19.7}$$

$$\frac{\partial u_2}{\partial t} = p_2(\alpha)\frac{\partial^2 u_2}{\partial x^2}, \tag{19.8}$$

for all $(x,t) \in I_1 \times I_2$ and all $\alpha \in [0,1]$. The fuzzy boundary conditions are $\tilde{U}(0,t) = \tilde{T}_0$ and $\tilde{U}(l,t) = \tilde{T}_l$ and fuzzy initial condition is $\tilde{U}(x,0) = \tilde{f}(x)$. Then, we write boundary conditions in terms of α-level sets as

$$u_1(0,t,\alpha) = t_{01}(\alpha),\ \ u_2(0,t,\alpha) = t_{02}(\alpha) \tag{19.9}$$

$$u_1(l,t,\alpha) = t_{l1}(\alpha),\ \ u_2(l,t,\alpha) = t_{l2}(\alpha). \tag{19.10}$$

The initial condition

$$u_1(x, 0, \alpha) = \tilde{f}_1(x, \alpha), \ u_2(x, 0, \alpha) = \tilde{f}_2(x, \alpha). \tag{19.11}$$

Let $u_i(x, t, \alpha)$ solve Eqs. (19.7) and (19.8) with boundary conditions (19.9) and (19.10) and initial conditions (19.11), $i = 1, 2$. If

$$[u_1(x, t, \alpha), u_2(x, t, \alpha)] \tag{19.12}$$

defines the α-level set of a fuzzy number, for each $(x, t) \in I_1 \times I_2$, then $\tilde{U}(x, t)$ is the S-solution.

Case (ii): $\frac{\partial^2 u_1}{\partial x^2} < 0, \ \frac{\partial^2 u_2}{\partial x^2} > 0$, the Eqs. (19.5) and (19.6) simplified as

$$\frac{\partial u_1}{\partial t} = p_2(\alpha) \frac{\partial^2 u_1}{\partial x^2}, \tag{19.13}$$

$$\frac{\partial u_2}{\partial t} = p_1(\alpha) \frac{\partial^2 u_2}{\partial x^2}, \tag{19.14}$$

for all $(x, t) \in I_1 \times I_2$ and all $\alpha \in [0, 1]$. We solve this system with specified boundary and initial conditions same as first case.

Case (iii): $\frac{\partial^2 u_1}{\partial x^2} < 0, \ \frac{\partial^2 u_2}{\partial x^2} < 0$, the Eqs. (19.5) and (19.6), we simply the system same as (19.13) and (19.14).

19.4 Illustration and Analysis

Let $I_1 = [0, 1]$ and $I_2 = [0, 1]$. Consider a fuzzy heat equation

$$\frac{\partial \tilde{U}}{\partial t} = \tilde{P} \otimes \frac{\partial^2 \tilde{U}}{\partial x^2}, \tag{19.15}$$

where $\tilde{P}$ is a fuzzy diffusivity, $\tilde{U}(x, t)$ is fuzzy temperature at $(x, t) \in I_1 \times I_2$ and $\otimes$ is fuzzy multiplication operator. We have specific fuzzy boundary conditions $\tilde{U}(0, t) = \tilde{U}(1, t) = \tilde{0}$ and fuzzy initial condition $\tilde{U}(x, 0) = \tilde{C} \odot \cos(\pi x - \pi/2)$, where $\tilde{C}$ is a fuzzy number (An operator $\odot$ defines multiplication of a fuzzy number with a real number), and $\tilde{0}(r) = 1$ at $r = 0$ and $\tilde{0}(r) = 0$ for $r \neq 0$.

We proceed to look for a S-solution. As the fuzzy initial condition involves cosine function, $\frac{\partial^2 u_1}{\partial x^2} < 0, \ \frac{\partial^2 u_2}{\partial x^2} < 0$, the Eqs. (19.5) and (19.6) can be simplified as

$$\frac{\partial u_1}{\partial t} = p_2(\alpha) \frac{\partial^2 u_1}{\partial x^2}, \tag{19.16}$$

$$\frac{\partial u_2}{\partial t} = p_1(\alpha)\frac{\partial^2 u_2}{\partial x^2}, \tag{19.17}$$

for all $(x, t) \in I_1 \times I_2$ and all $\alpha \in [0, 1]$. subject to

$$u_i(0, t, \alpha) = u_i(1, t, \alpha) = 0 \tag{19.18}$$

$$u_i(x, 0, \alpha) = c_i(\alpha)\cos(\pi x - \pi/2) \tag{19.19}$$

for $i = 1, 2$. The solution is

$$u_1(x, t, \alpha) = c_1(\alpha)e^{-p_2(\alpha)\pi^2 t}\cos(\pi x - \pi/2), \tag{19.20}$$

and

$$u_2(x, t, \alpha) = c_2(\alpha)e^{-p_1(\alpha)\pi^2 t}\cos(\pi x - \pi/2), \tag{19.21}$$

for $(x, t) \in I_1 \times I_2$ and $\alpha \in [0, 1]$.

If $[u_1(x, t, \alpha), u_2(x, t, \alpha)]$ defines α-level sets of a fuzzy number for each $x \in I_1$ and $t \in I_2$, then fuzzy solution of (19.15) exist with specified fuzzy boundary and initial conditions. Since $u_i(x, t, \alpha)$ are continuous and $u_1(x, t, 1) = u_2(x, t, 1)$, what we need to check is $\frac{\partial u_1}{\partial \alpha} > 0$ and $\frac{\partial u_2}{\partial \alpha} < 0$. Hence the S-solution exists if

$$\frac{\partial u_1}{\partial \alpha} = e^{-p_2(\alpha)\pi^2 t}\cos(\pi x - \pi/2)(c_1'(\alpha) - c_1(\alpha)p_2'(\alpha)\pi^2 t) > 0 \tag{19.22}$$

for all $\alpha \in [0, 1]$, and

$$\frac{\partial u_2}{\partial \alpha} = e^{-p_1(\alpha)\pi^2 t}\cos(\pi x - \pi/2)(c_2'(\alpha) - c_2(\alpha)p_1'(\alpha)\pi^2 t) < 0, \tag{19.23}$$

for all $\alpha \in [0, 1]$ and $(x, t) \in I_1 \times I_2$.

Analysis:

Now take fuzzy diffusivity constant as a fuzzy number $\tilde{P} = (1.9, 2, 2.1)$ a triangular fuzzy number (see Definition 19.3) with $p_1(\alpha) = 1.9 + 0.1\alpha$ and $p_2(\alpha) = 2.1 - 0.1\alpha$, $\alpha \in [0, 1]$. Let $\tilde{C} = \tilde{2} = (1, 2, 3)$ as a coefficient in the fuzzy initial condition $\tilde{U}(x, 0) = \tilde{C} \odot \cos(\pi x - \pi/2)$. So that $\tilde{U}(x, 0) = \tilde{2} \odot \cos(\pi x - \pi/2)$ where $c_1(\alpha) = 1 + \alpha$ and $c_2(\alpha) = 3 - \alpha$. By substituting $p_i(\alpha), c_i(\alpha)$, $i = 1, 2$ in (19.20) and (19.21), we get u_i, $i = 1, 2$ as

$$u_1(x, t, \alpha) = (1 + \alpha)e^{-(2.1-0.1\alpha)\pi^2 t}\cos(\pi x - \pi/2), \tag{19.24}$$

and

$$u_2(x, t, \alpha) = (3 - \alpha)e^{-(1.9+0.1\alpha)\pi^2 t} \cos(\pi x - \pi/2). \tag{19.25}$$

If $[u_1(x, t, \alpha), u_2(x, t, \alpha)]$ defines α-level sets of a fuzzy number for each $x \in I_1$ and $t \in I_2$, then it is a fuzzy solution of Eq. (19.15) with fuzzy boundary conditions $\tilde{U}(0, t) = \tilde{U}(1, t) = \tilde{0}$ and fuzzy initial condition $\tilde{U}(x, 0) = \tilde{2} \odot \cos(\pi x - \pi/2)$.

We see that $u_i(x, t, \alpha)$ are continuous, $i = 1, 2$ and $u_1(x, t, 1) = u_2(x, t, 1)$. Then, for a S-solution we need

$$\frac{\partial u_1}{\partial \alpha} = e^{-(2.1-0.1\alpha)\pi^2 t} \cos(\pi x - \pi/2)(1 + 0.1(1 + \alpha)\pi^2 t) > 0 \tag{19.26}$$

and

$$\frac{\partial u_2}{\partial \alpha} = e^{-(1.9+0.1\alpha)\pi^2 t} \cos(\pi x - \pi/2)(-1 - 0.1(3 - \alpha)\pi^2 t) < 0, \tag{19.27}$$

for each fixed (x, t) and for all $\alpha \in [0, 1]$. Since $e^{-(1+\alpha)\pi^2 t} > 0$ for each $t \in I_2 = [0, 1]$ and for all α, and $\cos(\pi x - \pi/2) > 0$ for each $x \in I_1 = [0, 1]$, we need to check $(1 + 0.1(1 + \alpha)\pi^2 t) > 0$ for each t and all α in (19.26) and $(-1 - 0.1(3 - \alpha)\pi^2 t) < 0$ for each t and all α in (19.27). By analysis, we see that $(1 + 0.1(1 + \alpha)\pi^2 t) > 0$ in (19.26) and $(-1 - 0.1(3 - \alpha)\pi^2 t) < 0$ in (19.27) for each $t \in [0, 1]$ and all α. Therefore, $u_1(x, t, \alpha)$ is increasing and $u_2(x, t, \alpha)$ is decreasing with respect to α and for all $(x, t) \in I_1 \times I_2$. Hence, we say that Seikkala solution for the given fuzzy heat equation exists for $(x, t) \in I_1 \times I_2$. The solution surfaces of u_1 and u_2 are shown in Fig. 19.1. To visualize u_1 and u_2 more clearly, we draw the surfaces in separate figures. See Figs. 19.2 and 19.3. We observed from the figures that u_2 values lies in upper surface and u_1 values lies in lower surface.

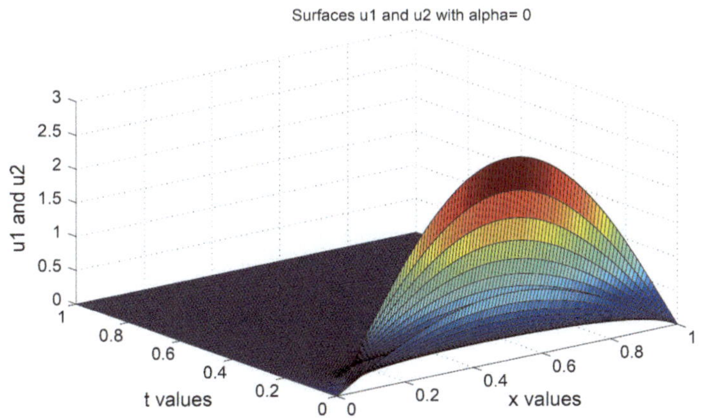

Fig. 19.1 Surfaces of u_1 and u_2 for $(x, t) \in [0, 1] \times [0, 1]$

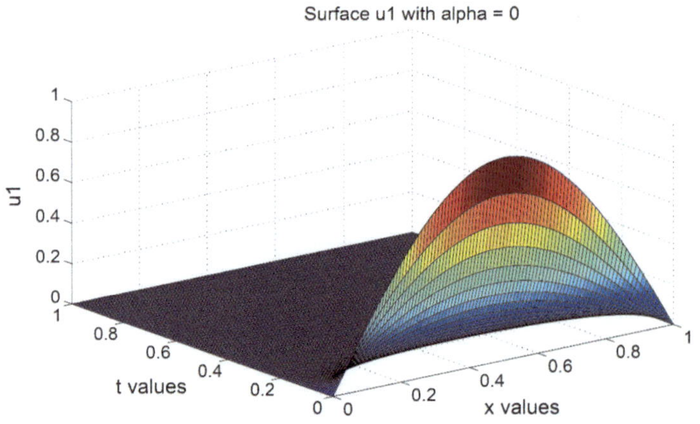

Fig. 19.2 Surface of u_1 for $(x, t) \in [0, 1] \times [0, 1]$

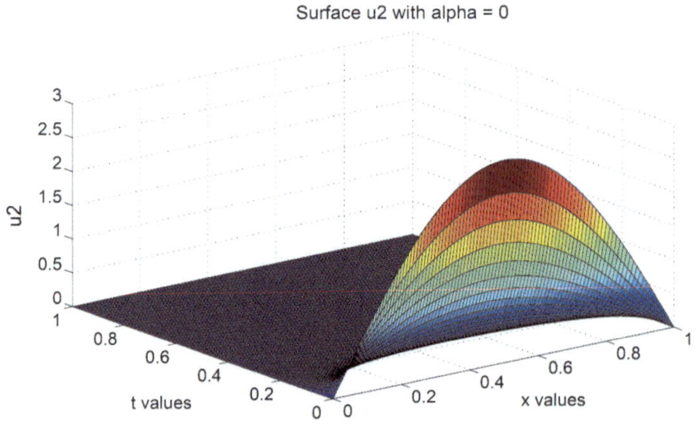

Fig. 19.3 Surface of u_2 for $(x, t) \in [0, 1] \times [0, 1]$

19.5 Conclusions

We have presented the solution of fuzzy heat equation with fuzzy temperature variable, fuzzy diffusivity and fuzzy boundary, and initial conditions. The solution is based on Seikkala derivatives of fuzzy-valued function. Our study allows us to select a flexible value for fuzzy thermal diffusivity, which can vary in certain range with different membership grades. We have analyzed the concept of solution by providing an appropriate illustration.

Acknowledgements This research work is supported by National Board for Higher Mathematics (NBHM), Department of Atomic Energy (DAE), India. The authors are thankful to Prof. V. D. Pathak for his fruitful discussions.

References

1. Allahviranloo, T.: Difference methods for fuzzy partial differential equations. Comput. Methods Appl. Math. **2**(3), 233–242 (2002)
2. Allahviranloo, T., Taheri, N.: An analytic approximation to the solution of fuzzy heat equation by Adomian decomposition method. Int. J. Contemp. Math. Sci. **4**(3), 105–114 (2009)
3. Allahviranloo, T., Afshar, K.M.: Numerical methods for fuzzy linear partial differential equations under new definition for derivative. Iran. J. Fuzzy Syst. **7**(3), 33–50 (2010)
4. Allahviranloo, Tofigh, et al.: On fuzzy solutions for heat equation based on generalized Hukuhara differentiability. Fuzzy sets Syst. (2014). doi:10.1016/j.fss.2014.11.009
5. Bertone, A.M., Jafelice, R.M., de Barros, L.C., Bassanezi, R.C.: On fuzzy solutions for partial differential equations. Fuzzy Sets Syst. **219**, 68–80 (2013)
6. Bede, B., Gal, S.G.: Generalization of the differentiability of fuzzy-number-valued functions with applications to fuzzy differential equations. Fuzzy Sets Syst. **151**, 581–599 (2005)
7. Buckley, J., Feuring, T.: Introduction to fuzzy partial differential equations. Fuzzy Sets Syst. **105**, 241–248 (1999)
8. Chen, Y.-Y., Chang, Y.-T., Chen, B.-S.: Fuzzy solutions to partial differential equations: adaptive approach. IEEE Trans. Fuzzy Syst. **17**(1), 116–127 (2009)
9. George, A.A.: Fuzzy Ostrowski type inequalities. Comput. Appl. Math. **22**, 279–292 (2003)
10. George, A.A.: Fuzzy Taylor formulae. CUBO, A Math. J. **7**, 1–13 (2005)
11. Goetschel, R., Voxman, W.: Elementary fuzzy calculus. Fuzzy Sets Syst. **18**, 31–43 (1986)
12. Faybishenko, B.A.: Introduction to modeling of hydrogeologic systems using fuzzy partial differential equation. In: Nikravesh, M., Zadeh, L., Korotkikh, V. (eds.) Fuzzy Partial Differential Equations and Relational Equations: Reservoir Characterization and Modeling. Springer (2004)
13. Mahmoud, M.M., Iman, J.: Finite volume methods for fuzzy parabolic equations. J. Math. Comput. Sci. **2**(3), 546–558 (2011)
14. Pirzada, U.M., Vakaskar, D.C.: Solution of fuzzy heat equations using adomian decomposition method. Int. J. Adv. Appl. Math. Mech. **3**(1), 87–91 (2015)
15. Seikkala, S.: On the fuzzy initial value problem. Fuzzy Sets Syst. **24**, 319–330 (1987)

Chapter 20
Chaos in Nanofluidic Convection of CuO Nanofluid

Rashmi Bhardwaj and Saureesh Das

Abstract This paper deals with the nonlinear stability dynamics of nanofluid convection under magnetic and temperature variation for Copper Oxide (CuO) nanofluid, which is used as coolant in heat transfer applications. The system comprises a cavity in which the fluid layer is subjected to external magnetic field and heat exposure. The partial differential equations of conservation of momentum and energy are the governing equations, which are converted to a system of nonlinear differential equations. Using stability, phase portrait and time series analysis, the effect of magnetic field and temperature variation through Hartmann number and Rayleigh number on the chaotic CuO nanofluid convection is studied. It is observed that as the value of Hartman number increases, then the system enters into a stable phase. However, on increasing the Rayleigh number system becomes chaotic. Also, it is observed that by controlling the Rayleigh number chaos cannot be controlled but only on increasing the applied field the chaotic state in nanofluid convection can be controlled, which indicates towards a kind of magnetic cooling. It is concluded that as temperature varies the nanofluid convection exhibit chaotic motion which can be stabilized by applying magnetic field which has many applications in drug delivery, nano technology, environmental engineering, industrial engineering and in pharmaceutical industry.

Keywords Stability analysis · Atmospheric interactions · Phase portrait
Time series · MATLAB simulation

R. Bhardwaj (✉) · S. Das
University School of Basic and Applied Sciences, Non-linear Dynamics Research Lab, Guru Gobind Singh Indraprastha University, New Delhi, India
e-mail: rashmib22@gmail.com

S. Das
e-mail: saureeshdas@gmail.com

P. Manchanda et al. (eds.), *Industrial Mathematics and Complex Systems*,
Industrial and Applied Mathematics, DOI 10.1007/978-981-10-3758-0_20

20.1 Introduction

The interest in understanding chaotic behavior of dynamical system is growing in the past few decades. In nature and laboratory, chaotic convection plays a significant role with extended application in understanding the evolution of dynamical systems. The effect of magnetic field on chaotic convection for magneto-convection with stress-free boundary conditions is studied by Bekki and Moriguchi [1]. Nanofluids are the mixture of very small amount of nanoparticles whose dimensions varies from 1 to 100 nm with water or ethylene-glycol as base fluid is proposed by Choi [2]. Garandet et al. [3] obtained analytical solution for the equations of magneto hydrodynamics to transverse the effect of a magnetic field on buoyancy-driven convection in a two-dimensional cavity and showed that for high Hartmann number the velocity gradient lies outside the two Hartmann layers at the vicinity of the walls normal to magnetic field in the core is constant. Idris and Hashim [4] demonstrated that the convective motion of a fluid in saturated porous layer can be delayed using magnetic field and observed transitions from steady state convection to chaotic via Hopf bifurcation and discussed the effect of magnetic field on the route to chaos of fluid in saturated porous layer. The uniform internal heating can enhance the onset of chaotic convection in porous medium, which is studied by Jawdat and Hashim [5]. Kimura et al. [6] used a pseudo-spectral numerical scheme and showed the evolution of convection of fluid in saturated porous layer from steady to chaotic phase with increase in Rayleigh number. A lot of studies are based upon the work of Lorenz [7] who studied the chaos in fluid layer for unpredicted weather behavior and discussed Rayleigh–Benard problem of two-dimensional fluid cells which are cooled and heated from above and below respectively. A set of three-dimensional partial differential equations known as model of fluid convection are obtained. The chaotic synchronization through linear control of two identical systems is discussed by Odibat et al. [8] who showed that synchronization of two different fractional order chaotic systems is feasible by active control of parameters. Pecora and Carroll [9] discussed the problems on synchronization of fractional chaotic systems.

Vadasz and Olek [10] discussed the transition in a porous layer from steady state convection to non-periodic state at a sub-critical value of Rayleigh number and observed the spatially coherent and temporally chaotic rolls in the long-term behavior of magneto-convection in contrast to that of highly turbulent fluids for analyzing the characteristic of transition during natural convection in porous media with sudden change in nature. Dynamical systems research by Yan [11] attracted much attention with the consistent improvement in models with fractional order differential structure.

The study finds its relevance in the utilization of nanofluids as coolants in reactors where exposure to high temperatures can lead to chaotic fluid convection and cause damage to the cooling system and the reactor. To prevent the damages, the stability of the steady state is essential for the nanofluid convection with an alternate mechanism to control it in chaotic state. Another importance of the study is in the Wiendemann–Franz Law, which states that the ratio of the thermal conductivity to the electrical conductivity of a metal is proportional to the temperature.

Qualitatively, this relationship is based upon the fact that the heat and electrical transport both involve the free electrons in the metal. The thermal conductivity increases with the average particle velocity as it increases the forward transport of energy. However, the electrical conductivity decreases with particle velocity increases because the collisions divert the electrons from forward transport of charge. When the fluid is heated particle velocity increases which lead to more collisions, disorder and chaos, thus electrical conductivity decreases.

20.2 Mathematical Modeling

Let us consider an infinitesimal cavity in Cartesian coordinate system through which electrically conductive CuO nanofluid passes. During its flow through cavity, the horizontal nanofluid layer is subjected to heat and magnetic exposure. The two long walls are maintained at temperature T_H and T_C, respectively, while the short end walls are thermally insulated. The vertical axis z is collinear with gravity and a uniform constant magnetic field B is applied normally to the heated side of the cavity as discussed in Fig. 20.1. Garandet et al. [3] discussed the mechanism of buoyancy is experienced by the fluid.

Due to heat transfer the fluid density changes with variation in temperature and the interaction of the magnetic field with the convective motion. The induced magnetic field is negligible as smaller magnetic Reynolds number being considered.

In Darcy's equation, the time derivative terms cannot be neglected for low values of Prandtl number. Darcy's law is assumed to govern the fluid flow. Boussinesq approximation is applied in momentum equation to study the effects of density variations in the gravity term only.

The set of equations governing the conservation of mass, momentum, energy and electric charge transfer for laminar flow [4] are given by:

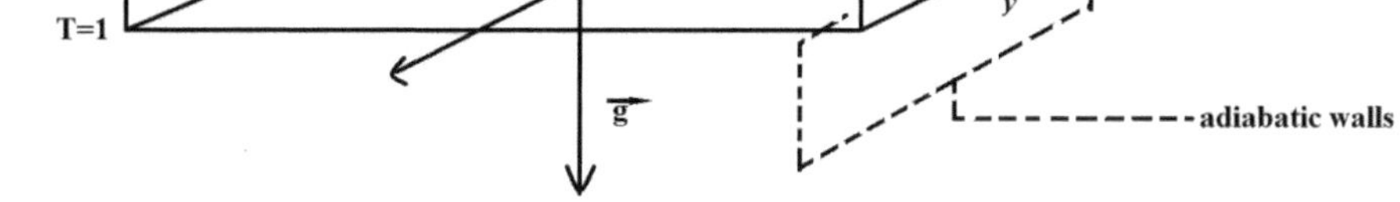

Fig. 20.1 The schematic diagram of the cavity with CuO nanofluid

$$\nabla.\mathbf{V} = 0 \tag{20.1}$$

$$\frac{\partial \mathbf{V}_*}{\partial t} + \mathbf{V}_*.\nabla\mathbf{V}_* = \frac{-1}{\rho_{nf}}\nabla\mathbf{p}_* + \nu_{nf}\nabla^2\mathbf{V}_* + \mathbf{J}\times\mathbf{B} - \frac{(\rho\beta)_{nf}}{\rho_{nf}}\mathbf{g}(T_* - T_C) \tag{20.2}$$

$$\nabla.\mathbf{J} = 0;\ J = \sigma(-\nabla\sigma + \mathbf{V}_* \times \mathbf{B}_*) \tag{20.3}$$

$$\frac{\partial T}{\partial t_*} + \mathbf{V}_*.(\nabla T) = \alpha_{nf}\nabla^2 T \tag{20.4}$$

where $\mathbf{V}_*$ = velocity, T = temperature, $\mathbf{p}_*$ = pressure, β = thermal expansion coefficient, ϕ = electric potential, ν = fluid viscosity, $\mathbf{J}$ = electric current density, σ = electric conductivity, $\mathbf{B}_*$ = magnetic field, ρ_{nf} = effective density, α_{nf} = thermal diffusivity, $\mathbf{g}$ = gravity, $(\rho\beta)_{nf}$ = thermal expansion coefficient, $(T_* - T_C)$ = temperature difference.

The following non-dimensional transformations are used for Eqs. (20.1)–(20.4) $\mathbf{V} = \frac{H_*}{\alpha_f}\mathbf{V}_*$; $\mathbf{p} = \frac{H_*^2}{\alpha_f^2}\mathbf{p}_*$; $t = \frac{\alpha_f}{H_*^2}t_*$; $(x, y, z) = \frac{1}{H_*}(x_*, y_*, z_*)$; $T\triangle T_C = (T_* - T_C)$; $\mathbf{B} = \frac{\mathbf{B}_*}{H_*}$
where $V_* = (u_*, v_*, z_*)$ is the velocity component, $\triangle T_C = (T_H - T_C)$ is the characteristic temperature difference, α_f is the effective thermal diffusivity and H_* is the scaling factor. The fluid layer with horizontal boundaries and stress-free condition is considered. The solution must follow the impermeability condition $\mathbf{V}.\hat{e_n} = 0$ and the stress-free condition $\frac{\partial u}{\partial z} = \frac{\partial v}{\partial z} = \frac{\partial^2 w}{\partial z^2} = 0$ on these boundaries, where $\hat{e_n}$ is a unit vector normal to the boundary. The temperature boundary conditions are: $T = 0$ at $z = 1$ and $T = 1$ at $z = 0$. In terms of stream function, the convective rolls with axes parallel to y axes, when $v = 0$ is defined by $u = \frac{-\partial\psi}{\partial z}$ and $w = \frac{\partial\psi}{\partial x}$. Applying curl on Eq. (20.2) the following partial differential equations are obtained:

$$\left[\frac{1}{P_r}\left(\frac{\partial}{\partial t} - \frac{\partial\psi}{\partial Z}\frac{\partial}{\partial X} - \frac{\partial\psi}{\partial X}\frac{\partial}{\partial Z}\right) - \bar{\nu}\nabla^2 + \bar{\gamma}\right]\nabla^2\psi = \bar{\beta}R_a\frac{\partial T}{\partial X} \tag{20.5}$$

$$\frac{\partial T}{\partial t} - \frac{\partial\psi}{\partial Z}\frac{\partial T}{\partial X} + \frac{\partial\psi}{\partial X}\frac{\partial T}{\partial Z} = \bar{\alpha}\left(\frac{\partial^2 T}{\partial X^2} + \frac{\partial^2 T}{\partial Z^2}\right) \tag{20.6}$$

where $P_r = \frac{\bar{\nu}}{\alpha_f}$; $\bar{\nu} = \nu_{nf}/\nu_f$; R_a(Rayleigh Number) $= \frac{[(\rho\beta)_{nf}\mathbf{g}\triangle H_*^3}{[\rho_{nf}\alpha_f^2]}$; $\bar{\beta} = -\frac{(\rho\beta)_{nf}}{\rho_{nf}\beta_f}$; H_a (Hartmann Number) $= BL\sqrt{\frac{\sigma}{\mu_{nf}}}$; L (Characteristic length) $= 1\ unit$; $\bar{\alpha} = \frac{\alpha_{nf}}{\alpha_f}$

The boundary condition for stream function on horizontal boundaries of the stream function is given as $\psi = \frac{\partial\psi}{\partial z} = 0$. The nonlinear coupled system formed by Eqs. (20.5) and (20.6) with the boundary conditions gives the basic motionless conduction solution. The following stream function and temperature function, which represent Galerkin expansion of the solution in both x and y direction, is considered:

$$\psi = A_{11} \sin(kx) \sin(\pi z) \tag{20.7}$$

$$T = 1 - z + B_{11} \cos(kx) \sin(\pi z) + B_{02} \sin(2\pi z) \tag{20.8}$$

The time and amplitude are rescaled with respect to convective fixed points and given as follows:

$$X = \frac{\tilde{A_{11}}}{\sqrt{\frac{\bar{\beta}\bar{\alpha}\lambda}{\bar{\upsilon}\pi^2}\left(\frac{\bar{\alpha}\bar{\upsilon}}{\bar{\beta}} - \frac{R}{L}\right)}}; \tag{20.9}$$

$$Y = \frac{-\tilde{B}_{11}\sqrt{\frac{\bar{\beta}\bar{\alpha}\lambda}{\bar{\upsilon}\pi^2}\left(\frac{\bar{\alpha}\bar{\upsilon}}{\bar{\beta}} - \frac{R}{L}\right)}}{L\sqrt{\frac{\bar{\upsilon}\bar{\alpha}\lambda}{\bar{\beta}}}}; \tag{20.10}$$

$$Z = \frac{-\tilde{B}_{02}}{\frac{L}{2}\left(\frac{\bar{\alpha}\bar{\upsilon}}{\bar{\beta}} - \frac{R}{L}\right)}; \tag{20.11}$$

Thus, the following system of differential equation is obtained:

$$\dot{X} = P_r \bar{\upsilon} L \left[\frac{\pi}{\left(\frac{\bar{\alpha}\bar{\upsilon}}{\bar{\beta}} - \frac{R}{L}\right)} Y - X\right] \tag{20.12}$$

$$\dot{Y} = \left[\frac{R\bar{\beta}}{\bar{\upsilon}L\pi}\right]\left(\frac{\bar{\alpha}\bar{\upsilon}}{\bar{\beta}} - \frac{R}{L}\right) X - \bar{\alpha}Y + \left(\frac{\bar{\beta}}{\bar{\upsilon}}\right)\left(\frac{\bar{\alpha}\bar{\upsilon}}{\bar{\beta}} - \frac{R}{L}\right)^2 XZ \tag{20.13}$$

$$\dot{Z} = \bar{\alpha}\lambda\left[\left(\frac{\bar{\alpha}\bar{\upsilon}}{\bar{\beta}} - \frac{R}{L}\right)^{-1} XY - Z\right] \tag{20.14}$$

where $L = \left(1 + \frac{\bar{\gamma}}{\pi^2 + k^2}\right)$.

20.3 Stability Analysis

The system of Eqs. (20.12)–(20.14) is expressed as a system of following equations:

$$\dot{X} = c[NY - X] \tag{20.15}$$

$$\dot{Y} = TX - aY - MXZ \tag{20.16}$$

$$\dot{Z} = s(FXY - Z) \tag{20.17}$$

where $L = (1 + \frac{\bar{\gamma}}{\pi^2+k^2})$; $c = LP_r\bar{\nu}$; $T = \left[\frac{R\bar{\beta}}{\bar{\nu}L\pi}\right]\left(\frac{\bar{\alpha}\bar{\nu}}{\bar{\beta}} - \frac{R}{L}\right)$; $M = \left(\frac{\bar{\beta}}{\bar{\nu}}\right)\left(\frac{\bar{\alpha}\bar{\nu}}{\bar{\beta}} - \frac{R}{L}\right)^2$; $a = \bar{\alpha}$; $F = \left(\frac{\bar{\alpha}\bar{\nu}}{\bar{\beta}} - \frac{R}{L}\right)^{-1}$; $s = a\lambda$; $N = \frac{\pi}{\left(\frac{\bar{\alpha}\bar{\nu}}{\bar{\beta}} - \frac{R}{L}\right)}$.

The fixed points $(0, 0, 0)$; $(\sqrt{\frac{a-TN}{MF}}, \sqrt{\frac{a-TN}{MF}}, \frac{a-TN}{MF})$; $(-\sqrt{\frac{a-TN}{MF}}, -\sqrt{\frac{a-TN}{MF}}, \frac{a-TN}{MF})$ are obtained.

- The point$(0, 0, 0)$ is stable for $NT < a$, unstable for $NT > a$ and critical for $NT = a$.
- The point $(\sqrt{\frac{a-NT}{MF}}, \frac{1}{N}\sqrt{\frac{a-NT}{MF}}, \frac{a-NT}{MN})$ and $(-\sqrt{\frac{a-NT}{MF}}, -\frac{1}{N}\sqrt{\frac{a-NT}{MF}}, \frac{a-NT}{MN})$ are stable for $NT < NT_c$, critical for $NT = NT_c$ and chaotic for $NT > NT_c$ where $NT_c = c(3a + c + s)/(c - a - s)$ which is obtained from characteristic equation of the following Jacobian: $J = \begin{bmatrix} -c & cN & 0 \\ \frac{a}{N} & -a & \sqrt{\frac{M(TN-a)}{F}} \\ \frac{s}{N}\sqrt{\frac{F(TN-a)}{M}} & \sqrt{\frac{F(TN-a)}{M}} & -s \end{bmatrix}$.

20.4 Result and Discussion

In order to observe the effect of magnetic field and temperature on CuO nanofluid convection, the system of equations are numerically solved using MATLAB. The thermophysical properties of the fluid and CuO nanoparticle are given in Tables 20.1 and 20.2 respectively mentioned as follows: The three stages of the dynamics of CuO nanofluid convection as determined through stability analysis and observed through numerical simulation on MATLAB as shown in Figs. 20.2, 20.3 and 20.4 (Table 20.3).

The transition from stable to chaotic stage on varying R is shown and transition from chaotic stage to stable stage on varying Ha are shown in Fig. 20.5. In Fig. 20.6

Table 20.1 Thermophysical properties of fluid

Fluid	ρ (kgm^{-3})	k (Wm^{-1}K^{-1})	C_p (Jkg^{-1}K^{-1})	$\beta \times 10^5$ (K^{-1})
Water	997.1	0.613	4179	21

Table 20.2 Thermophysical properties of nanoparticles

Nanoparticle	ρ (kgm^{-3})	k (Wm^{-1}K^{-1})	C_p (Jkg^{-1}K^{-1})	$\beta \times 10^5$ (K^{-1})	$\bar{\nu}$	$\bar{\beta}$	$\bar{\alpha}$
CuO	6500	18	540	0.85	0.890	0.7548	1.150

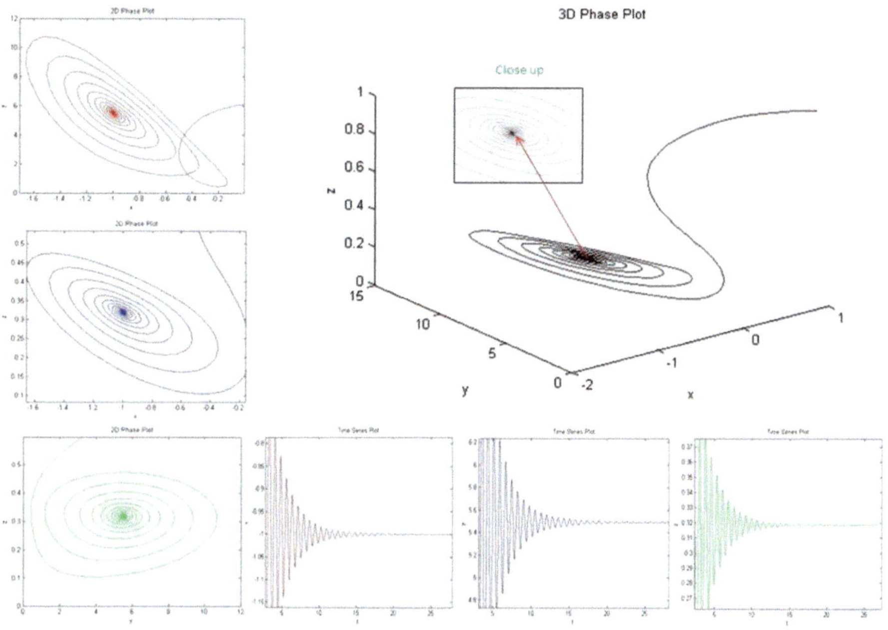

Fig. 20.2 Stable phase of CuO nanofluid at R = 20 and Ha = 0.5

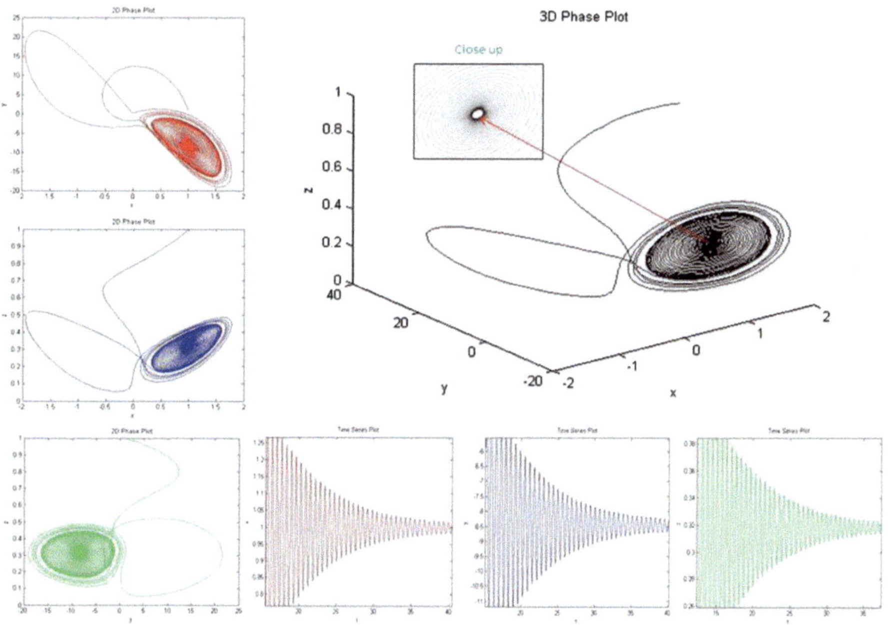

Fig. 20.3 Stable phase of CuO nanofluid at R = 30 and Ha = 0.5

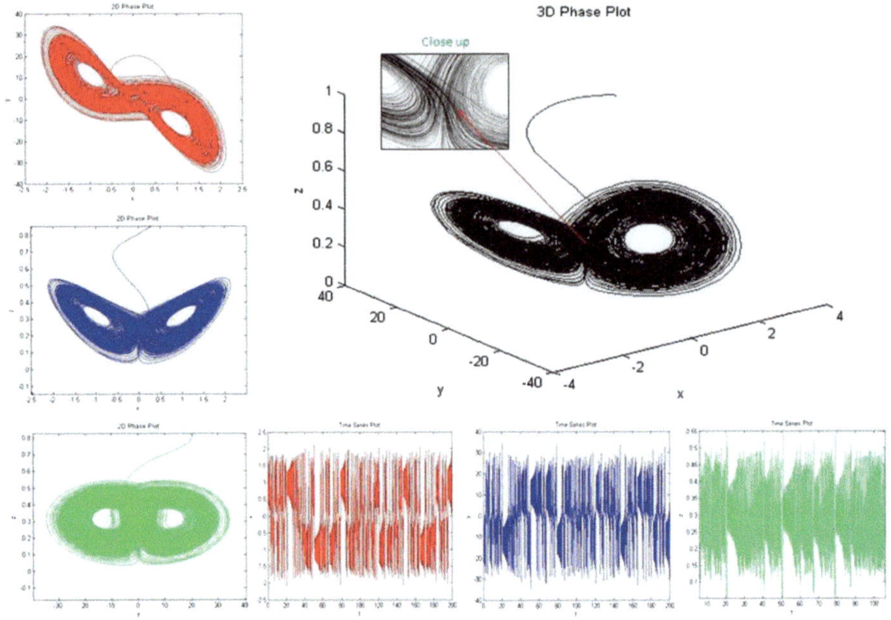

Fig. 20.4 Stable phase of CuO nanofluid at R = 40 and Ha = 0.5

Table 20.3 Effect of Temperature and Magnetic field variations

State	Variation of R at Ha = 0.5	Variation of Ha at R = 46.4
Stable	≤22	≥1.3
Critical	23–32	1.1–1.2
Chaotic	≥33	≤1.0

the Lyapunov plots for both the transitions are shown. In Table 20.4 the Lyapunov exponents, Hurst exponent(H)and fractal dimension are listed for different phases of the system for whom the critical value $NT_c = 28.18$ as per the parameter values listed in Tables 20.1 and 20.2. The bifurcation diagram is shown in Fig. 20.7. The chaotic stage is observed only when the NT exceeds the critical value of NT_c whose relation has been derived from stability analysis.

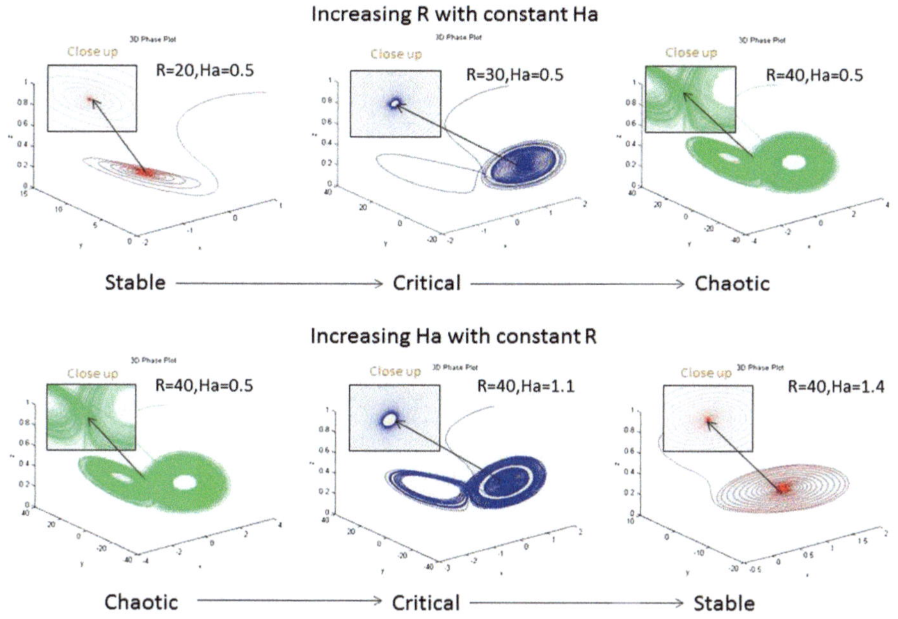

Fig. 20.5 Phase Transition of CuO nanofluid for variation in R and Ha

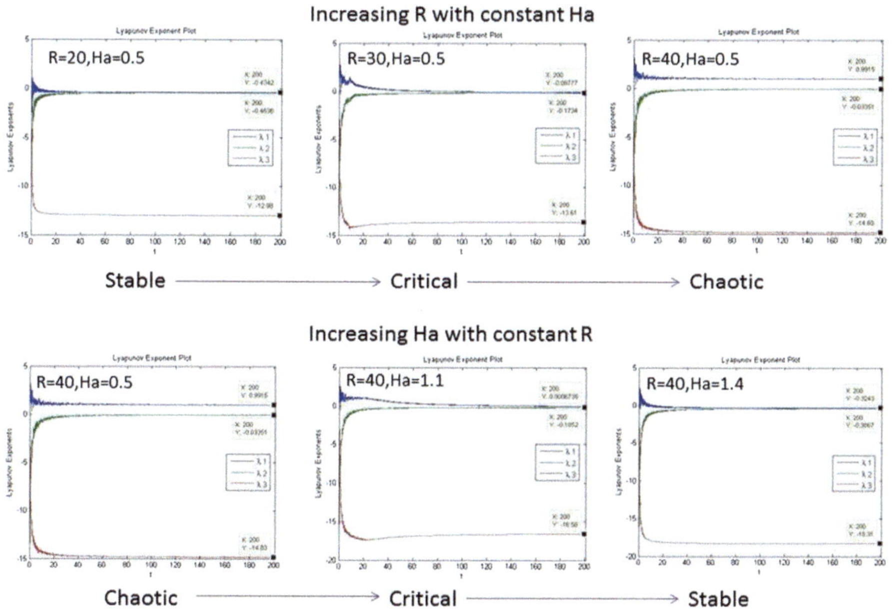

Fig. 20.6 Lyapunov plots of CuO nanofluid for variation in R and Ha

Table 20.4 Lyapunov Exponent, Hurst Exponent(H) and Fractal Dimension(D) for different stages

R	Ha	NT	Phase	λ_1	λ_2	λ_3	Hurst exponent	Fractal dimension
20	0.5	15.8	Stable	−0.43	−0.46	−12.91	0.7	1.3
30	0.5	23.7	Critical	−0.09	−0.18	−13.52	0.6	1.4
40	0.5	31.5	Chaotic	1.08	−0.03	−14.76	0.5	1.5
40	1.1	24.9	Critical	0.00	−0.18	−16.58	0.7	1.3
40	1.4	22.5	Stable	−0.32	−0.38	−18.31	0.7	1.3

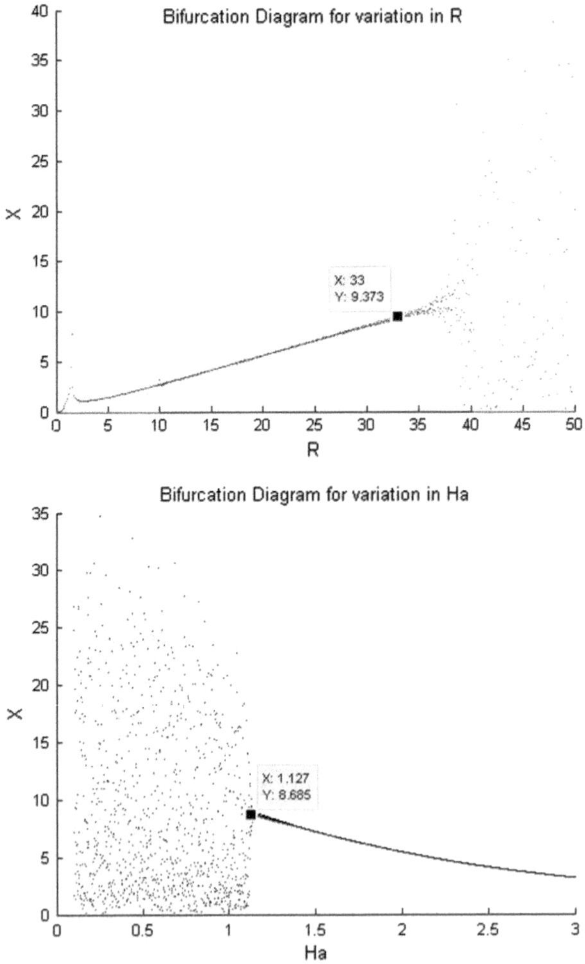

Fig. 20.7 Bifurcation plots of CuO nanofluid for variation in R and Ha

20.5 Conclusion

In this paper, the convection of CuO nanofluid is modeled and effect of magnetic field and temperature on the nanofluid convection is studied using stability analysis and numerical simulation on MATLAB. The three stages, stable state, critical state and chaotic state of nanofluid convection are observed. The transition of the nanofluid convection from stable to chaotic state on increasing Rayleigh number clearly indicates that increase in temperature leads the convection system to chaotic from where it cannot be restored to stability again until magnetic field is applied externally. On increasing the Hartmann number, the system restore to stable state of nanofluid convection indicating towards magnetic cooling. Thus in case of transition of nanofluid to chaotic state of convection is obtained by increasing the applied magnetic field intensity which provides an alternate mechanism to bring the system back to stable state.

Acknowledgements The author is thankful to Guru Gobind Singh Indraprastha University, Delhi (India) for providing research facilities.

References

1. Bekki, N., Moriguchi, H.: Temporal chaos in Boussinesq magnetoconvection. Phys. Plasmas **14**, Art. no. 012306 (2007)
2. Choi, S.U.S.: Enhancing thermal conductivity of fluids with nanoparticles. Proc. ASME Fluids Eng. Div. 231 (1995)
3. Garandet, J.P., Alboussiere, T., Moreau, R.: Buoyancy-driven convection in a rectangular enclosure with a transverse magnetic field. Int. J. Heat Mass Transf. **35**, 741–748 (1992)
4. Idris, R., Hashim, I.: Effects of a magnetic field on chaos for low Prandtl Number convection in porous media. Nonlinear Dyn. **62**, 905–917 (2010)
5. Jawdat, J.M., Hashim, I.: Low Prandtl number chaotic convection in porous media with uniform internal heat generation. Int. Commun. Heat Mass Transf. **37**, 629–636 (2010)
6. Kimura, S., Schubert, G., Straus, J.M.: Route to chaos in porous-medium thermal convection. J. Fluid Mech. **166**, 305–324 (1986)
7. Lorenz, E.N.: Deterministic non-periodic flow. J. Atmos. Sci. **20**, 130–141 (1963)
8. Odibata, Z.M., Corsonb, N., Aziz-Alaouib, M.A., Bertellec, C.: Synchronization of chaotic fractional-order systems via linear control. Int. J. Bifurc. Chaos **20**(1), 81–97 (2010)
9. Pecora, L.M., Carroll, T.L.: Synchronization in chaotic systems. Phys. Rev. Lett. **64**, 821–824 (1990)
10. Vadasz, P., Olek, S.: Transitions and chaos for free convection in a rotating porous layer. Int. J. Heat Mass Transf. **41**, 1417–1435 (1998)
11. Yan, J., Li, C.: On chaos synchronization of fractional differential equations. Chaos Soliton Fractal, **32**(2), 725–735 (2007)

Chapter 21
Study of the Seasonal Variability of Plankton and Forage Fish in Chilika Lagoon Using NPZF Model: A Case Study

Bhanumati Panda, Anumeha Dube and Sushil Kumar

Abstract A four-compartment, Nutrient (N), Phytoplankton (P), Zooplankton (Z) and Forage fish (F), nonlinear mathematical model is used to study the seasonal variability of plankton and forage fish in the Chilika lagoon (19°28′ N–19°54′ N, 85°06′ E–85°36′ E), the largest brackish water lagoon with estuarine character on the east coast of India. It is a highly biological productive and ideal system for aquaculture study. Almost every component at each tropic level of an aquatic food web are dependent on phytoplankton and the availability of nutrient in the study domain. The coupled ordinary differential equations with four state variables represent the interaction of the biological and chemical processes in a marine ecosystem. The main objective of the study is to obtain a set of parameters which can be used in a mathematical model to simulate the ecology of a shallow water lagoon. The model which is presently used in this study enables to bring significant changes in planktonic distribution in the lagoon.

Keywords *NPZF* model · Plankton · Forage fish · Chilika lagoon

21.1 Introduction

The shallow coastal lagoons are highly productive systems with large concentration of species like algae, plankton and fish, etc. The dynamical behaviour of coastal lagoon ecosystem is unstable because of high variability of physical, biological and

B. Panda (✉)
I.T.S Engineering College, 46, Knowledge Park III, Greater Noida, India
e-mail: jhununeel@gmail.com

A. Dube
NCMRWF, Ministry of Earth Sciences, Govt. of India, Sec-62, Noida, India
e-mail: anumeha.dube@gmail.com

S. Kumar
Department of Applied Mathematics, School of Vocational Studies and Applied Sciences, Gautam Buddha University, Greater Noida, India
e-mail: sushil.kumar@gbu.ac.in

P. Manchanda et al. (eds.), *Industrial Mathematics and Complex Systems*, Industrial and Applied Mathematics, DOI 10.1007/978-981-10-3758-0_21

chemical parameters. Since the lagoon exhibit estuarine characters hence the primary productivity is high in these regions. The water characteristics of the lagoons are also influenced by fresh water influx from river run off and saline influx from the ocean. Hence the spatial as well as temporal variations observed in these ecosystems are high as compared to purely fresh water or saline water bodies. Hence these lagoons are regions of high fertility and fish productivity which is the main source of food and income for the population around this area. The *NPZ* (Nutrient, Phytoplankton and Zooplankton) model is one of the conventional model for researchers and oceanographers to describe the oceanic plankton dynamics. The simulation of the model becomes complicated when new state variables are added to the system. Many mathematical models are used to study plankton dynamics of marine ecosystem (Franks et al. [9]; Wroblewski et al. [22]; Fasham et al. [7]; Marra and Ho [15]; Dippner [2]; Felip and Chatalan [8]; Edwards [5]; Sarkar et al. [20]). Evans and Parslow [6] formulated an *NPZ* model to understand the factors controlling different annual plankton cycles in the Atlantic and Pacific ocean basins by using vertical mixing cycles. Through this study they were able to show that the occurrence of an algal bloom requires a low rate of primary production in winter and does not require shallowing of the mixed layer. Dube and Jayaraman [3, 4] used four (nutrient, fresh water plankton, marine plankton and zooplankton) and three compartment (*NPZ*) ecological models to study the seasonal variability of plankton in different sectors of Chilika lagoon. They studied the effect of increased salinity influx on the production of fresh water and marine phytoplankton in the lagoon. Earlier studies (Lehodey et al. [14]; Isoda [11]; Pikitch et al. [18]; Naithani et al. [16]; Ghosh and Kar [10]) have explained successfully by including forage fish or tuna fish to marine environment. Turner et al. [21] observed that *NPZ* dynamics must extend to couple benthic nutrient cycling to the basic *NPZ* interactions. Kumar and Kumari [13] used *NPZF* (nutrient, phytoplankton, zooplankton and forage fish), a nonlinear mathematical model to explain the variability of plankton and forage fish in the Gulf of Kutch.

In the current paper, a four-compartment nonlinear mathematical model *NPZF* (Fig. 21.1) is used as a case study to understand the seasonal variations in Chilika lagoon. In Sect. 21.2, the mathematical formulation of the model and the details of the parameters are elaborated. Section 21.3 gives the study area, where the model is

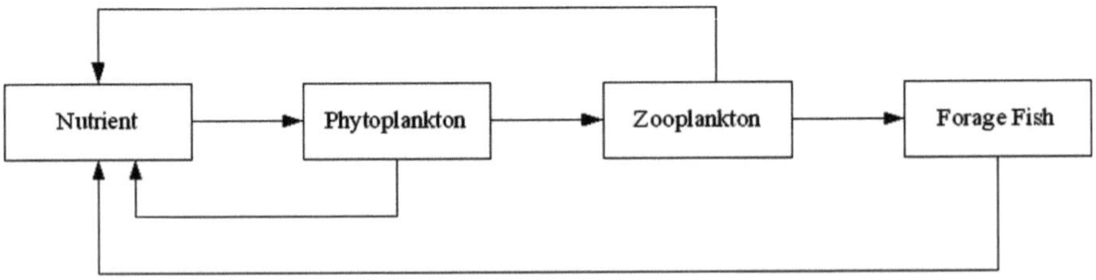

Fig. 21.1 Schematic diagram of NPZF Model. *Source* The End-to-End Approach to Marine Ecosystem Modelling (Mandy Bunke, Department of Biology, University of York, March 2010)

used as a case study. Section 21.4 gives the model validation along with sensitivity analysis and followed by Sects. 21.5 and 21.6 which gives the result, discussion and conclusion of this study.

21.2 Mathematical Formulation of the Model

The governing equations of a coupled physical, biological quantity are same as Kumar and Kumari [13], given by

$$\frac{dC_i}{dt} = S_i + D_i, i = 1, 2, 3, \ldots\ldots\ldots \tag{21.1}$$

where S_i is the source term and D_i is the decay term of the ith biological or chemical tracer of concentration C_i.

The four-compartment *NPZF* model is nonlinear and the governing system of equations are as follows:

$$\frac{dN}{dt} = -\left(\frac{\alpha(\phi, H, t)N}{K_N + N} - r\right)P + \frac{m_0}{H}(N_0 - N) \tag{21.2}$$

$$\frac{dP}{dt} = \left(\frac{\alpha(\phi, H, t)N}{K_N + N} - r\right)P - \frac{c_0(P - P_0)Z}{K_Z + (P - P_0)} - \frac{q_1 P}{B}\frac{c_1(B - B_0)F}{K_F + (B - B_0)} - \frac{m_1 P}{H} \tag{21.3}$$

$$\frac{dZ}{dt} = \frac{e_0 c_0(P - P_0)Z}{K_Z + (P - P_0)} - g_1 Z - \frac{q_2 Z}{B}\frac{c_1(B - B_0)F}{K_F + (B - B_0)} \tag{21.4}$$

$$\frac{dF}{dt} = \frac{e_1 c_1(B - B_0)F}{K_F + (B - B_0)} - g_2 F \tag{21.5}$$

where $B = q_1 P + q_2 Z$, represents the total food perceived by forage fish, q_1 and q_2 are the preferences of forage fish for phytoplankton and zooplankton respectively.

The state variables *N*,*P*,*Z* and *F* are measured in (mg/l), 't' is the time, α is the photosynthetic growth rate of phytoplankton, r (d^{-1}) is the mortality loss rate of phytoplankton, K_N, K_Z and K_F which are measured as (mg/l) are the uptake, grazing, predation half saturation coefficient of phytoplankton, zooplankton and forage fish respectively. $m_0(m\, d^{-1})$ is the vertical mixing rate and H is the depth of the lagoon in metres. N_0 (mg/l) is the nutrient source and m_1 is the settling rate of phytoplankton in $(m\ d^{-1})$. P_0 and B_0 are the threshold value of phytoplankton and the threshold value of B respectively. $g_1(d^{-1})$ and $g_2(d^{-1})$ are the mortality rate of zooplankton and forage fish respectively. c_0 and c_1 are the maximum grazing and predation rate of zooplankton and forage fish measured as (d^{-1}) respectively. e_0 and e_1 are the grazing and predation efficiency of zooplankton and forage fish respectively. q_1 and q_2 are palatability coefficient of zooplankton and forage fish respectively.

21.2.1 Photosynthetic Growth Rate (α)

The photosynthetic growth rate of phytoplankton depends on the surface light and mixed layer depth. The growth rate of phytoplankton is averaged over the course of a day (Evans and Parslow [6]; Dube and Jayaraman [4]; Kumar and Kumari [13]) is as follows:

$$\alpha(\phi, H, t) = \frac{2\,Q}{k_1 H} \int_0^{\tau} \int_{\beta}^{\beta e^{k_1 H}} \frac{t\,dy\,dt}{y(y^2 + t^2)^{1/2}} \tag{21.6}$$

where $\beta = \frac{Q\,\tau}{k_2 J}$, $\tau = \frac{1}{2}\cos^{-1}(-\tan\delta\ \tan\phi)$, $\delta = 23.45\sin(2\pi(t-81)/365)$, $J\tau = \frac{R}{\pi}(\tau\sin\delta\sin\phi + \cos\delta\cos\phi\sin\tau)$, $R = \frac{3}{8}(1-a_0)\ S_0$.

Where a_0 is the average albedo of earth and $S_0 = 1.375$ kWm^{-2} is the solar constant.

21.2.2 Nutrient (N)

The Eq. (21.2) represents the rate of change of nutrients where the first term in Eq. (21.2), $-(\frac{\alpha(\phi,H,t)N}{K_N+N})$ represents the loss of nutrients concentration due to the growth of phytoplankton and $(\frac{N}{K_N+N})$ is the nutrient limited growth rate of phytoplankton. The second term rP represents the increase of nutrients due to the mortality loss rate of phytoplankton, $\frac{m_0}{H}(N_0)$ represents the addition of nutrients due to the vertical diffusion.

21.2.3 Phytoplankton (P)

The Eq. (21.3) represents the rate equation of phytoplankton. The first term $(\frac{\alpha(\phi,H,t)N}{K_N+N})$ of the Eq. (21.3) represents the growth rate of phytoplankton due to loss of nutrients and the terms $-\frac{c_0(P-P_0)Z}{K_Z+(P-P_0)}Z$ and $-\frac{q_1 P}{B}\frac{c_1(B-B_0)F}{K_F+(B-B_0)}$ represent the loss of phytoplankton due to grazing and predation by zooplankton and forage fish respectively and the loss term $-\frac{m_1 P}{H}$ is due to the settling of phytoplankton.

21.2.4 Zooplankton (Z)

The model Eq. (21.4) represents the rate of change of zooplankton and the term $\frac{e_0 c_0(P-P_0)Z}{K_Z+(P-P_0)}$ represents the growth of zooplankton due to the loss by phytoplankton and $-\frac{q_2 Z}{B}\frac{c_1(B-B_0)F}{K_F+(B-B_0)}$ represents the decrease of zooplankton due to the predation by forage fish and the loss term $g_1 Z$ is due to the mortality loss of zooplankton.

21.2.5 Forage Fish (F)

The Eq. (21.5) represents the rate equation of forage fish and the term $\frac{e_1 c_1 (B-B_0)F}{K_F+(B-B_0)}$ represents the growth rate of forage fish due to the loss of phytoplankton and zooplankton. The term $-g_2F$ represents the mortality loss of forage fish.

21.3 Study Area (Chilika Lagoon)

Chilika lagoon (19°28′ N – 19°54′ N, 85°06′ E – 85°36′ E) (Fig. 21.2) on the east coast of India in the state of Odisha is one of the largest tropical lagoon in the world and the largest brackish water lagoon in Asia with estuarine character and designated as Ramsar site of Wetland in 1981. The average depth of the lagoon is about 2 m and of length 65 km and spreading over an area of 960 km^2 during summer and 1,165 km^2 during monsoon. The lake is divided into four different sectors i.e., southern, central, northern sectors and outer channel area. A 32 km long outer channel is connected with the lake at Arakhuda village with the Bay of Bengal.

Phytoplankton community of Chilika lagoon consists of mixture of marine, brackish and fresh water. The four group of algae, diatoms (Bacillariphyceae), dinoflagellates (Phyrrophyceae and Dinophyceae), blue-green algae (Chlorophyceae) and green algae are mainly found in the lagoon. During summer season phytoplankton of the lagoon are dominated by diatoms, certain dinoflagellates and blue-green algae (Adhikary and Sahu [1]) and during winter it is mainly dominated by blue-green algae and green algae (Panigrahi et al. [17]). The entire lagoon is dominated by diatoms except the northern region.

The different species of zooplankton which are the dominant group of the mainly copepods, chaetognaths, cladocerans, mysids, lucifers, euphasids, siphonophores and sergestids (Sarkar et al. [19]).

Forage fish are small, pelagic fish which are a prey to many large fishes, seabirds and marine mammals. They include particularly fishes of the family

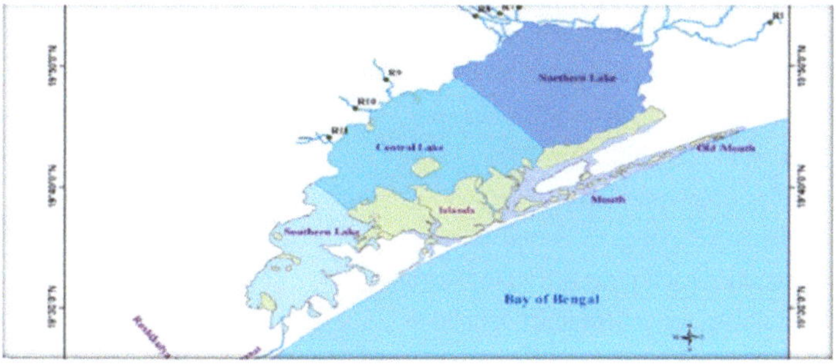

Fig. 21.2 Chilika Lagoon: The study area

clupeidai (herring, sardines, shad, hilsa, menhaden, anchovies and sparts). In Chilika lagoon, hilsa is the dominant group of forage fish (Jones and Sujansingani [12]).

21.4 Numerical Experiments and Sensitivity Analysis

The model discussed in Sect. 21.2 is used to study the effect of additional state variable forage fish to the plankton dynamics in Chilika lagoon. Earlier the model is used by Kumar and Kumari [13] to study the dynamics of plankton and forage fish in the Gulf of Kutch. The model equations are solved by using fourth-order Runge Kutta method. There are total 22 parameters involved in the model equations. During model simulation the sensitivity analysis is performed for stability of the model and some parameters, K_N, K_Z, r, P_0, B_0, q_2, m_1, e_0, c_0, g_1 and g_2 responds more sensitive to the model and the parameters are obtained from sensitivity analysis whereas the other parameters Q, k_1, k_2, a_0, S_0 are constants and collected from the literature (Dube and Jayaraman, [3, 4]). Table 21.1 gives the description of the parameters and their values that are fixed for the current study.

Table 21.1 Description of parameters and its values used for model simulation

Symbols	Parameters	Units	Values as per stability criteria
N_0	Nutrient Source	mg/l	100.25
K_N	Half saturation coefficient of phytoplankton	mg/l	1.5
K_Z	Half saturation coefficient of zooplankton	mg/l	280
K_F	Half saturation coefficient of forage fish	mg/l	90
P_0	Threshold value of Phytoplankton	mg/l	50.1
B_0	Threshold value of B	mg/l	50
r	Mortality loss rate of phytoplankton	d^{-1}	0.1
m_0	Vertical mixing rate	md^{-1}	5.7
m_1	Settling rate of phytoplankton	md^{-1}	0.54
H	Depth of Chillika lagoon	m	4
g_1	Mortality rate of Zooplankton	d^{-1}	0.01
g_2	Mortality rate of Forage fish	d^{-1}	0.01
c_0	Maximum grazing rate of Zooplankton	d^{-1}	0.45
c_1	Maximum predation rate of Forage fish	d^{-1}	0.1
e_0	Grazing efficiency of Zooplankton	–	0.44
e_1	Predation efficiency of Forage fish	–	0.15
q_1	Palatability coefficient of Phytoplankton	–	50.0
q_2	Palatability coefficient of Zooplankton	–	10

[a]Parameters description and values used in simulations
Source Kumar and Kumari [13], Dube and Jayaraman [4]

21.5 Result and Discussion

The model simulated results of phytoplankton and zooplankton are depicted in (Figs. 21.3, 21.4, 21.5 and 21.6). The model results of phytoplankton are compared with the observed value of Adhikary and Sahu [1] as well as Dube and Jayaraman [4]. The inclusion of extra component forage fish results in variations in the mortality of phytoplankton and zooplankton population due to the predation by forage fish. So it effects the temporal distribution of phytoplankton and zooplankton population. In Fig. 21.5, it is seen that the increase in the concentration of the zooplankton is due to the loss of phytoplankton. For the comparison of the model with the available data and stability criteria, the parameters involved in the model equations are required to be perfectly tuned. It was found through sensitivity analysis that the effective parameter, which controls the plankton distribution of the system, is the photosynthetic growth

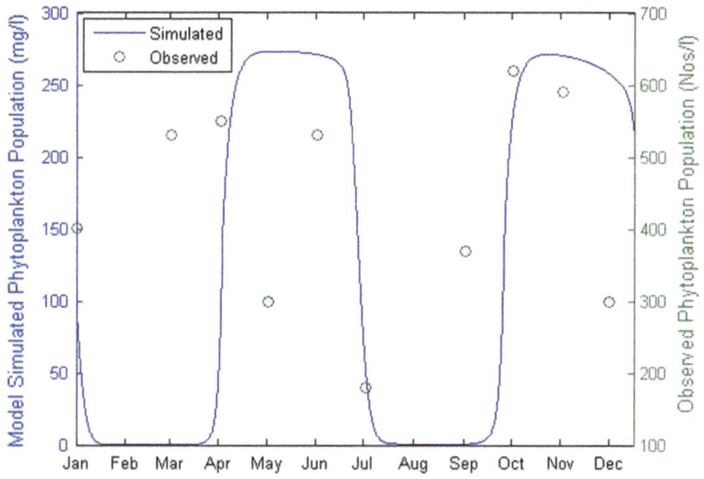

Fig. 21.3 Model simulated and observed phytoplankton

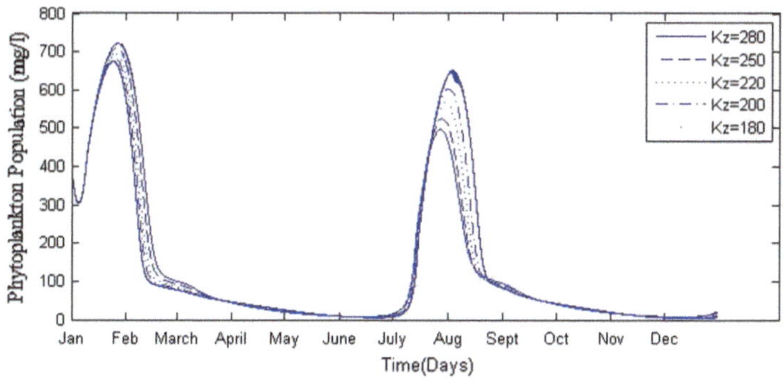

Fig. 21.4 Sensitivity of K_Z for phytoplankton

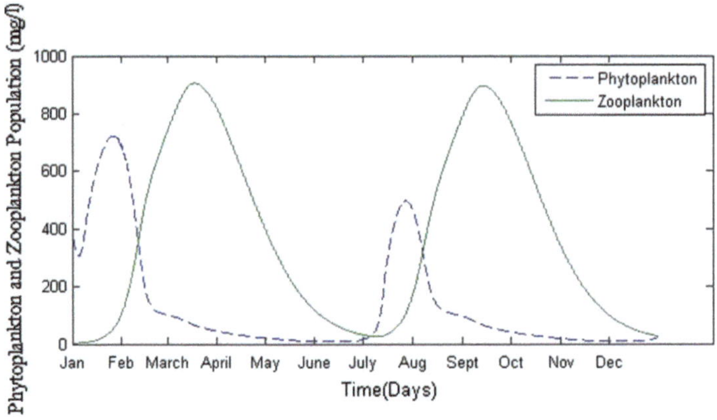

Fig. 21.5 Model simulated phytoplankton and zooplankton

Fig. 21.6 Sensitivity of K_Z for zootoplankton

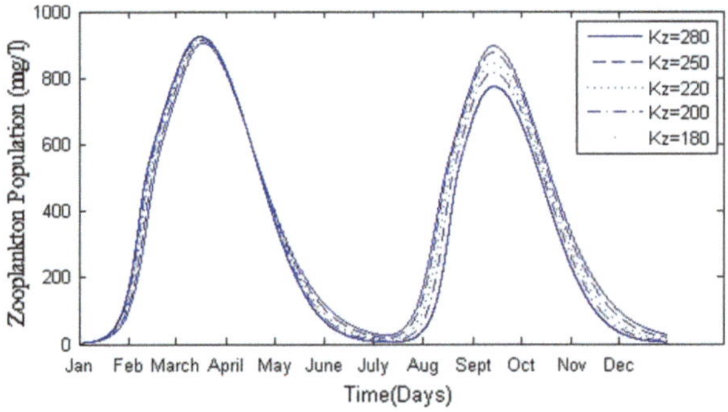

rate. Figures 21.4 and 21.6 show the effect of K_Z through sensitivity analysis for the distribution of phytoplankton and zooplankton. Beyond certain range (<180 mg/l), the model becomes numerically unstable. Another critical parameter is the mortality loss rate r and for value ($< 0.5\ d^{-1}$) the model becomes numerically unstable. The other parameters which are case sensitive to the model have been discussed in Sect. 21.4.

21.6 Conclusion

The simulation of the model shows that addition of forage fish to *NPZ* model changes the system dynamics significantly and for stability of the model, parameter estimation is done through sensitivity analysis and the range of the values are fixed. The trend of temporal variation of plankton is also seen through model-simulated results. The agreement between the model and observations is not perfect which may be attributed to many factors like imperfection in formulation of model, non-inclusion of climatic variables and their effect on the ecosystem, imperfect initial conditions, etc. All these factors cause a difference between the simulations and observations. Since the system of equations is highly nonlinear, it is difficult to set the parameters in order to match the model results with the observed data exactly. Currently the work is going on to extend the study to include indepth sensitivity and phase space analysis.

Acknowledgements The author would like to acknowledge Mr. Vijay Kumar, (Senior research fellow, IIT Delhi) for his suggestion and discussion to implement the NPZF model in Chilika Lagoon. Authors are grateful to the anonymous reviewers for their constructive comments which improved the quality and readability of the manuscript to a great extent.

References

1. Adhikary, S.P., Sahu, J.K.: Distribution and seasonal abundance of algal forms in Chilika lake, east coast of India. Japanese J. Limnol. **53**, 197–205 (1992)
2. Dippner, J.W.: A lagrangian model of phytoplankton growth dynamics for the Northern Adriatic Sea. Cont. Shelf Res. **13**, 331–355 (1993)
3. Dube, A., Jayaraman, G.: Proceedings of the 8th WSEAS International Conference on Mathematics and Computers in Biology and Chemistry, Vancouver, Canada, June 19–21, pp. 103–111 (2007)
4. Dube, A., Jayaraman, G.: Mathematical modelling of the seasonal variability of plankton in a shallow lagoon. Nonlinear Anal. **69**, 850–865 (2008)
5. Edwards, A.M.: Adding detritus to a nutrient-phytoplankton-zooplankton model: a dynamical-systems approach. J. Plankton Res. **23**, 389–413 (2001)
6. Evans, G.T., Parslow, J.S.: A models of annual plankton cycles. Biol. Oceanogr. **3**, 327–347 (1985)
7. Fasham, M.J.R., Ducklow, H.W., Mckelive, S.M.: A nitrogen based model of plankton dynamics in the ecosystem in the oceanic mixed layer. J. Mar. Res. **48**, 591–639 (1990)
8. Felip, M., Chatalan, J.: The relationship between phytoplankton biovolume and chlorophyll in a deep oligotrophic lake: decoupling in their spatial and temporal maxima. J. Plank. Res. **22**, 91–105 (2000)
9. Franks, P.J.S., Wroblewski, J.S., Flierl, G.R.: Behaviour of a simple plankton model with food level accimilation by herbivores. Marine Biol. **91**, 121–129 (1986)
10. Ghosh, B., Kar, T.K.: Sustainable use of prey species in a prey–predator system: jointly determined ecological thresholds and economic trade-offs. Ecol. Model. **272**, 49–58 (2014)
11. Isoda, Y.: Decadal response of marine food-chain to bottom-up and top-down controls. Bull. Fish. Sci. Hokkaido Univ. **54**(3), 29–41 (2003)
12. Jones, S., Sujansingani, K.H.: Fish and fisheries of the Chilka Lake with statistics of fish catches for the years 1948–1950. Indian J. Fish **1**, 256–344 (1954)
13. Kumar, V., Kumari, B.: Mathematical modelling of seasonal variability of plankton and forage fish in the Gulf of Kachchh. Ecol. Model. **313**, 237–250 (2015)
14. Lehodey, P., Andre, J.M., Bertignac, M., Hampton, J., Stones, A., Menkes, C., Memery, L., Grima, N.: Predicting skipjack tuna forage distributions in the equatorial Paciic using a coupled dynamical bio-geochemical model. Fish. Oceanogr. **7**, 317–325 (1998)
15. Marra, J., Ho, C.: Initiation of the spring bloom in the northeast Atlantic (470 N, 200 W): a numerical simulation. Deep-Sea Res. **II**(40), 55–73 (1993)
16. Naithani, J., Darchambeaub, F., Deleersnijder, E., Descy, J.P., Wolanski, E.: Study of the nutrient and plankton dynamics in Lake Tanganyika using a reduced-gravity model. Ecol. Model. **200**, 225–233 (2007)

17. Panigrahi, S., Wikner, J., Panigrahi, R.C., Satpathy, K.K., Acharya, B.C.: Variability of nutrients and phytoplankton biomass in a shallow brackish water ecosystem (Chilika Lagoon, India). Limnology **10**, 73–85 (2009)
18. Pikitch, E.K., Rountos1, K.J., Essington, T.E., Santora, C., Pauly, D., Watson, R., Sumaila, U.R., Boersma, P.D., Boyd5, I.L., Conover, D.O., Cury, P., Heppell, S.S., Houde, E.D., Mangel, M., Plagányi, E., Sainsbury, K., Steneck, R.S., Geers, T.M., Gownaris, N., Munch, S.B.: The global contribution of forage fish to marine fisheries and ecosystems. Fish Fish **15**, 43–64 (2014)
19. Sarkar, S.K., Bhattacharya, B.D., Mohanty, A.K., Panigrahi, S.: Encyclopedia of Lakes and Reservoirs, pp. 148–155. Springer, Berlin (2012)
20. Sarkar, R.R., Pal, J., Das, K.P., Chattopadhyay, J.: Control of harmful algal blooms through input of competitive phytoplankton and the effect of environmental variability. J. Cal. Math. Soc. **4**, 1–8 (2008)
21. Turner, E.L., Bruesewitz, D.A., Mooney, R.F., Montagna, P.A., McClelland, J.W., Sadovski, A., Buskey, E.J.: Comparing performance of five nutrient phytoplankton zooplankton (NPZ) models in coastal lagoons. Ecol. Model. **277**, 13–26 (2014)
22. Wroblewski, J.S., Sarmiento, J.L., Flierk, G.R.: An ocean basin scale model of plankton dynamics in the North Atantic 1. Solution for the climatological oceanographic conditions in May. Glob. Biogeochem. **2**, 199–218 (1988)

Chapter 22
Effect of Glycerol Kinetics and Mass Transfer During Enzymatic Biodiesel Production from Jatropha Oil

Fahad Al Basir, Xianbing Cao, Sushil Kumar and Priti Kumar Roy

Abstract Enzymatic transesterification for biodiesel production from *Jatropha curcas* oil has gained favorable attention due to high selectivity and mild reaction conditions. Beside this, mass transfer limitation is a barrier for maximum biodiesel yield. Stirring and enzyme regulates the mass transfer in transesterification of Jatropha oil. Moreover, the effects of enzyme and stirring have been considered by many researchers but the effect of glycerol was neglected while studying mass transfer kinetics. In this article, the aim is to study the mass transfer resistance due to immiscibility of alcohol and oil as well as to reduce glycerol inhibition to increase enzyme activity by formulating a mathematical model. Optimal control approach has been applied on mixing intensity to avoid mass transfer limitations in both the phases which minimizes the glycerol effect and gives cost effective production of biodiesel. Simulation results of the model system are in a good agreement with experimental results available in the literature.

Keywords Biodiesel · Enzymatic transesterification · Stirring · Mass transfer Enzyme inhibition · Glycerol kinetics · Optimization

F. Al Basir · P.K. Roy (✉)
Centre for Mathematical Biology and Ecology, Department of Mathematics,
Jadavpur University, Kolkata 700032, West Bengal, India
e-mail: pritiju@gmail.com

S. Kumar
Department of Applied Mathematics, School of Vocational Studies
and Applied Sciences, Gautam Buddha University, Gr. Noida, India
e-mail: sushil.kumar@gbu.ac.in

X. Cao
College of Science, Beijing Technology and Business University,
Beijing 100048, China

P. Manchanda et al. (eds.), *Industrial Mathematics and Complex Systems*,
Industrial and Applied Mathematics, DOI 10.1007/978-981-10-3758-0_22

22.1 Introduction

Biodiesel, the most appropriate alternative fuel for diesel engines, is gaining the enormous importance as diminishing petroleum sources and the environmental consequences due to exhaust gases from petroleum based engines. To reduce the cost of biodiesel production, less expensive feedstock such as Jatropha oil is used as feed stock [1]. *Jatropha curcas* (Linn), a multipurpose plant, contains high amount of oil in its seeds which is used to produce biodiesel. The fuel properties of Jatropha biodiesel are similar to diesel and can be used as an alternative fuel in diesel engine [2]. Moreover, *Jatropha curcas* can be cultivated on semiarid and barren land to supply raw material for biodiesel production and thereby reduces production cost [3–5]. Hence, Jatropha oil is chosen as raw material for biodiesel production.

Generally, biodiesel is produced by transesterification of triglycerides by homogeneous alkaline catalysts [6, 7]. But this process is not suitable for transesterification of Jatropha oil since it contains high amount of free fatty acid (FFA). FFA reacts with the alkaline catalysts to form soap as a by-product and deactivates the catalyst [8]. This process requires high temperature and stirring. But higher temperature and stirring enhances saponification during the reaction process. Instead of using chemical process, enzymatic production features more safe, eco-friendly, and cost-effective alternative to generate biodiesel [9]. Enzymatic transesterification has also attracted much attention for biodiesel production as it produces high purity product. Separation from the by-product, glycerol, is very easy. But the cost of enzyme remains a barrier for its industrial implementation. In order to increase the cost effectiveness of the process, enzyme deactivation by glycerol is reduced and thus it is reused by immobilizing in a suitable biomass supported particle [10].

In enzymatic transesterification reaction, Jatropha oil is immiscible with alcohol due to polar and nonpolar nature of alcohol and oil respectively which causes mass transfer resistance problem [11]. Mechanical stirring is applied to overcome this problem. It enhances the rate of reaction by reducing mass transfer resistance [12–14]. Stirring reduces mass transfer resistance by producing higher diffusion and low effectiveness factor. Moreover, glycerol is formed during the transesterification of Jatropha oil. It decreases the enzyme activity by forming a layer on it. The glycerol inhibition is due to mass transfer limitation in immobilized enzymes [15]. Glycerol forms a hydrophilic layer that is not completely miscible with oil. This hydrophilic layer serves as a partition between alcohol and oil and decreases in the conversion rate of oil to biodiesel. Glycerol accumulates in the mixture to such an extent that the reactivity of the enzyme is decreased during the transesterification process [16]. To overcome this problem, stirring is introduced in the system which removes the glycerol-enzyme layer thereby increasing the activity of the enzyme. Thus, optimization of mechanical agitation for evaluation of the mass transfer resistance and glycerol phase is essential for maximum biodiesel production and to make the process cost effective.

Mathematical modeling is an important tool to determine suitable reaction conditions for chemical or biochemical reactive system [17–20]. There are few research articles available on modeling considering enzymatic biodiesel production. Deng et al. [21] established a kinetic model for lipase-catalyzed biodiesel production from waste cooking oil and determines reaction conditions using the mathematical model. Halim et al. [12] have experimentally shown that biodiesel yield is initially mass transfer controlled, and mathematical models for biodiesel production have also been developed for the cases of mass transfer [22–24]. Roy et al., 2014 [22] and Basir et al., 2015 [17] have established mathematical models transesterification of Jatropha oil and shown that mass transfer limitation can be avoided by controlling stirrer rotation in an optimum level. But, there are no such mathematical models considering mass transfer limitation between the polar methanol/glycerol phase and nonpolar oil phase in biodiesel production.

Here, the aim is to reduce mass transfer resistance due to immiscibility of alcohol and oil, as well as to remove glycerol phase to increase enzyme activity by optimizing stirring. With this view, a mathematical model is developed considering the glycerol effect and mass transfer resistance in different phases of biodiesel production. Using optimal control theory, an optimal stirring profile is derived to avoid mass transfer limitation and glycerol-related problem, and to make the production process cost effective. Results obtained from numerical simulation of the model system are in a good agreement with experimental results.

22.2 The Mathematical Model

Here, we have considered the mathematical model formulated by Basir, Datta, and Roy [17], and developed the following model by some additional assumptions. The following assumptions are taken to develop the mathematical model for the enzymatic transesterification of Jatropha oil.

Enzymatic transesterification of Jatropha curcas oil with an alcohol (AL) can be described as two-step process. The first step is the hydrolysis of Jatropha curcas oil or TG to produce acylated enzyme (F) and release of glycerol through a complex C_1 (i.e., complex $[E.TG]$); and the second step is the esterification of methanol (AL) with F to form the desired product, i.e., biodiesel with the release of free enzyme (E) through a second complex C_2 $[F.AL]$, [25]. Moreover, Glycerol reacts with enzyme and forms a hydrophilic layer (complex C_3, i.e., [E.G]) that serves as a partition between enzyme and oil [15, 16].

All the mechanistic steps for the biodiesel production can be represented by the following sequence of reactions:

$$
\begin{aligned}
E + TG &\underset{k_{-1}}{\overset{k_1}{\rightleftharpoons}} [E.TG] \underset{k_{-2}}{\overset{k_2}{\rightleftharpoons}} F + GL \\
F + AL &\underset{k_{-3}}{\overset{k_3}{\rightleftharpoons}} [F.AL] \underset{k_{-4}}{\overset{k_4}{\rightleftharpoons}} E + BD \\
E + GL &\underset{k_{-5}}{\overset{k_5}{\rightleftharpoons}} [E.GL]
\end{aligned}
\tag{22.1}
$$

Here, k_1, k_2, k_3 and k_4 are the forward rate constants and k_{-1}, k_{-2}, k_{-3}, and k_{-4} are the backward rate constants.

Mixing intensity directs the mass transfer between reaction phases, so mechanical stirring has an effect on biodiesel yield. Here, we use k_s as the mass transfer rate constant and the term has been defined as below [17, 22]

$$k_s = \frac{a}{1 + e^{-b(N-c)}}, \tag{22.2}$$

where N is the speed of stirrer and a, b, and c are constants. Here B_{max} represents maximum production of biodiesel in an ideal reaction conditions. We denote the concentration of TG, E, F, $C1$, $C2$, AL, BD, $C3$ and GL as x_T, x_E, x_F, x_{C1}, x_{C2}, x_A, x_B, x_G and x_{C3} respectively. Now from the above assumptions with the above reaction mechanism followed by law of mass action, we can formulate the set of differential equations given below:

$$\begin{aligned}
\frac{dx_E}{dt} &= -k_1 x_T x_E + k_{-1} x_{C1} + k_4 x_{C2} - k_{-4} x_E x_B \\
&\quad -k_5 x_E x_G + k_{-5} x_{C3}, \\
\frac{dx_T}{dt} &= -k_1 x_T x_E + k_{-1} x_{C1}, \\
\frac{dx_F}{dt} &= k_2 x_{C1} - k_{-2} x_F x_G - k_3 x_F x_A + k_{-3} x_{C2}, \\
\frac{dx_B}{dt} &= k_4 x_{C2} - k_{-4} x_E x_B + k_s x_B \left(1 - \frac{x_B}{B_{max}}\right), \\
\frac{dx_A}{dt} &= -k_3 x_F x_A + k_{-3} x_{C2}, \\
\frac{dx_{C1}}{dt} &= k_1 x_T x_E - k_{-1} x_{C1} - k_2 x_{C1} + k_{-2} x_F x_G, \\
\frac{dx_{C2}}{dt} &= k_3 x_F x_A - k_{-3} x_{C2} - k_4 x_{C2} + k_{-4} x_E x_B, \\
\frac{dx_G}{dt} &= k_2 x_{C1} - k_{-2} x_F x_G - k_5 x_E x_G + k_{-5} x_{C3}, \\
\frac{dx_{C3}}{dt} &= k_5 x_E x_G - k_{-5} x_{C3} - k_s x_{C3} \left(1 - \frac{x_{C3}}{C_{3max}}\right).
\end{aligned} \tag{22.3}$$

The initial conditions are as follows:

$$\begin{aligned}
&x_E(0) = x_{E_0},\ x_{C1}(0) = 0,\ x_F(0) = 0, \\
&x_A(0) = x_{A_0},\ x_T(0) = x_{T_0},\ x_{C2}(0) = 0, \\
&x_B(0) = 0,\ x_G(0) = 0, x_{C3}(0) = 0.
\end{aligned} \tag{22.4}$$

22.3 The Optimal Control Problem

The objective of giving optimal control on stirring is to find an optimal profile of stirring to get maximum and cost-effective production of biodiesel. We use the control variable $u(t)$, which represents the stirring activator input at time t satisfying $0 \leq u(t) \leq 1$ [22].

Incorporating the control $u(t)$, the system (22.3) becomes,

$$\begin{aligned}
\frac{dx_E}{dt} &= -k_1 x_T x_E + k_{-1} x_{C1} + k_4 x_{C2} - k_{-4} x_E x_B - k_5 x_E x_G + k_{-5} x_{C3}\\
\frac{dx_T}{dt} &= -k_1 x_T x_E + k_{-1} x_{C1},\\
\frac{dx_F}{dt} &= k_2 x_{C1} - k_{-2} x_F x_G - k_3 x_F x_A + k_{-3} x_{C2},\\
\frac{dx_B}{dt} &= k_4 x_{C2} - k_{-4} x_E x_B + u(t) k_s x_B \left(1 - \frac{x_B}{B_{max}}\right),\\
\frac{dx_A}{dt} &= -k_3 x_F x_A + k_{-3} x_{C2},\\
\frac{dx_{C1}}{dt} &= k_1 x_T x_E - k_{-1} x_{C1} - k_2 x_{C1} + k_{-2} x_F x_G,\\
\frac{dx_{C2}}{dt} &= k_3 x_F x_A - k_{-3} x_{C2} - k_4 x_{C2} + k_{-4} x_E x_B,\\
\frac{dx_G}{dt} &= k_2 x_{C1} - k_{-2} x_F x_G - k_5 x_E x_G + k_{-5} x_{C3},\\
\frac{dx_{C3}}{dt} &= k_5 x_E x_G - k_{-5} x_{C3} - u(t) k_s x_{C3} \left(1 - \frac{x_{C3}}{C_{3max}}\right).
\end{aligned} \tag{22.5}$$

with initial conditions as given by (22.4).

The above state system can be written in a compact form as

$$\frac{dx}{dt} = f(x, u, t), \tag{22.6}$$

$x = (x_1, x_2, \ldots, x_9)^T$ and $f = (f_1, f_2, \ldots, f_9)^T$, f_i, i = 1, 2, ..., 9, 3 are right sides of the above system. The cost function is thus formulated as

$$J[u(t)] = \int_{t_0}^{t_f} [Pu^2(t) - Qx_B^2(t) + Rx_{C3}^2]dt. \tag{22.7}$$

The parameter P (> 0) is the weight constant on the benefit of the cost of production and $Q > 0$, $R > 0$ are the penalty multiplier. Thus, we have to find out the optimal control $u^*(t)$ such that

$$J(u^*) = \min\ \{J(u) : u \in U\},$$

where U is the admissible control set defined as:

$$U = \{u(t) : u(t) \text{ is measurable}, 0 \leq u(t) \leq 1,\ t \in [t_i, t_f]\}. \tag{22.8}$$

Pontryagin Minimum Principle [27, 28] is used to find the optimal stirring in term of $u^*(t)$. For this, the Hamiltonian is formulated as,

$$H = Pu^2(t) - Qx_B^2(t) + Rx_{C3}^2 + \xi^T f. \tag{22.9}$$

Theorem 22.3.1 *If the given optimal control $u^*(t)$ and the solution $(x_E^*, x_T^*, x_F^*, x_B^*, x_A^*, x_{C1}^*, x_{C2}^*, x_G)$ of the corresponding system (22.3) minimize $J(u)$ over U, then there exists adjoint variables ξ_1 - ξ_9 which satisfying the following equations:*

$$\begin{aligned}
\frac{d\xi_1}{dt} &= k_1 x_T(\xi_1 - \xi_6) + k_{-4} x_B(\xi_4 - \xi_7) + k_5 x_G(\xi_1 + \xi_8 - \xi_9),\\
\frac{d\xi_2}{dt} &= k_1 x_E(\xi_1 + \xi_2 - \xi_6),\\
\frac{d\xi_3}{dt} &= k_3 x_A(\xi_5 - \xi_7) + k_{-2} x_G(\xi_3 - \xi_6),\\
\frac{d\xi_4}{dt} &= 2Qx_B - \xi_4\left[k_{-4}x_E + \frac{u(t)k_s x_B}{B_{max}} + u(t)k_s\left(1 - \frac{x_B}{B_{max}}\right)\right] - \xi_7 k_{-4} x_E,\\
\frac{d\xi_5}{dt} &= k_3 x_F(\xi_5 - \xi_7),\\
\frac{d\xi_6}{dt} &= k_{-1}(\xi_6 - \xi_2) + k_2(\xi_6 - \xi_3),\\
\frac{d\xi_7}{dt} &= -k_4\xi_4 - k_{-3} + k_{-3}\xi_7,\\
\frac{d\xi_8}{dt} &= k_{-2}(\xi_8 - \xi_6) + k_5 x_E(\xi_1 + \xi_8 - \xi_9),\\
\frac{d\xi_9}{dt} &= k_{-5}(-\xi_1 - \xi_8 + \xi_9) + \xi_9\left[\frac{u(t)k_s x_{C3}}{C_{3max}} + u(t)k_s\left(1 - \frac{x_{C3}}{C_{3max}}\right)\right],
\end{aligned} \tag{22.10}$$

along with the boundary conditions $\xi_i(t_f) = 0$ for $i = 1$ to 9. Further, $u^*(t)$ can be written as,

$$u^*(t) = max\left(0,\ min\left(1,\ \frac{uk_s[\xi_9 x_{C3}(1 - \frac{x_{C3}}{C3_{max}}) - \xi_4 x_B(1 - \frac{x_B}{B_{max}})]}{2P}\right)\right).$$

Proof The Hamiltonian (22.9) can be written as

$$\begin{aligned}
H = {} & Pu^2(t) + uk_s\left[\xi_4 x_B(1 - \frac{x_B}{B_{max}}) - \xi_9 x_{C3}(1 - \frac{x_{C3}}{C3_{max}})\right]\\
& + \text{terms without } u(t).
\end{aligned} \tag{22.11}$$

According to the Pontryagin Minimum Principle, the unconstrained optimal control variable $u^*(t)$ satisfies

$$\frac{\partial H}{\partial u^*} = 0. \tag{22.12}$$

Thus from (22.11) and (22.12), we have

$$\frac{\partial H}{\partial u^*} = 2Pu^* + \xi_4 k_s x_B \left(1 - \frac{x_B}{B_{max}}\right) - k_s \xi_9 x_{C3} \left(1 - \frac{x_{C3}}{C3_{max}}\right) = 0.$$

Solving we get,

$$u^*(t) = \frac{k_s[\xi_9 x_{C3}(1 - \frac{x_{C3}}{C3_{max}}) - \xi_4 x_B(1 - \frac{x_B}{B_{max}})]}{2P}. \tag{22.13}$$

Due to the boundedness of the standard control, the compact form of $u^*(t)$ is

$$u^*(t) = max\left(0,\ min\left(1,\ \frac{k_s[\xi_9 x_{C3}(1 - \frac{x_{C3}}{C3_{max}}) - \xi_4 x_B(1 - \frac{x_B}{B_{max}})]}{2P}\right)\right). \tag{22.14}$$

According to Pontryagin Minimum Principle, adjoint variables satisfy the following equation:

$$\frac{d\xi}{dt} = -\frac{\partial H}{\partial x}, \tag{22.15}$$

where $x = (x_E, x_T, x_F, x_B, x_A, x_{C1}, x_{C2}, x_G, x_{C3})^T$ and $\xi = (\xi_1, \xi_2, \ldots, \xi_9)^T$ and the necessary condition satisfying the optimal control $u(t)$ is given by

$$H\left(x(t), u^*(t), \xi(t), t\right) = \min_{u \in U}\left(H(x_i(t), u(t), \xi(t), t)\right). \tag{22.16}$$

So, adjoint equations (22.10) can be determined by Eq. (22.15) with boundary conditions $\xi(t_f) = 0$.

The adjoint system (22.5) together with state (22.14) and optimal control (22.10) represents the optimality system. Thus, the optimal system consists of state system, adjoint system, and the optimal control. Moreover, the optimal profiles for stirring (N^*) can be obtained by the following relations:

$$u^* k_s = \frac{a}{1 + e^{-b(N^* - c)}}. \tag{22.17}$$

22.4 Numerical Simulation

In this section, numerical simulation of the model equations are shown in the following figures. To study the effect of glycerol phase on mass transfer resistance, reaction parameters such as enzyme loading and stirring on are varied. The optimal control problem is solved to find the optimal stirring profile to minimize enzyme deactivation by glycerol and cost-effective production of biodiesel.

To study the possible mass transfer effect on the biodiesel productivity, simulation of the model has been performed using different agitating speeds (150–600 rpm) keeping all other variables fixed. In Fig. 22.1, raising the stirring rate from 150 to 600 rpm, biodiesel yield is plotted at the fixed molar ratio (4:1). We have seen that, stirrer rotation reduces the mass transfer resistance and increases overall reaction rate. It is also seen that, the conversion of oil to biodiesel could not be enhanced by further increment of agitation speed above 300 rpm. The stirring over 350 rpm decreases the yield, which is possibly due to the shearing of the lipase molecule or inactivation of the lipase.

Influence of glycerol on enzyme is shown in Fig. 22.2. Enzyme activity is paralyzed by glycerol due to the formation of complex C3. As stirring increases, formation of complex C3 is decreased and hence inhibition of enzyme by glycerol is reduced by this process. Stirrer speed removes the hydrodynamic boundary layer near the enzyme surface which enhances the activity of the enzyme.

Figure 22.3 shows that effect of enzyme amount has significant effect on biodiesel yield. Here, final concentration of biodiesel is plotted by varying the amount of enzyme. It is established that 0.25 mol/L enzyme loading is the best for biodiesel production in presence of 200–250 rpm stirring. The addition of larger lipase quantity

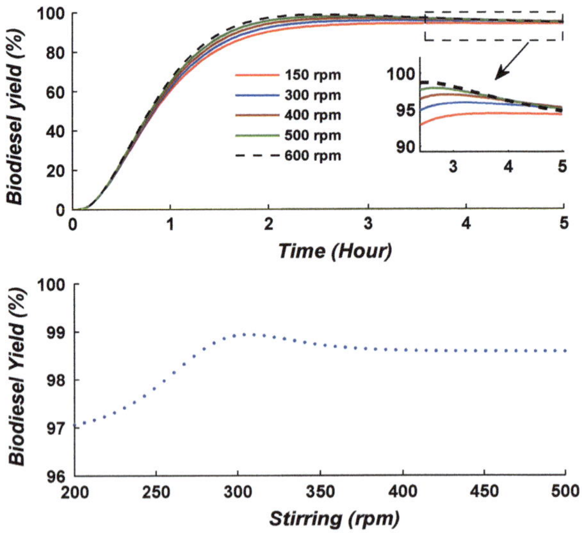

Fig. 22.1 Effect of parameters such as stirring on biodiesel production is shown with parameter as given in Table 22.1

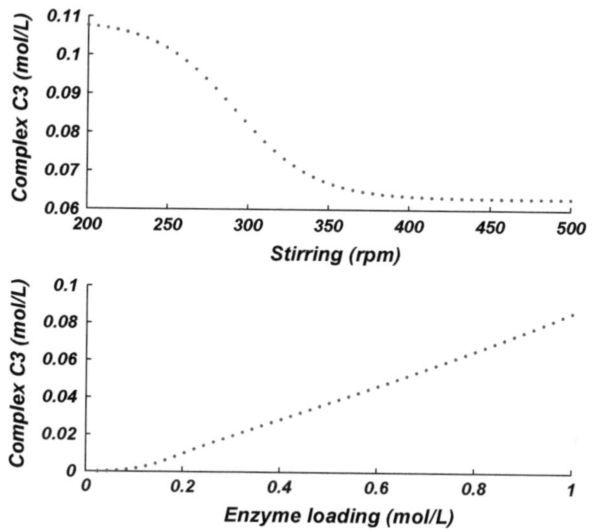

Fig. 22.2 Concentration of C3 as function stirring (N) is shown using the parameters as given in Table 22.1

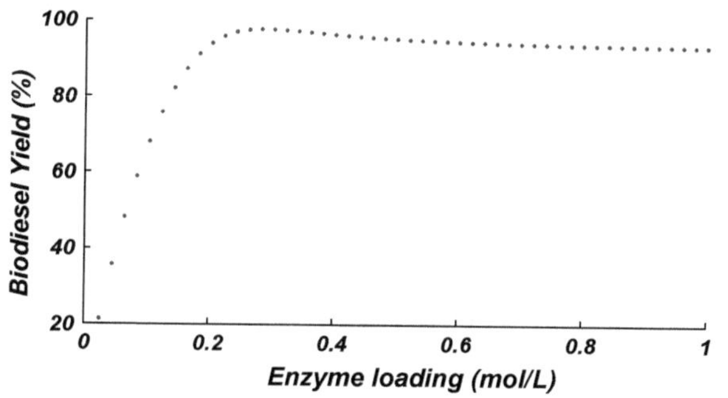

Fig. 22.3 Yield of biodiesel is plotted with varying enzyme concentration (*lower panel*) and combined effect of stirring and initial enzyme concentration on biodiesel yield is shown (*upper panel*) with 4:1 methanol to oil molar ratio and other parameters as given in Table 22.1 and 4 h of reaction time

is not practical due to the formation of excess amount of complex $C3$ (see Fig. 22.2) and also the raw materials and lipase together make the solution extremely viscous [13]. Thus, higher amount of lipase will not help to increase of biodiesel yield further.

Figure 22.4 shows that optimal stirring produces the highest biodiesel yield using a lower enzyme loading ($x_E(0) = 0.25$ mol/L). Initial mass transfer rate is increased and reaction time is also reduced. Thus, production of biodiesel is more favorable in enzymatic transesterification reaction using optimum stirring profile. It reduces time and cost of production. This figure also shows that optimal stirring is needed for three hours from the beginning of the reaction. Also from Fig. 22.5, it is seen that

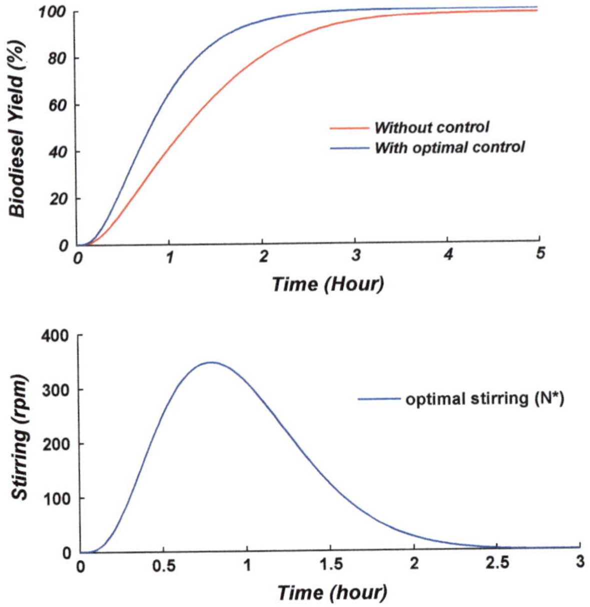

Fig. 22.4 Control profile of biodiesel and optimal stirring ($N^*(t)$) is plotted as a function of time and for two cases, with control and without control with parameters as given in Table 22.1

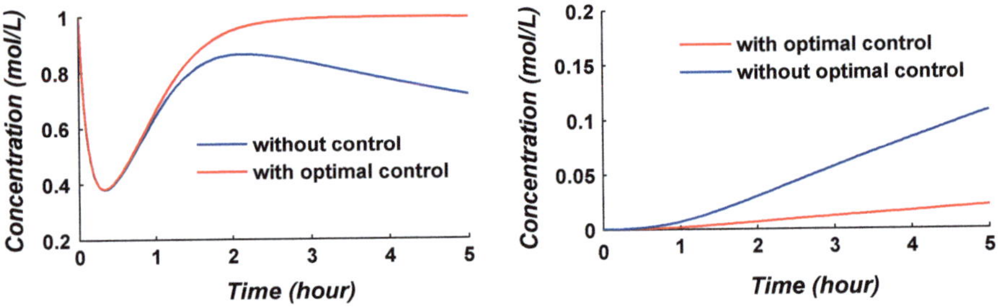

Fig. 22.5 Control profile of enzyme and complex C_3 are plotted respectively as function of time for two cases, with control and without control with parameters as given in Table 22.1

inhibition of enzyme by glycerol is minimized, enzyme is reverted back at the end of the reaction which can be reused further.

22.5 Discussion and Conclusion

In this research article, a mathematical model has been developed for enzymatic biodiesel production from Jatropha oil considering mass transfer resistance in both oil phase and glycerol phase. The main focus is on maximum production of biodiesel

Table 22.1 Values of rate constants at temperature 40 °C and other parameters used for numerical simulation of the model system [17, 26]

Parameters	Value	Unit
k_1	7.5128	mol L^{-1} h^{-1}
k_{-1}	0.1147	h^{-1}
k_2	0.1032	h^{-1}
k_{-2}	0.0988	mol L^{-1} h^{-1}
k_3	1.937	mol L^{-1} h^{-1}
k_{-3}	0.0323	h^{-1}
k_4	1.9230	h^{-1}
k_{-4}	0.0011	mol L^{-1} h^{-1}
r_5	0.1213	mol L^{-1} h^{-1}
r_{-5}	0.03887	h^{-1}
a	0.320	–
b	0.003	rpm^{-1}

and therefore mass transfer limitation is minimized in both the phases. Influences of glycerol on mass transfer resistance on biodiesel productivity, in terms of stirring are studied using the mathematical model. For cost-effective production of biodiesel through enzyme catalyzed reaction, we have applied control approach on mixing intensity. It has been shown that control measures have a great impact on reaction system and gives maximum production of biodiesel. Mass transfer is significantly influenced by agitation speed, enzyme amount, and glycerol. Finally, using control theory, an optimal profile for stirring is obtained to minimize mass transfer resistance, hence cost-effective maximum glycerol-free biodiesel can be produced within 3 h.

Acknowledgements The research work is supported by UGC-DRS Programe, Dept. of Mathematics, Jadavpur University and DST PURSE, Govt. of India.

References

1. Knothe, G., Gerpen, J.V., Krahl, J.: The Biodiesel Handbook. AOCS Press, Champaign (2005)
2. Tapanes, N.C.O., Aranda, D.A.G., Carneiro, J.W.D.M., Antunes, O.A.C.: Transesterification of Jatropha curcas oil glycerides: theoretical and experimental studies of biodiesel reaction. Fuel **87**, 2286–95 (2008)
3. Venturino, E., Roy, P.K., Al Basir, F., Datta, A.: A model for the control of the mosaic virus disease in Jatropha curcas plantations. Energy, Ecology and Environment, pp. 1–10 (2016)
4. Roy, P.K., Li, X.-Z., Al Basir, F., Datta, A., Chowdhury, J.: Effect of insecticide spraying on Jatropha curcas plant to control mosaic virus: a mathematical study. Commun. Math. Biol. Neurosci. **2015**, 1–19 (2015)
5. Al Basir, F., Venturino, E., Roy, P.K.: Effects of awareness program for controlling mosaic disease in Jatropha curcas plantations. Math. Methods App. Sci. (2016). doi:10.1002/mma.4149

6. Tapanes, N.O.: Transesterification of Jatropha Curcas oil glycerides: theoretical and experimental studies of biodiesel reaction. Fuel **87**, 2286–95 (2008)
7. Freedman, B., Pryde, E.H., Mounts, T.L.: Variables affecting the yields of fatty esters from transesterified vegetable oils. JAOCS **61**, 1638–43 (1984)
8. Berchmans, H.J., Morishita, K., Takarada, T.: Kinetic study of hydroxide-catalyzed methanolysis of Jatropha curcas-waste food oil mixture for biodiesel production. Fuel **104**, 46–52 (2013)
9. Fjerbaek, L., Christensen, K.V., Norddahl, B.: A review of the current state of biodiesel production using enzymatic transesterification. Biotech. Bioenerg. **102**, 1298–315 (2009)
10. Ko, C., Yeh, K., Wang, Y., Wu, C., Chang, F., Cheng, M., Liou, C.: Impact of methanol addition strategy on enzymatic transesterification of Jatropha oil for biodiesel processing. Energy **48**, 375–379 (2012)
11. Marjanovíc, A.V., Stamenkovíc, O.S., Todorovíc, Z.B., Lazíc, M.L., Veljkovíc, V.B.: Kinetics of the base-catalyzed sunflower oil ethanolysis. Fuel **89**(3), 665–671 (2010)
12. Halim, S.F.A., Kamaruddin, A.H.: Catalytic studies of lipase on FAME production from waste cooking palm oil in a tert-butanol system. Process Biochem. **43**, 1436–1439 (2008)
13. Sim, J.H., Harun, K.A., Bhatia, S.: Effect of mass transfer and enzyme loading on the biodiesel yield and reaction rate in the enzymatic transesterification of crude palm oil. Energy Fuel **23**, 4651–4659 (2009)
14. Yadav, G.D., Trivedi, A.H.: Trivedi, Kinetic modeling of immobilized-lipase catalyzed transesterification of n-octanol with vinyl acetate in non-aqueous media. Enzyme Microb. Technol. **30**, 783–789 (2003)
15. Hong, W.P., Park, J.Y., Min, K., Ko, M.J., Park, K., Yoo, Y.J.: Kinetics of glycerol effect on biodiesel production for optimal feeding of methanol. Korean J. Chem. Eng. **28**(9), 1908–1912 (2011)
16. Dossat, V., Combes, D., Marty, A.: Continuous enzymatic transesterification of high oleic sunflower oil in a packed bed reactor: influence of the glycerol production. Enzyme Microb. Technol. **25**, 194 (1999)
17. Al Basir, F., Datta, S., Roy, P.K., Studies on biodiesel production from Jatropha Curcas oil using chemical and biochemical methods–a mathematical approach. Fuel **158**, 503–511 (2015)
18. Zhao, X., Qi, F., Yuan, C., Du, W., Liu, D.: Lipase-catalyzed process for biodiesel production: enzyme immobilization, process simulation and optimization. Renew. Sustain. Energy Rev. **44**, 182–197 (2015)
19. Likozar, B., Levec, J.: Effect of process conditions on equilibrium, reaction kinetics and mass transfer for triglyceride transesterification to biodiesel: experimental and modeling based on fatty acid composition. Fuel Process. Technol. **122**, 30–41 (2014)
20. Al Basir, F., Roy, P.K.: Production of biodiesel using enzymatic transesterification of Jatropha Curcas oil: a mathematical study. J. MESA **5(2)**, 175–184 (2014)
21. Liu, S., Nie, K., Zhang, X., Wang, M., Deng, L., Ye, Z.: Kinetic study on lipasecatalyzed biodiesel production from waste cooking oil. J. Mol. Catal. B: Enzym. **99**, 43–50 (2014)
22. Roy, P.K., Datta, S., Nandi, S., Basir, F.A.: Effect of mass transfer kinetics for maximum production of biodiesel from Jatropha Curcas oil: a mathematical approach. Fuel **134**, 39–44 (2014)
23. Likozar, B., Pohar, A., Levec, J.: Transesterification of oil to biodiesel in a continuous tubular reactor with static mixers: modelling reaction kinetics, mass transfer, scale-up and optimization considering fatty acid composition. Fuel Process. Technol. **142**, 326–336 (2016)
24. Aniya, V.K., Muktham, R.K., Alka, K., Satyavathi, B.: Modeling and simulation of batch kinetics of non-edible karanja oil for biodiesel production: a mass transfer study. Fuel **161**, 137–145 (2015)
25. Fjerbaek, L., Knud, V.C., Birgir, N.: A Review of the Current State of Biodiesel Production Using Enzymatic transesterification. Wiley InterScience (2009)
26. Liu, S., Nie, K., Zhang, X., Wang, M., Deng, L., Ye, Z., Wang, F., Tan, T.: Kinetic study on lipase-catalyzed biodiesel production from waste cooking oil. J. Mol. Catal. B: Enzym. **99**, 43–50 (2014)

27. Pontryagin, L.S., Boltyanskii, V.G., Gamkarelidze, R.V., Mishchenko, E.F.: Mathematical Theory of Optimal Process. Gordon and Breach Science Publishers, New York (1986)
28. Swan, G.M.: Application of optimal control theory in biomedicine. Marcel Dekker Inc., New York (1984)

Chapter 23
Role of Bio-Pest Control on Theta Logistic Populations: A Case Study on Jatropha Curcus Cultivation System

Jahangir Chowdhury, Sourav Rana, Sabyasachi Bhattacharya and Priti Kumar Roy

Abstract Renewable crops are the most demanding source for biodiesel production. *Jatropha sp.* oil has shown promising features in generating renewable energy source. Presently, the crop cultivation faces severe damage to the seed production during pest invasion. We consider a nonlinear mathematical system with biomass of *Jatropha sp.*, susceptible pest population, infected pest population and virus population. The biomass of *Jatropha sp.* and susceptible pest population follows theta logistic growth as theta logistic growth curve is a more natural choice in comparison with the classical logistic growth curve model. Introduction of pest control by the application of Nuclear Polyhedrosis Virus (NPV) was applied through foliar spraying to arrest the pest invasion. The values of θ have to depend on the process of interaction at different densities. In this research communication we observe how various values of theta can affect crop survival during pest invasion which makes the model biologically more realistic. Stability and bifurcation analyses have been worked out for the system. Analytically and numerically we find out the threshold value of $\theta = 0.74$. We have seen that for $\theta < 0.74$ the system is stable and for $\theta \geq 0.74$, the system shows limit cycle oscillation which holds upto $\theta = 1$. Analytical and numerical results based on simulated findings validate our mathematical model.

Keywords Biodiesel · Biological control · Pest · Mathematical modeling · Virus

---

J. Chowdhury · P.K. Roy (✉)
Centre for Mathematical Biology and Ecology, Department of Mathematics, Jadavpur University, Kolkata 700032, India
e-mail: pritiju@gmail.com

J. Chowdhury
e-mail: jahangirchowdhury.ju@gmail.com

S. Rana
Department of Statistics, Visva-Bharati University, Santiniketan 731235, West Bengal, India

S. Bhattacharya
Agricultural and Ecological Research Unit, Indian Statistical Institute, Kolkata, India

P. Manchanda et al. (eds.), *Industrial Mathematics and Complex Systems*, Industrial and Applied Mathematics, DOI 10.1007/978-981-10-3758-0_23

23.1 Introduction

The growing consumption of oil resource can deplete the present natural fossil oil reserve in near future. In order to meet the growing oil demand, we need to develop alternative resource. Recently, biodiesel evolves as a promising alternative fuel to replace traditional petroleum diesel. Biodiesel is best obtained from the plant *Jatropha sp.* among many alternative crops. *Jatropha sp.* is a renewable non-edible oil producing plant [1]. *Jatropha sp.* is not pest and disease resistant [2]. The main obstacle of *Jatropha sp.* seed oil production is the pest attack. The major pests and diseases affecting Jatropha are: (1) the leaf miner *Stomphastis thraustica*, (2) the leaf and stem miner *Pempelia morosalis*, and (3) the shield-backed bug *Calidea panaethiopica*, which can cause flower and fruit abortion.

Pest management models through control strategies have been designed and analyzed by many researchers mathematically oftentimes [3–5]. The practical evidence of the use of virus against insect pests is practised in North America and European countries [6]. The experimental and field use of pathogenic viruses in Europe is listed by Falcon et al. [7]. Roy et al. [8] studied the effect of insecticide spraying on *Jatropha sp.* and using mathematical model they showed how to control mosaic virus by adopting appropriate strategy. In this system, the growth profile of the pest is the primary source of this attack. However, the logistic growth of the pest population is considered in most of the recent studies [3, 4].

Note that in a recent study Sibly et al. [9] suggested density regulation in species growth for most of the taxonomic species (e.g., fish, birds, mammals, and insects) is concave. The pest is a member of this insect family and it is reasonable to assume theta logistic structure in their growth to incorporate this density regulation. So we consider the theta logistic growth of insect population instead of logistic growth which is more appropriate. Theta logistic equation is followed in the form of density dependence with extra liberty, where the parameter θ indicates the curvature of relationship. There are concave relationships between abundance and per capita growth rate (PGR) if $\theta < 1$ and convex relationship if $\theta > 1$ [10]. When $\theta = 1$, the term is equivalent to the logistic growth term.

Recently, researchers [11–16] are paying attention to explore ecological and epidemiological outbreak in discrete-time setup. There are three definite causes behind using of discrete time models. Firstly, it is very much appropriate and perfect to explain pest control scenario by applying discrete-time models as statistical information is collected at discrete time. Secondly, numerical simulations of continuous models are attained through discretization. Finally, too many critical dynamical behavioral patterns are furnished for discrete-time models [16, 17]. However, no substantial and systematic work has been done on bio-pest control with theta logistic growth profile. We have studied the impact of theta logistic growth on this cultivation system by mathematical analysis and numerical simulations.

The NPV belongs to the family of Baculoviruses. NPV is known for high epizootic levels, self-perpetuating and safe to natural enemies due to host specificity. NPV is being one of the important biopesticides, as it is eco-friendly, having less residual toxicity, compatible with many chemical pesticides. Hence, NPV can be implemented as one of major components in IPM programme.

Our goal is to study whether the density dependent theta logistic growth of pest has a substantial impact on the survival of *Jatropha sp.*. Through mathematical modeling we want to study how these type of eco-epidemiological system minimizes the damage of *Jatropha sp.*. We also study the impact of NPV to minimize pest population within the system dynamics.

The research article is organized as follows, in the first section we discuss the ecological backdrop of the problem and a general introduction. In the next section we present the formulation of the mathematical model. In Sect. 23.3 we perform local stability analysis of different equilibriums. In Sect. 23.4, we study the dynamics of the system, namely stability of the system. Numerical simulations are shown in Sect. 23.5. Finally, we discuss our results and conclude our findings.

23.2 Formulation of the Mathematical Model

In our formulated model, we consider four populations to analyze the system:

(i) Biomass of Jatropha plant, $j(t)$,
(ii) The susceptible pest population, $s(t)$,
(iii) The infected pest, $i(t)$ and
(iv) The virus, $v(t)$.

We formulate a four-species mathematical model, which contains biomass of *Jatropha sp.*, susceptible pest, infected pest and virus. Biomass of *Jatropha sp.* grows in a logistic fashion with carrying capacity $k_1 (k_1 > 0)$ and with an intrinsic growth rate constant r_1 for pest-free system. *Jatropha sp.* is affected by pest and that is why plant biomass is abolished with a simple mass action βjs, where β represents effective per capita pest contact rate with plant. Susceptible pest infected by virus with mass action λsv, where λ represents effective per capita pest contact rate with virus. We assume that only susceptible pest is capable of reproducing with logistic growth term, i.e., infected class of pest is removed by lysis before having possibility of reproducing. The infected individuals fail to contribute the reproduction process due to their inability to compete for resources. However, they still contribute with susceptible pest to population growth towards the carrying capacity. The lysis of infected pest largely produces virus, on average per insect is called the virus replication number. The virus population $v(t)$ has natural mortality due to temperature changes, enzymatic attack, pH dependence, etc.

Based on the above assumptions, we can formulate the following mathematical model:

$$\frac{dj}{dt} = jr_1\left(1 - \frac{j}{k_1}\right) - \beta js,$$

$$\frac{ds}{dt} = sr_2\left(1 - \frac{s+i}{k_2}\right) + \beta js - \lambda sv,$$
$$\frac{di}{dt} = \lambda sv - \xi i,$$
$$\frac{dv}{dt} = \pi_v + \kappa\xi i - \mu v, \qquad (23.1)$$

where r_2 is the per capita growth rate of susceptible pest population, k_2 is the per capita carrying capacity of susceptible pest population, ξ is the mortality rate of infected pest, κ is the virus replication number and π_V is the constant rate of reproduction of free virus.

Logistic growth of a population follows an S-shaped sigmoid curve when the population increases its density. But in nature relationship between density and per capita growth are concave for most of the species (e.g., fish, birds, mammals and insects) as observed by Sibly (2005) [9]. Incorporation of this density regulation as theta logistic equation is appropriate, because pest is a member of insect family. Hence, theta logistic growth curve is more realistic than the classical logistic growth model. Here we incorporate the discrete version of the model (23.1), since the species abundance data for eco-epidemiological study are generally observed in discrete time setup.

Based on the above perception along with the theta logistic growth in susceptible pest using the Forward Euler Scheme for discretization [16], we have revised the Eq. (23.2) as

$$j_{t+1} = j_t + l\left[j_t r_1\left[1 - \left(\frac{j_t}{k_1}\right)\right] - \alpha s_t j_t\right],$$
$$s_{t+1} = s_t + l\left[s_t r_2\left[1 - \left(\frac{s_t + i_t}{k_2}\right)^{\theta}\right] + \alpha s_t j_t - \beta s_t v_t\right],$$
$$i_{t+1} = s_t + l\left[\beta s_t v_t - \xi i_t\right],$$
$$v_{t+1} = v_t + l\left[\pi_v + \kappa\xi i_t - \mu v_t\right], \qquad (23.2)$$

where $l(>0)$ denotes the step size and $\theta(>0)$ describes the curvature of the relationship.

23.3 Dynamics of the System

23.3.1 Equilibria and Stability

The above system (23.2) has four equilibrium points, viz.

(a) The axial equilibrium point $E_0(0, 0, 0, \frac{\pi_v}{\mu})$,
(b) Pest free equilibrium point $E_1(k_1, 0, 0, \frac{\pi_v}{\mu})$,

(c) Virus free equilibrium point $E_2(j_2, s_2, 0, 0)$ here $\pi_v = 0$ and
(d) The interior equilibrium point $E^*(j^*, s^*, i^*, v^*)$.

where,

$j^* = k_1\left(1 - \frac{\alpha\mu\xi i^*}{r_1\beta(\kappa\xi i^* + \pi_v)}\right)$,
$s^* = \frac{\mu\xi i^*}{\beta(\kappa\xi i^* + \pi_v)}$,
$v^* = \frac{\kappa\xi i^* + \pi_v}{\mu}$ and i^* is the positive root of the following equation:

$$r_2\left[1 - \left(\frac{s^* + i^*}{k_2}\right)^\theta\right] + \alpha j^* - \beta v^* = 0.$$

Proposition 23.1 *The system always unstable around axial equilibrium point $E_0(0, 0, 0, \frac{\pi_v}{\mu})$. Again at disease free equilibrium point $E_1(k_1, 0, 0, \frac{\pi_v}{\mu})$ the system is stable around E_1 if $\mid 1 - lr_1\theta_1 \mid < 1$, $\mid 1 + lr_2 + l\alpha k_1 - \frac{\beta\pi_v l}{\mu} \mid < 1$, $\mid 1 - \xi l \mid < 1$ and $\mid 1 - l\mu \mid < 1$, critically stable if at least one of above inequality hold equality. Otherwise the system is unstable.*

The Jacobian matrix of the system at vanishing equilibrium point $E_0(0, 0, 0, \frac{\pi_v}{\mu})$ is given by:

$$J(E_0) = \begin{bmatrix} 1 + lr_1 & 0 & 0 & 0 \\ 0 & 1 + lr_2 - \frac{\beta\pi_v l}{\mu} & 0 & 0 \\ 0 & \frac{\beta\pi_v l}{\mu} & 1 - \xi l & 0 \\ 0 & 0 & \kappa\xi l & 1 - l\mu \end{bmatrix}.$$

$1 + lr_1$, $1 + lr_2 - \frac{\beta\pi_v l}{\mu}$, $1 - \xi l$ and $1 - l\mu$ are eigenvalues of $J(E_0)$. Here, l and r_1 are always positive and thus the axial equilibrium point is unstable.

The Jacobian matrix of the system at pest free equilibrium point $E_1(k_1, 0, 0, \frac{\pi_v}{\mu})$ is given by:

$$J(E_1) = \begin{bmatrix} 1 - lr_1 & -l\alpha k_1 & 0 & 0 \\ 0 & 1 + lr_2 + l\alpha k_1 - \frac{\beta\pi_v l}{\mu} & 0 & 0 \\ 0 & \frac{\beta\pi_v l}{\mu} & 1 - \xi l & 0 \\ 0 & 0 & \kappa\xi l & 1 - l\mu \end{bmatrix}.$$

The eignevalue of $J(E_1)$ are $1 - lr_1$, $1 + lr_2 + l\alpha k_1 - \frac{\beta\pi_v l}{\mu}$, $1 - \xi l$ and $1 - l\mu$.

Thus, by Jury Condition the system is stable around E_1 if modulus of eigenvalues are less than one (Fig. 23.1).

The Jacobian matrix of the system at virus free equilibrium point $E_2(j_2, s_2, 0, 0)$ is given by:

$s_2 = \frac{r_1}{\alpha}(1 - (\frac{j_2}{k_1})^{\theta_1})$ and j_2 satisfy the equation $r_2(1 - (\frac{s_2}{k_2})^\theta) + \alpha j_2 = 0$.

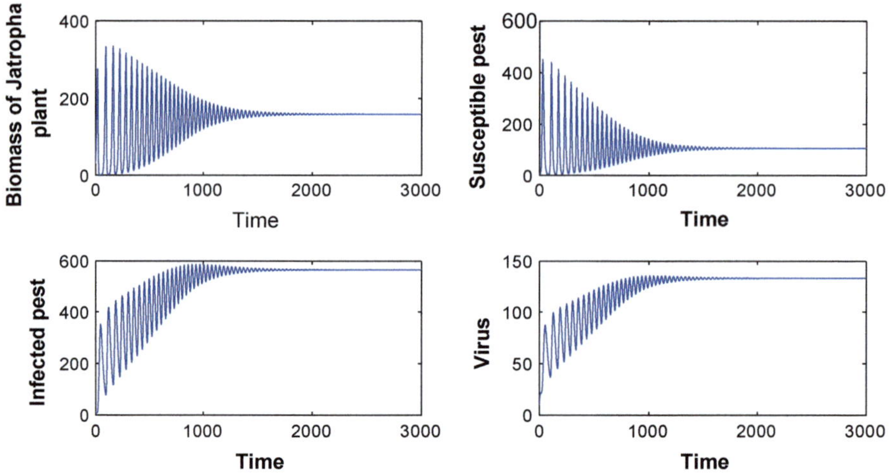

Fig. 23.1 Trajectory portrait of model system (23.2). Which shows that the system stable around pest free equilibrium point corresponds to $\theta = 0.1$ other parameter values given in Table 23.1

$$J(E_2) = \begin{bmatrix} B_{11} & -l\alpha j_2 & 0 & 0 \\ l\alpha s_2 & B_{22} & -\frac{ls_2 r_2 \theta}{k_2}\left(\frac{s_2}{k_2}\right)^{\theta-1} & -l\beta s_2 \\ 0 & 0 & 1-\xi l & l\beta s_2 \\ 0 & 0 & -\kappa\xi l & 1-l\mu \end{bmatrix}.$$

Where

$$B_{11} = 1 + lr_1 - \frac{2lr_1 j_2}{k_1} - l\alpha s_2 \tag{23.3}$$

$$B_{22} = 1 + lr_2\left[1 - \left(\frac{s_2}{k_2}\right)^{\theta}\right] - \frac{lr_2 s_2 \theta}{k_2}\left(\frac{s_2}{k_2}\right)^{\theta-1} + l\alpha j_2 \tag{23.4}$$

has the following characteristic equation:

$$(\lambda^2 + d_1\lambda + d_2)(\lambda^2 + d_3\lambda + d_4) = 0. \tag{23.5}$$

Where,

$$d_1 = l\mu + l\xi - 2,\ d_2 = (1-\xi l)(1-l\mu) + l^2\kappa\xi\beta s_2$$

$$d_3 = \frac{lr_2 s_2 \theta}{k_2}\left(\frac{s_2}{k_2}\right)^{\theta-1} - \left\{lr_1\left[1 - \left(\frac{2j_2}{k_1}\right)\right] + \right.$$

$$\left. lr_2\left[1 - \left(\frac{s_2}{k_2}\right)^{\theta_2}\right]\right\} + l\alpha s_2 - l\alpha j_2 - 2,$$

$$d_4 = \left(1 + lr_1 - \frac{2lr_1 j_2}{k_1} - l\alpha s_2\right)\left(1 + lr_2\left[1 - \left(\frac{s_2}{k_2}\right)^{\theta}\right] - \frac{lr_2 s_2 \theta}{k_2}\left(\frac{s_2}{k_2}\right)^{\theta-1} + l\alpha j_2\right) + l^2\alpha^2 j_2 s_2.$$

Then eigenvalues of the system satisfy the equation $P_1(\lambda) = \lambda^2 + d_1\lambda + d_2 = 0$ and $P_2(\lambda) = \lambda^2 + d_3\lambda + d_4 = 0$, by Jury Condition the system is asymptotically stable around $E_2(j_2, s_2, 0, \frac{\pi_v}{\mu})$ if,

$$P_1(1) > 0,\ P_1(-1) > 0 \text{ and } d_2 < 1,$$
$$P_2(1) > 0,\ P_2(-1) > 0 \text{ and } d_4 < 1.$$

23.3.2 Stability of Interior Equilibrium

The Jacobian matrix of the system at interior equilibrium point $E^*(j^*, s^*, i^*, v^*)$ is given by:

$$J_{E^*} = \begin{bmatrix} C_{11} & C_{12} & 0 & 0 \\ C_{21} & C_{22} & C_{23} & C_{24} \\ 0 & 0 & C_{33} & C_{34} \\ 0 & 0 & C_{43} & C_{44} \end{bmatrix}.$$

Where,

$$\begin{aligned}
C_{11} &= 1 + lr_1\left[1 - \left(\frac{2j^*}{k_1}\right)\right] - l\alpha s^* \\
C_{22} &= 1 + lr_2\left[1 - \left(\frac{s^* + i^*}{k_2}\right)^{\theta}\right] - \frac{lr_2 s^* \theta}{k_2}\left(\frac{s^* + i^*}{k_2}\right)^{\theta-1} + l\alpha j^* - \beta l v^* \\
C_{12} &= -l\alpha j^* \\
C_{21} &= l\alpha s^* \\
C_{23} &= -\frac{ls^* r_2 \theta}{k_2}\left(\frac{s^* + i^*}{k_2}\right)^{\theta-1} \\
C_{24} &= -l\beta s^* \\
C_{33} &= 1 - \xi l \\
C_{34} &= l\beta s^* \\
C_{43} &= -\kappa \xi l \\
C_{44} &= 1 - l\mu.
\end{aligned} \tag{23.6}$$

Now we choose the Lyapunov function as follows:

$$\psi(j, s, i, v) = \frac{1}{2}(c_1 j^2 + c_2 s^2 + c_3 i^2 + c_4 v^2),$$

where $c_i > 0$; $i = 1, 2, 3, 4$, is to be chosen suitably. Obviously ψ is positive definite. Derivative of ψ along the solution of the equation $\dot{X}(t) = J_{E^*} X(t)$, where $X(t) = (j(t), s(t), i(t), v(t))^T$ is as follows:

$$\begin{aligned}
\dot{\psi} &= c_1 j\dot{j} + c_2 s\dot{s} + c_3 i\dot{i} + c_4 v\dot{v}, \\
&= c_1 j[jC_{11} + sC_{12}] + c_2 s[jC_{21} + sC_{22} + iC_{23} + vC_{24}] \\
&\quad + c_3 i[iC_{33} + vC_{34}] + c_4 v[iC_{43} + vC_{44})], \\
&= c_1 C_{11} j^2 + c_2 C_{22} s^2 + c_3 C_{33} i^2 + c_4 C_{44} v^2 + \\
&\quad (c_1 C_{12} + c_2 C_{21}) js + c_2 C_{23} si + c_2 C_{24} sv + (c_3 C_{34} + c_4 C_{43}) iv.
\end{aligned}$$

Thus symmetric matrix corresponding to $\dot{\psi}$ is given as:

$$M = \frac{1}{2}\begin{bmatrix} 2c_1 C_{11} & (c_1 C_{12} + c_2 C_{21}) & 0 & 0 \\ (c_1 C_{12} + c_2 C_{21}) & 2c_2 C_{22} & c_2 C_{23} & c_2 C_{24} \\ 0 & c_2 C_{23} & 2c_3 C_{33} & (c_3 C_{34} + c_4 C_{43}) \\ 0 & c_2 C_{24} & (c_3 C_{34} + c_4 C_{43}) & 2c_4 C_{44} \end{bmatrix}.$$

The positive equilibrium E^* is locally asymptotically stable if $\dot{\psi}$ is negative definite. Which in turns follows if the symmetric matrix M is negative definite. But M is negative definite if odd rank principal minor in order is negative and the even rank principal minor in order is positive, which in turn follows if,

(i) $2c_1 C_{11} < 0$,
(ii) $4c_1 c_2 C_{11} C_{22} - (c_1 C_{12} + c_2 C_{21})^2 > 0$,
(iii) $2c_1 C_{11}[4c_2 c_3 C_{22} C_{33} - (c_2 C_{23})^2] - 2c_3 C_{33}(c_1 C_{12} + c_2 C_{21})^2 < 0$,
(iv) $2c_4 C_{44}$[LHS of expression of inequality (iii)]
$-(c_3 C_{34} + c_4 C_{43})$[Minor with respect to (4, 3) element of matrix M]
$+c_2 C_{24}$[Minor with respect to (4, 2) element of matrix M] > 0.

Now, we choose c_1, c_2, c_3, c_4 such as

$$(c_1 C_{12} + c_2 C_{21}) = 0, \tag{23.7}$$

$$(c_3 C_{34} + c_4 C_{43}) = 0, \tag{23.8}$$

$$c_4 C_{44}[4c_2 c_3 C_{22} C_{33} - (c_2 C_{23})^2] - c_2 c_3 (C_{24})^2 = 0, \tag{23.9}$$

$$4c_2 c_3 C_{22} C_{33} - (c_2 C_{23})^2 > 0. \tag{23.10}$$

Then the above four inequalities are satisfied if,

$$1 + lr_1\left[1 - \left(\frac{2j^*}{k_1}\right)\right] - l\alpha s^* < 0 \quad (23.11)$$

$$1 + lr_2\left[1 - \left(\frac{s^* + i^*}{k_2}\right)^{\theta}\right] - \frac{lr_2 s^* \theta}{k_2}\left(\frac{s^* + i^*}{k_2}\right)^{\theta - 1} + l\alpha j^* - \beta l v^* < 0 \quad (23.12)$$

$$1 - l\mu < 0. \quad (23.13)$$

We summarize the above results with the following theorem:

Theorem 23.2 *The system is globally asymptotically stable around* $E^*(j^*, s^*, i^*, v^*)$ *if* $1 + lr_1\left[1 - \left(\frac{2j^*}{k_1}\right)\right] - l\alpha s^* < 0$, $1 + lr_2\left[1 - \left(\frac{s^*+i^*}{k_2}\right)^{\theta}\right] - \frac{lr_2 s^* \theta}{k_2}\left(\frac{s^*+i^*}{k_2}\right)^{\theta-1} + l\alpha j^* - \beta l v^* < 0$ *and* $1 - l\mu < 0$.

23.4 Bifurcation Analysis

The characteristic equation of $J_{E^*} = Cij$ is

$$\lambda^4 + \sigma_1\lambda^3 + \sigma_2\lambda^2 + \sigma_3\lambda + \sigma_4 = 0. \quad (23.14)$$

Where,

$$\begin{aligned}
\sigma_1 &= -\sum C_{11},\\
\sigma_2 &= \sum C_{11}C_{22} - C_{12}C_{21} - C_{34}C_{43}\\
\sigma_3 &= -\sum C_{11}C_{22}C_{33} + C_{12}C_{21}(C_{33} + C_{44}) + C_{34}C_{43}(C_{11} + C_{22})\\
\sigma_4 &= C_{11}C_{22}C_{33}C_{44} - C_{12}C_{21}C_{33}C_{44} - C_{11}C_{22}C_{34}C_{43} + C_{12}C_{21}C_{34}C_{43}.
\end{aligned} \quad (23.15)$$

Here, C_{ij} are given in (23.6).

Now, we shall find out the conditions for which E^* enters Hopf bifurcation as θ varies over the interval (0, 1).

Routh–Hurwitz criterion and Hopf bifurcation: Let $\Psi : (0, \infty) \to \mathbb{R}$ be the following continuously differentiable function of θ:

$$\Psi(\theta) := \sigma_1(\theta)\sigma_2(\theta)\sigma_3(\theta) - \sigma_3^2(\theta) - \sigma_4(\theta)\sigma_1^2(\theta).$$

Hopf bifurcation can occur if the following conditions are satisfied:

(A) There exists $\theta^* \in (0, 1)$, at which a pair of complex eigenvalues $\lambda(\theta^*), \bar{\lambda}(\theta^*) \in \sigma(\theta)$ are such that

$$Re\lambda(\theta^*) = 0, \quad Im\lambda(\theta^*) = \omega_0 > 0,$$

and the transversality condition

$$\frac{dRe\lambda(\theta)}{d\theta}|_{\theta^*} \neq 0;$$

(B) All other elements of $\sigma(\theta)$ have negative real parts, where $\sigma(\theta) = \{\rho : D(\rho) = 0\}$ is the spectrum of the characteristic equation (23.14).

Theorem 23.3 *The system (5) around the interior equilibrium E^* enters Hopf bifurcation at $\theta = \theta^* \in (0, 1)$ if and only if*

i. $\Psi(\theta^*) = 0$
ii. $\sigma_1^3\sigma_2'\sigma_3(\sigma_1 - 3\sigma_3) > 2(\sigma_2\sigma_1^2 - 2\sigma_3^2)(\sigma_3'\sigma_1^2 - \sigma_1'\sigma_3^2)$,

and all other eigenvalues are of negative real parts, where $\lambda(\theta)$ is purely imaginary at $\theta = \theta^*$.

Proof The existence of θ^* can be obtained by solving $\Psi(\theta^*) = 0$. By the condition $\Psi(\theta^*) = 0$, the characteristic equation can be written as

$$\left(\lambda^2 + \frac{\sigma_3}{\sigma_1}\right)\left(\lambda^2 + \sigma_1\lambda + \frac{\sigma_1\sigma_4}{\sigma_3}\right) = 0.$$

If it has four roots, say λ_i, (i=1,2,3,4) with the pair of purely imaginary roots at $\theta = \theta^*$ as $\lambda_1 = \bar{\lambda}_2$, then we have

$$\begin{aligned} \lambda_3 + \lambda_4 &= -\sigma_1, \\ \omega_0^2 + \lambda_3\lambda_4 &= \sigma_2, \\ \omega_0^2(\lambda_3 + \lambda_4) &= -\sigma_3, \\ \omega_0^2\lambda_3\lambda_4 &= \sigma_4, \end{aligned} \tag{23.16}$$

where $\omega_0 = Im\lambda_1(\theta^*)$. By above $\omega_0 = \sqrt{\frac{\sigma_3}{\sigma_1}}$. Now, if λ_3 and λ_4 are complex conjugate, then from (23.16), it follows that $2Re\lambda_3 = -\sigma_1$; if they are real roots, then by (23.14) and (23.16) $\rho_3 < 0$ and $\rho_4 < 0$. To complete the discussion, it remains to verify the transversality condition.

As $\Psi(\theta^*)$ is a continuous function of all its roots, so there exists an open interval $\theta \in (\theta^* - \varepsilon, \theta^* + \varepsilon)$ where λ_1 and λ_2 are complex conjugate for θ. Suppose, their general forms in this neighborhood are

$$\begin{aligned} \lambda_1(\theta) &= \chi(\theta) + i\nu(\theta), \\ \lambda_2(\theta) &= \chi(\theta) - i\nu(\theta). \end{aligned}$$

Now, we shall verify the transversality condition

$$\frac{dRe(\lambda_j(\theta))}{d\theta}|_{\theta=\theta^*} \neq 0, \ j = 1, \ 2.$$

Substituting $\lambda_j(\theta) = \chi(\theta) \pm i\nu(\theta)$, into (23.14) and calculating the derivative, we have

$$K(\theta)\chi'(\theta) - L(\theta)\nu'(\theta) + M(\theta) = 0,$$
$$L(\theta)\chi'(\theta) + K(\theta)\nu'(\theta) + N(\theta) = 0.$$

Here,

$$\begin{aligned}
K(\theta) &= 4\chi^3 - 12\chi\nu^2 + 3\sigma_1(\chi^2 - \nu^2) + 2\sigma_2\chi + \sigma_3,\\
L(\theta) &= 12\chi^2\nu + 6\sigma_1\chi\nu - 4\chi^3 + 2\sigma_2\chi,\\
M(\theta) &= \sigma_1\chi^3 - 3\sigma_1'\chi\nu^2 + \sigma_2'(\chi^2 - \nu^2) + \sigma_3'\chi,\\
N(\theta) &= 3\sigma_1'\chi^2\nu - \sigma_1'\nu^3 + 2\sigma_2'\chi\nu + \sigma_3'\chi.
\end{aligned}$$

Solving for $\chi'(\theta^*)$ we have

$$\left[\frac{dRe(\lambda_j(\theta))}{d\theta}\right]_{\theta=\theta^*} = \chi'(\theta)_{\theta=\theta^*} = -\frac{L(\theta^*)N(\theta^*)+K(\theta^*)M(\theta^*)}{K^2(\theta^*)+L^2(\theta^*)}$$
$$= \frac{\sigma_1^3\sigma_2'\sigma_3(\sigma_1-3\sigma_3)-2(\sigma_2\sigma_1^2-2\sigma_3^2)(\sigma_3'\sigma_1^2-\sigma_1'\sigma_3^2)}{\sigma_1^4(\sigma_1-3\sigma_3)^2+4(\sigma_2\sigma_1^2-2\sigma_3^2)^2} > 0,$$

$$\text{if } \sigma_1^3\sigma_2'\sigma_3(\sigma_1 - 3\sigma_3) > 2(\sigma_2\sigma_1^2 - 2\sigma_3^2)(\sigma_3'\sigma_1^2 - \sigma_1'\sigma_3^2).$$

Thus the transversality conditions hold and hence Hopf bifurcation occurs at $\theta = \theta^*$.

23.5 Numerical Simulation

The dynamics of the model system are analyzed using Mathworks MATLAB ver.2008. In this section we verify the analytical predictions obtained in the previous sections through numerical results of the system (23.2).

For our numerical studies, we have taken the step size parameter value $l = 0.4$. From Fig. 23.2 it is clear that the model variables $j(t)$, $s(t)$, $i(t)$ and $v(t)$ oscillate initially before the system moves toward its stable region as time increases for $\theta = 0.1$. From Figs. 23.3, 23.4 and 23.5 we have shown that system (23.2) is asymptotically stable at interior equilibrium point $E^*(j^*, s^*, i^*, v^*)$ for $\theta < 0.74$ and also it is seen that stability of the system took longer time as the value of theta increases. On the other side from Figs. 23.6 and 23.11 it is seen that for $\theta \geq 0.74$, the system is unstable and periodic which also holds for $\theta = 1$, i.e., the pest population follows logistic growth. Our result suggests that the assumption of theta logistic growths (Sibly et al. 2005) [9] for insect population are more realistic and natural. Thus theta logistic growth may help to sustain the biomass of *Jatropha sp.*-pest–virus population which is not the case for simple logistic growth.

In Fig. 23.7 and 23.8, we have shown the stability region by varying intrinsic growth rate parameter (r_1, r_2) and carrying capacity parameter (k_1, k_2) for biomass of *Jatropha sp.* and susceptible pest at pest free equilibrium point $E_1(k_1, 0, 0, \frac{\pi_v}{\mu})$.

Phase portraits of biomass of *Jatropha sp.*, healthy pest, infected pest and virus are drawn for $\theta = 0.1$ in Fig. 23.9 and for $\theta = 0.9$ in Fig. 23.10. Figure 23.9a is phase diagram for the virus, biomass of *Jatropha sp.* and susceptible pest. Figure 23.9b is

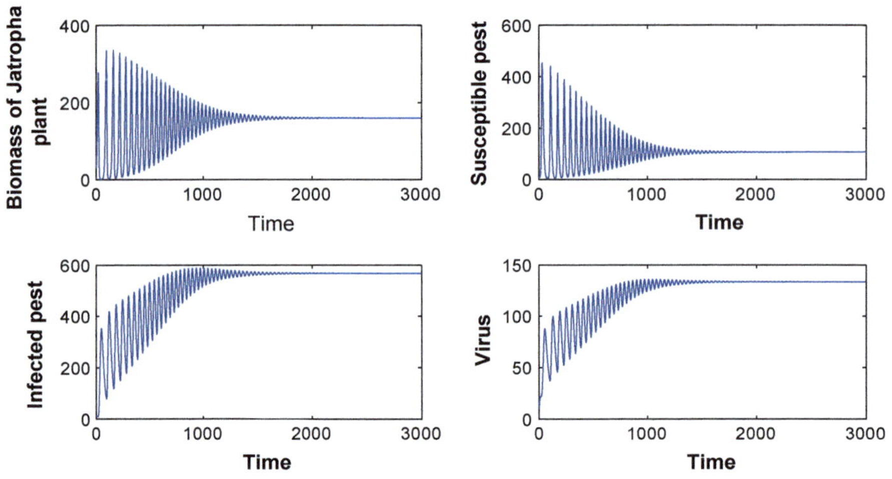

Fig. 23.2 Densities of biomass of *Jatropha sp.*, Susceptible pest, Infected pest and Virus population for $\theta = 0.1$ for model system (23.2), other parameter values given in Table 23.1

Table 23.1 Values of parameters used in numerical calculation for system (23.2).

Parameters	Definition	Values (Unit)
r_1	The growth rate of *Jatropha sp.*	$0.5\,\text{day}^{-1}$ [18]
k_1	The maximum density of *Jatropha sp.*	$500\,\text{ha}^{-1}$ [18]
β	The infection rate	$0.0032\,\text{pest}^{-1}\,\text{day}^{-1}$ [4, 19]
λ	Interaction rate of virus with pest	$0.004\,\text{pest}^{-1}\,\text{day}^{-1}$ [19]
ξ	The mortality rate of infected pest	$0.1\,\text{day}^{-1}$ [4]
r_2	The growth rate of pest	$0.1\,\text{day}^{-1}$ [19]
π_V	The acquisition rate of virus	$0.01\,\text{day}^{-1}$ [4]
k_2	The pest carrying capacity	$800\,\text{plant}^{-1}$ [20]
κ	The virus replication parameter	$1\,\text{pest}^{-1}\,\text{day}^{-1}$ (estimated)

phase diagram for the infected pest, biomass of *Jatropha sp.* and susceptible pest. Both figures show that the system (23.2) starting with the initial value (30, 5, 12, 10) converges to the interior equilibrium point $E^*(j^*, s^*, i^*, v^*)$. On the other hand Fig. 23.10a is phase diagram for the virus, susceptible pest and biomass of *Jatropha sp.* and Fig. 23.10b is phase diagram for the infected pest, susceptible pest and biomass of *Jatropha sp.* with the initial value (30, 5, 12, 10). Limit cycle oscillation occurs in both figures, i.e., the system (23.2) is unstable compare to logistic curve.

In Fig. 23.11 we have drawn time series plot corresponding to the Figs. 23.9 and 23.10.

In Fig. 23.12 we have drawn the bifurcation diagrams of biomass of *Jatropha sp.*, susceptible pest, infected pest and virus with respect to the parameter θ. From this figure it is clear that the system under goes Hopf bifurcation around $\theta = 0.74$.

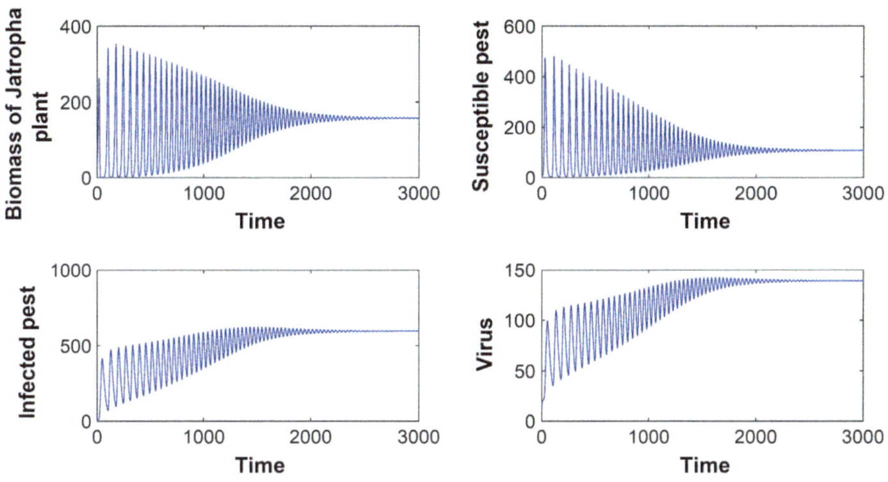

Fig. 23.3 Densities of biomass of *Jatropha sp.*, Susceptible pest, Infected pest and Virus population for $\theta = 0.3$ for model system (23.2), other parameter values given in Table 23.1

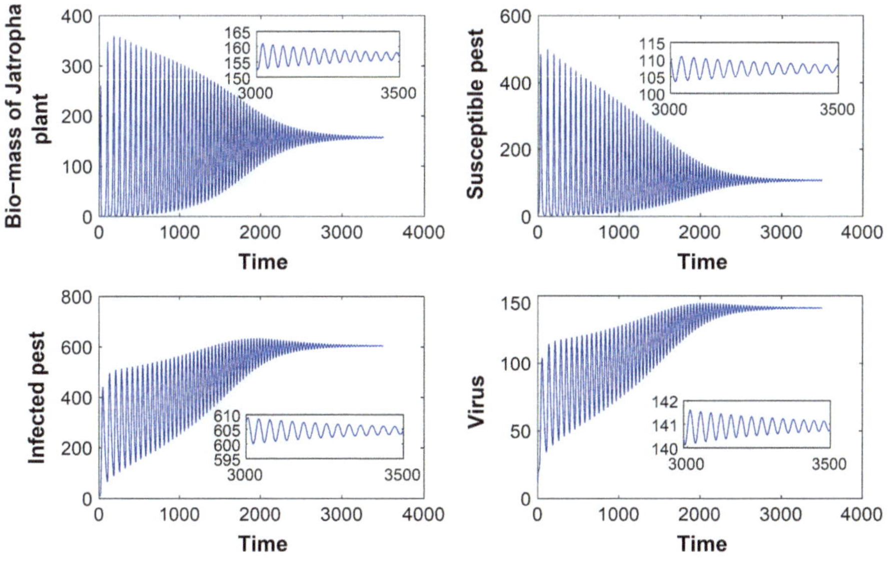

Fig. 23.4 Densities of biomass of *Jatropha sp.*, Susceptible pest, Infected pest and Virus population for $\theta = 0.4$ for model system (23.2), other parameter values given in Table 23.1

23.6 Discussion and Conclusion

Jatropha sp. is a renewable non-edible plant, and it can be used as a potential resource for biodiesel production. Unfortunately, the *Jatropha sp.* production is immensely

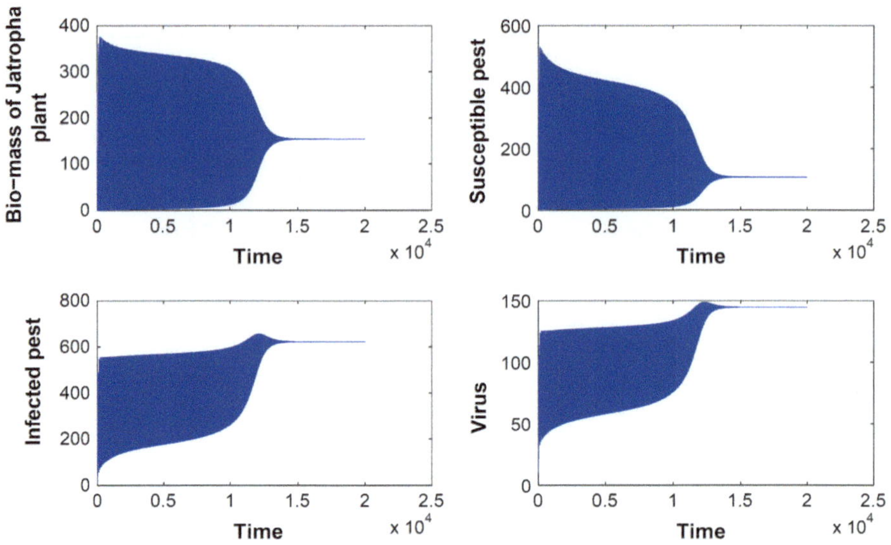

Fig. 23.5 Densities of biomass of *Jatropha sp.*, Susceptible pest, Infected pest and Virus population for $\theta = 0.71$ for model system (23.2), other parameter values given in Table 23.1

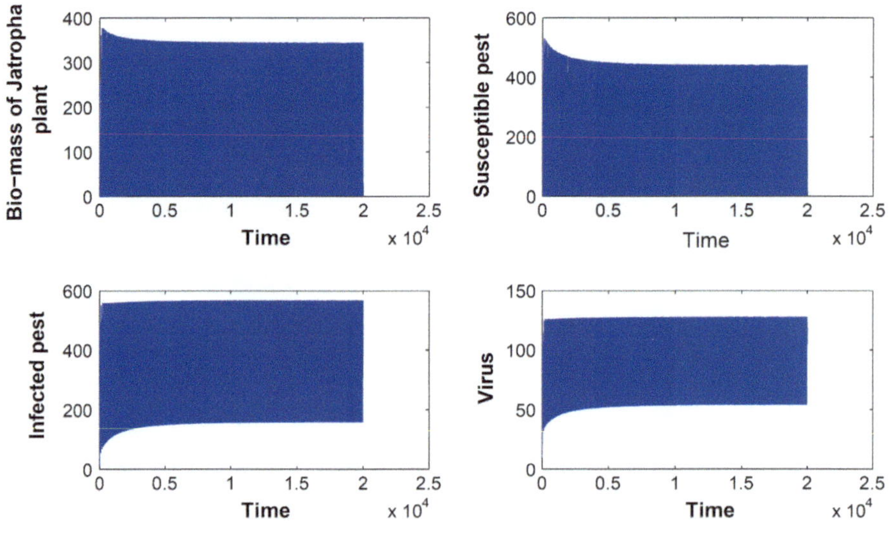

Fig. 23.6 Densities of biomass of *Jatropha sp.*, Susceptible pest, Infected pest and Virus population for $\theta = 0.74$ for model system (23.2), other parameter values given in Table 23.1

hampered due to serious pest attack. Therefore, the pest population needs to be controlled for better prospect of the biodiesel production. There are several ways to control this *Jatropha sp.* pest population, but the most elegant and modern strategy

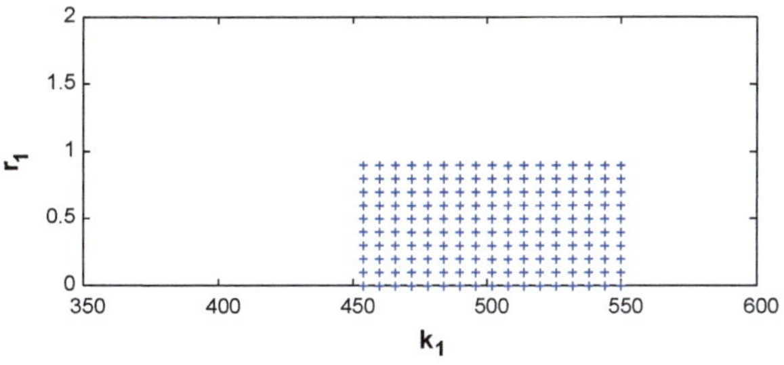

Fig. 23.7 Domain of stability region in which pest free equilibrium of model (23.2) is stable with respect to k_1 and r_1 corresponds to $\theta = 1$ other parameter values given in Table 23.1

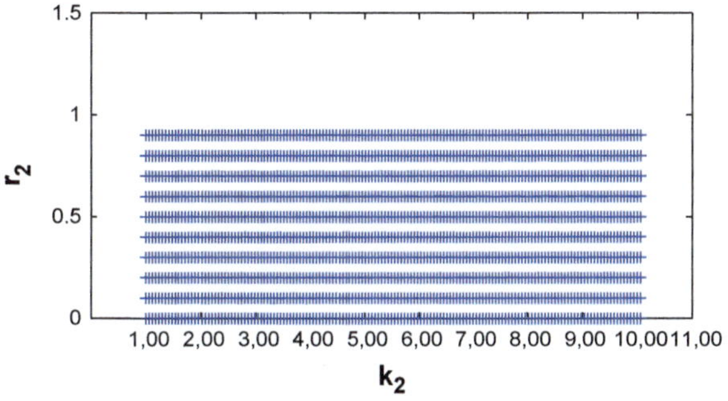

Fig. 23.8 Domain of stability region in which pest free equilibrium of model (23.2) is stable with respect to k_2 and r_2 corresponds to $\theta = 1$ other parameter values given in Table 23.1

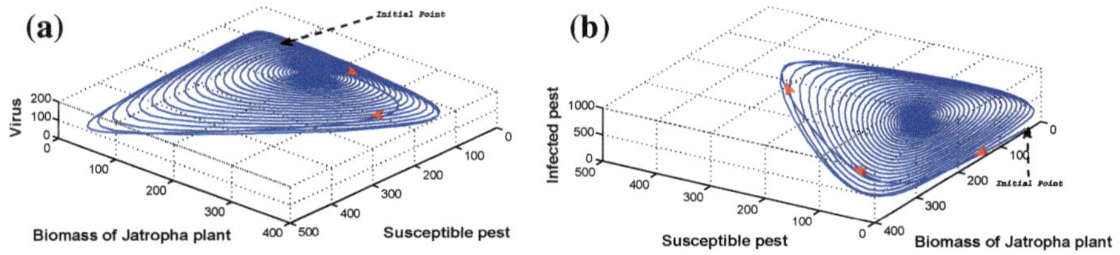

Fig. 23.9 Phase diagram of biomass of *Jatropha sp.*, susceptible pest, infected pest and virus population as a function of time shows that the system starting with the initial value (30, 5, 12, 10) converges to the interior equilibrium point $E^*(j^*, s^*, i^*, v^*)$. We have taken the parameter value $\theta = 0.1$ and the other parameter values are the same as in Table 23.1. **a** The phase diagram for the virus, susceptible pest and biomass of *Jatropha sp.*. **b** The phase diagram for the infected pest, biomass of *Jatropha sp.* and susceptible pest

is to introduce the viral population, viz. NPV, by which we can suppress the effects of the pest on *Jatropha sp.*.

To capture the entire dynamics, we proposed a mathematical model based on the three major populations viz., Biomass of *Jatropha sp.*–pest–virus. Recent study reveals that most of the insect population follows the theta logistic growth instead

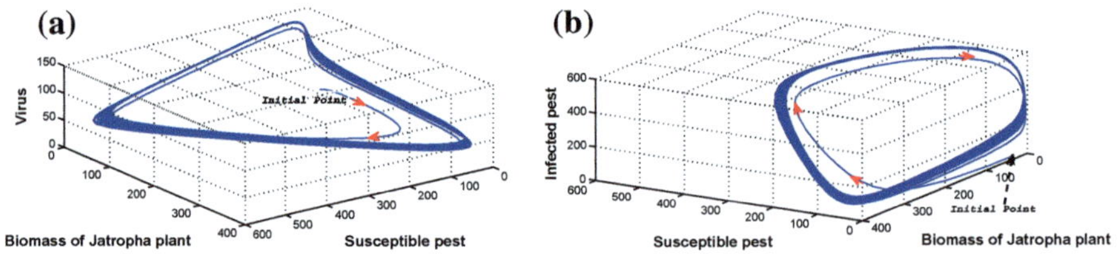

Fig. 23.10 Phase diagram of biomass of *Jatropha sp.*, susceptible pest, infected pest and virus population as a function of time shows that the system starting with the initial value (30, 5, 12, 10) exhibit limit cycle oscillation. We have taken the parameter value $\theta = 0.9$ and the other parameter values are the same as in Table 23.1. **a** The phase diagram for the virus, susceptible pest and biomass of *Jatropha sp.*. **b** The phase diagram for the infected pest, biomass of *Jatropha sp.* and susceptible pest

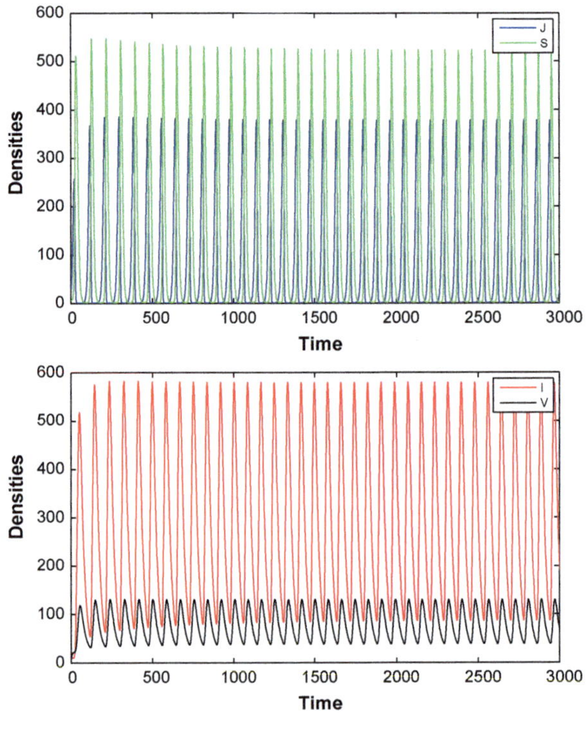

Fig. 23.11 Densities of biomass of *Jatropha sp.*, Susceptible pest, Infected pest and Virus population for $\theta = 0.9$ for model system (23.2). Clearly the system oscillate periodically after 500 days other parameter values given in Table 23.1

of logistic and the density - PGR (per capita growth rate) relationship is concave in nature (Sibly et al. 2005) [9]. So theta logistic must be a realistic and natural choice for explaining susceptible pest growth profile.

Assuming theta logistic growth for susceptible pest the proposed system with four populations exhibits stable behavior (i.e., all the populations co-exists) for low value of theta ($\theta < 0.74$). For $\theta > 0.74$, the system shows limit cycle oscillation.

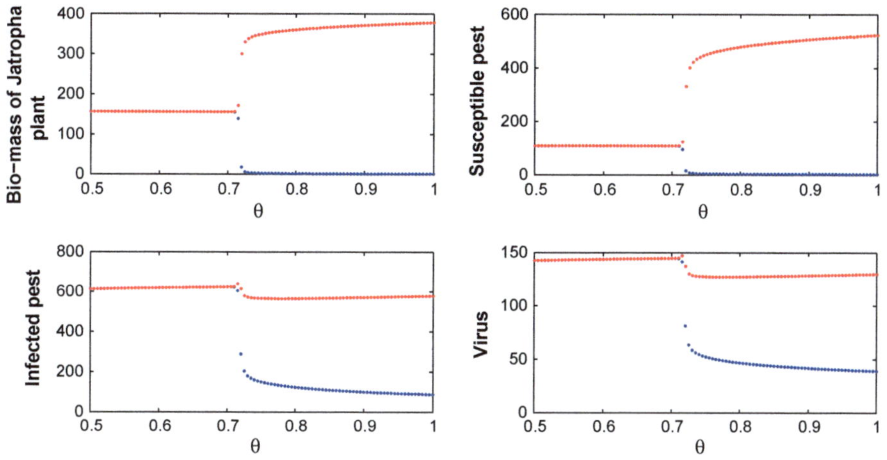

Fig. 23.12 Bifurcation diagrams of densities with respect to θ, assuming the parameters are as in Table 23.1

The same unstable behavior is observed even the theta reached the value one, which is the simple logistic case.

So we observed that for sustainability of the four populations we need to assume theta logistic growth in the susceptible prey. That is density regulation can be acted as a stabilizing agent for such four species interactive model. As nature supports (Sibly et al. 2005) [9] this theta logistic growth for the pest, so pest control by introducing viral population must a potential solution to retrieve *Jatropha sp.* from infection.

Acknowledgements The research is supported by State Government Fellowship, DST PURSE and UGC DRS Programme, Department of Mathematics, Jadavpur University.

References

1. Basir, F.A., Datta, S., Roy, P.K.: Studies on biodiesel production from Jatropha sp. oil using chemical and biochemical methods - a mathematical approach. Fuel **158**, 503–511 (2015). (3944)
2. Basir, F.A., Venturino, E., Roy, P.K.: Effects of awareness program for controlling mosaic disease in Jatropha curcas plantations. Math. Methods Appl. Sci. (2016). doi:10.1002/mma.4149
3. Ghosh, S., Bhattacharya, D.K.: Optimization in microbial pest control: an integrated approach. Appl. Math. Model. **34**, 1382–1395 (2010)
4. Bhattacharya, D.K., Karan, S.: On bionomic model of integrated pest management of a single pest population. J. Differ. Equ. Dyn. Syst. **12**(4), 301–330 (2004)
5. Vincent, T.L.: Pest management programs via optimal control theory. Biometrics **31**(1), 110 (1975)
6. Franz, J.M., Huber, J.: Entomophaga **24**, 333 (1979)
7. Falcon, L.A.: In: Conference on Viral Pesticide, Myrtle Beach, March, pp. 21–23 (1977)

8. Roy, P.K., Li, X.-Z., Basir, F.A., Datta, A., Chowdhury, J.: Effect of insecticide spraying on Jatropha curcas plant to control mosaic virus: a mathematical study. Commun. Math. Biol. Neurosci. **2015** (2015). Article ID 36
9. Sibly, R.M., Barker, D., Denham, M.C., Hone, J., Pagel, M.: On the regulation of populations of mammals, birds, fish, and insects. Science **309**, 607 (2005)
10. Clark, F., Brook, B.W., Delean, S., Akcakaya, H.R., Bradshaw, C.J.A.: The theta- logistic is unreliable for modelling most census data. Methods Ecol. Evol. **1**(3), 253–262 (2010)
11. Franke, J.E., Yakubu, A.A.: Disease-induced mortality in density-dependent discrete-time SIS epidemic models. J. Math. Biol. **57**, 755–790 (2008)
12. Castillo-Chavez, C., Yakubu, A.A.: Discrete-time SIS models with complex dynamics. Nonlinear Anal. **47**, 4753–4762 (2001)
13. Sekiguchi, M., Ishiwata, E.: Global dynamics of a discretized SIRS epidemic model with time delay. J. Math. Anal. Appl. **371**, 195–202 (2010)
14. Allen, L.J.S., Burgin, A.M.: Comparison of deterministic and stochastic SIS and SIR models in discrete time. Math. Biosci. **163**, 1–33 (2000)
15. Li, J., Ma, Z., Brauer, F.: Global analysis of discrete-time SI and SIS epidemic models. Math. Biosci. Eng. **4**, 699–710 (2007)
16. Hu, Z., Teng, Z., Giang, H.: Stability analysis in a class of discrete SIRS epidemic models. Nonlinear Anal. Real World Appl. **13**, 2017–2033 (2012)
17. Chen, L., Chen, L.: Permanence of a discrete periodic volterra model with mutual interference. Discrete Dyn. Nat. Soc. (2009). doi:10.1155/2009/205481
18. Venturino, E., Roy, P.K., Al Basir, F., Datta, A.: A model for the control of the mosaic virus disease in Jatropha curcas plantations. Energy Ecol. Environ. **1** (2016). doi:10.1007/s40974-016-0033-8
19. Roy, P.K., Chowdhury, J., Basir, F.A.: Renewable energy biodiesel: a mathematical approach from ecology to production. Imhotep Math. Proc. **3** (2015)
20. Terren, M., Mignon, J., De Clerck, C., Jijakli, H., Savery, S., de Haveskercke, P.J., Winandy, S., Mergeai, G.: Principal disease and insect pests of Jatropha curcas L. in the lower valley of the Senegal river. Tropicultura **30**(4), 222–229 (2012)

Chapter 24
Dynamics of Sirs Model with Single Time Delay

Sudipa Chauhan, Sumit Kaur Bhatia and Nidhi Purohit

Abstract In this paper, we have considered two models: SICRS model without and with single time delay in infective population. The global dynamics of both models have been carried out and the effect of carriers on transmission dynamics is discussed through the basic reproduction number.

Keywords Equilibrium point · Reproduction number · Local stability
Global stability

24.1 Introduction

The repeated threats of infectious disease have become an alarming issue in today's world. In the context of global health, control of such diseases has become mandatory. It has been recorded that in 2008, infectious diseases accounted for about sixteen percent of deaths worldwide. The infectious diseases are those which get transmitted from one person to another through various agents like, bacteria, virus (foodborne, waterborne, airborne) which are known as carriers. They play a vital role in transmission of the infection from one class of individuals to another but, they themselves do not exhibit the disease. Considering a closed population, i.e., (susceptible, infective, carriers and recovered) the introduction of an infective individual or an external vector can result in the spread of an infectious disease within the population [1]. Some carriers carry the disease on their respective gene known as genetic carriers, but here our concern is asymptotic carriers which particularly causes typhoid, hepatities B and diarrhea. In our models we will discuss these infectious diseases in general.

S. Chauhan (✉) · S.K. Bhatia · N. Purohit
Amity Institute of Applied Science, Amity University, Sector-125, Noida, India
e-mail: sudipachauhan@gmail.com

S.K. Bhatia
e-mail: sumit2212@gmail.com

N. Purohit
e-mail: nidhipurohit95@gmail.com

P. Manchanda et al. (eds.), *Industrial Mathematics and Complex Systems*,
Industrial and Applied Mathematics, DOI 10.1007/978-981-10-3758-0_24

Typhoid, hepatitis B, and diarrhea are most common infectious diseases that are transmitted through asymptotic carriers. These diseases remain an important public health problem in developing countries.Typhoid fever is a systemic infection caused by Salmonella enterica serotype Typhi (S. typhi). In 2000, it was estimated that over 2.16 million episodes of typhoid occurred worldwide, resulting in 216 000 deaths, and that more than 90% of this morbidity and mortality occurred in Asia [2]. It has been observed that the pathogens causing typhoid and diarrhea mainly occur in the areas surrounded by water bodies. As the inhabitants remain in contact with the natural reservoir in their daily routine and thus catch the disease easily and spread it among the individuals. The bacteria Salmonella is a bacteria genus that is closely related to diarrhea but recent data show that the mortality from diarrhea has declined over the past two decades from an estimated 5 million deaths among children under five to 1.5 million deaths in 2004 [3]. Another major infectious disease that causes long-term asymptomatic carriage is hepatitis B, a liver disease caused by the HBV virus of the Hepadna virus family. The population of infective increases as the individuals are generally not aware of the treatment methods and if at all, the treatment is given, they may not respond to it due to the unavailability of adequate facilities. Most people infected with HBV recover completely and develop a lifelong immunity and 15–25% of these develop liver disease. Clostridium difficile is a bacterium that causes Clostridium difficile-associated diseases (CDAD). CDAD remains the most common cause of acute hospital-acquired diarrhea, responsible for more than 300,000 cases of diarrhea annually in acute-care facilities in the United States. Asymptomatic carriage rates of up to 30 have been reported in long-term care facilities. It is believed that carriers are responsible for transmission and large outbreaks of CDAD in Europe and North America [4]. Keeping the public health significance, many studies have been done in forming the models in which vaccination has been given to cure the disease and then the work has been done to predict the further situation but models in which the role of vaccination is not taken into consideration, there the prominent role of reproduction number comes into the scene to analyze the system [5–8].

The study of this model is focused mainly on the impact of the effects of carriers of disease on population. Although infectious diseases are present in human populations at all times to some degree, the effects of epidemics are the most noticeable and spectacular. It is possible to mathematically model the progress of typhoid fever to discover the likely outcome of an epidemic [9]. Fever in KISII Town Kenya [9]. In the present paper, we have formulated two models showing the transmission of the infectious disease through asymptotic carriers. We have proposed two models, in Model I we have considered delay in time to come to recovered class from infected class and in Model II we have introduced a probability factor which is more realistic. Both the models are different from classical SEIR model as the growth of infective class is due to exposed class whereas in our model growth of infective class is due interaction between carrier and susceptible. Also, we have incorporated the interaction between susceptible and infective and the diagnosis rate of carrier. We have shown boundedness and existence of the endemic equilibrium point for both the models and their basic dynamical features.

This paper is organized in three sections. In the first section, we have formulated Model I and Model II and in the second section we obtained boundedness of both models. Further, we have obtained basic dynamical features in third section.

24.2 Model Development

In this section, we have developed the models under the following assumptions:

Let C(t) be the number of the carriers, S(t) be the number of susceptibles, I(t) be the number of infectives, R(t) be the number of recovered population. Models are appropriate to use under the following assumptions:

1. The population is fixed.
2. Age, sex, social status, and race do not affect the probability of being infected.
3. There is no inherited immunity.
4. The member of the population mix homogeneously (have the same interactions with one another to the same degree).
5. The natural birth and death rates are included.
6. All births are into the susceptible class.
7. The death rate is equal for members of all four classes, and it is assumed that the birth and death rates are equal so that the total population is stationary.

Model I

1. Susceptible population is growing at a rate α_1 and members of the population of susceptible class interact with members of population of carrier class at the rate α. Also, recovered is coming again into susceptible class at the rate δ and susceptible die naturally at the rate d_1. Thus, the equation is as follows:

$$\frac{dS}{dt} = \alpha_1 S - \alpha SC + \delta R - d_1 S \tag{24.1}$$

   where
   α_1 = constant rate at which the susceptible grows.
   δ = rate of loss of disease induced immunity (waning rate).
   d_1 = natural death of susceptible.
2. Members of infective class leave the class at the rate β with the time delay T. Also, infective dies naturally at the rate d_1. Thus, equation is as follows:

$$\frac{dI}{dt} = \alpha SC - \beta I(t-T) - d_1 I \tag{24.2}$$

   where
   d_1 = natural death of infective.

3. Members of infective class will enter in recovered class at the rate β. Also, recovered dies naturally at the rate d_1.Thus, equation is as follows:

$$\frac{dR}{dt} = \beta I - \delta R - d_1 R \tag{24.3}$$

$d_1 =$ natural death of recovered.

4. Carriers die naturally at the rate d_2. Thus, equation is as follows:

$$\frac{dC}{dt} = \gamma I - d_2 C \tag{24.4}$$

where, $d_2 =$ natural death of carriers.

Model II

1. The susceptible population S(t) at any time t is growing exponentially with growth rate of α_1, this respective population is converting to the carriers and infective with some conversation rate of α and β respectively. The recovered individuals have become susceptible again with a rate of π and dying with a natural death rate of d_1.

$$\frac{dS}{dt} = \alpha_1 S - \alpha SC - \beta SI - d_1 S + \delta R \tag{24.5}$$

2. The infective population I(t) at any time t can become a carrier with the probability (1 − p) and the carriers are getting prone to infection with the rate α_2. The population is dying with a natural and disease caused death rate of d_2.

$$\frac{dI}{dt} = (1-p)\alpha SC + (1-p)\beta SI + \alpha_2 C - m_1 I - d_2 I \tag{24.6}$$

3. The carriers population C(t) at any time t is dying with a natural and disease caused death rate of d_3.

$$\frac{dC}{dt} = p\alpha SC + p\beta SI - d_3 C - \alpha_2 C \tag{24.7}$$

4. The infective population is recovering with a rate π. The individuals are becoming susceptible again with the rate δ and dying with a natural rate of d_4.

$$\frac{dR}{dt} = m_1 I - \delta R - d_4 R \tag{24.8}$$

where
α_1: growth rate of susceptible.
α_2: conversion rate of susceptible into infective.
β: role of transmission of susceptible into infective and carriers.
m_1: recovery rate.
δ: rate at which the recovered transmit into susceptible.
d_1: natural deaths of susceptible.
d_2: natural and disease caused death of the infective.
d_3: natural and disease caused death of the carriers.
d_4: natural death of recovered population.
p: susceptible will become a carriers with probability p.

24.3 Boundedness

In this section, we shall obtain the boundedness of the system.

Proposition 24.1 *Model I is bounded.*

Proof Let W = S + I + C + R

$$\frac{dW}{dt} = \alpha_1 S + \gamma I + d_1 S - d_1 I - d_1 R - d_2 C$$

$$\frac{dW}{dt} \leq \alpha_1 S + \gamma I - dW$$

where $d = min(d_1, d_2)$

$$\frac{dW}{dt} + dW = A_1 W - AR - BC$$

where $A_1 = max(\alpha_1, \gamma)$

$$\frac{dW}{dt} + dW - A_1 W \leq 0$$
$$\frac{dW}{dt} + (d - A_1)W \leq 0$$
$$We^{(d-A_1)dt} \leq 0$$

Therefore the system is bounded.

Proposition 24.2 *Model II is bounded.*

Proof Let N = S + I + C + R

$$\frac{dN}{dt} = \frac{dS}{dt} + \frac{dI}{dt} + \frac{dC}{dt} + \frac{dR}{dt} \tag{24.9}$$

$$\frac{dN}{dt} = \alpha_1 S + (d_1 + d_2 + d_3 + d_4)(S + I + R + C) \tag{24.10}$$

Let $d = min(d_1, d_2, d_3, d_4)$

$$\frac{dN}{dt} = \alpha_1 S - dN$$

Since, S is bounded above by $S^* = \frac{\alpha_1}{d_1}$.

$\frac{dN}{dt} \leq \alpha_1 S^* - dN$

$\frac{dN}{dt} + dN \leq A$

Thus $N \leq \frac{A}{d}$.

Therefore, Model II is bounded.

24.4 Basic Dynamical Features

In this section, we will obtain the equilibrium points of Model I and Model II.

Equilibrium points: Model I

Equilibrium points for Model I are:

- Trivial equilibrium point: $E_0(S = 0, I = 0, R = 0, C = 0)$
- Endemic equilibrium point:

$$E^* \left(S^* = \frac{d_2(\beta + d_1)}{\alpha\gamma}, I^* = \frac{d_2(\beta + d_1)(\alpha_1 + d_1)(\delta + d_1)}{\alpha\gamma(d_1(\beta + d_1 + \delta))}, C^* = \frac{\gamma I^*}{d_2}, \right.$$
$$\left. R^* = \frac{\beta I^*}{(\delta + d_1)} \right)$$

Thus, disease-free equilibrium point does not exist.

Equilibrium Points: Model II

Equilibrium points for Model II are as follows:

- Disease-free equilibrium point
  $E_0(\frac{\alpha_1}{d_1}, 0, 0, 0)$
- Non-trivial equilibrium point
  $E^*(S^*, I^*, C^*, R^*)$

where
$I^* = \frac{(d_3 + \alpha_2 - p\alpha S^*)(\alpha_1 - d_1)S^*}{(d_3 + \alpha_2 - p\alpha S^*)(\beta S^* - \frac{\pi\delta}{\delta + d_4}) + p\beta S^*} > 0$ provided $d_3 + \alpha_2 > p\alpha S^*$ and $\beta S^* > \frac{\pi\delta}{\delta + d_4}$

$$S^* = \frac{(d_3+\alpha_2)(d_2+m_1)}{p\beta\alpha_2+(d_3+\alpha_2)(1-p)\beta+p\alpha(m_1+d_2)}$$

$$R^* = \frac{m_1 I^*}{\delta+d_4}$$

$$C^* = \frac{p\beta S^* I^*}{d_3+\alpha_2-p\alpha_2 S^*}$$

Now, we define,

$$R_0 = \frac{\alpha_1}{S^* d_1} = \frac{\alpha_1(p\beta\alpha_2+(d_3+\alpha_2)(1-p)\beta+p\alpha(m_1+d_2))}{(d_3+\alpha_2)(d_2+m_1)d_1}$$

which is a threshold parameter.

R_0 can be written as

$$R_0 = \left[\frac{(1-p)\beta}{(d_2+m_1)} + p(\frac{\alpha}{(d_3+\alpha_2)} + \frac{\beta\alpha_2}{(d_3+\alpha_2)(d_2+m_1)}\right]\frac{\alpha_1}{d_1}$$

We will show that R_0 is the basic reproduction number which is the average number of secondary infections caused by a single infective in an entire susceptible population during its entire infectious period. The probability of introduction of non-carrier into the system is $(1-p)$ which makes β effective contacts per unit time. We multiply β with (average infectious period) $\frac{1}{(d_2+m_1)}$ for non carriers. The infective is a carrier with probability p, which makes β effective contacts per unit time during the average period $\frac{1}{d_3+\alpha_2}$ it remains a carrier. Even though infective have become a non-carrier, it caused $\beta\frac{1}{(d_2+m_1)}$ number of infective with probability $\frac{\alpha_2\beta}{d_3+\alpha_2}$ so that it can survive in the carrier stage. Therefore $[\frac{(1-p)\beta}{(d_2+m_1)} + p(\frac{\alpha}{(d_3+\alpha_2)} + \frac{\beta\alpha_2}{(d_3+\alpha_2)(d_2+m_1)}]$ is the per capita average number of secondary infections. The product of this expression with the number of susceptible at the disease-free equilibrium, $(\frac{\alpha_1}{d_1})$, gives R_0.

Since R_0 have the parameter α_2, β and p all are related to the carrier class and all are present in the reproduction number from Eq. (24.7), therefore carriers have great effect on R_0. We can analyze that R_0 increases as β increases.

The effect of p and R_0.

$$\frac{\partial R_0}{\partial p} = \left[\frac{-\beta}{(d_2+m_1)} + \frac{\alpha}{(d_3+\alpha_2)} + \frac{\beta}{(d_2+m_1)}\frac{\alpha_2}{(d_3+\alpha_2)}\right]\frac{\alpha_1}{d_1} = \frac{\alpha_1}{d_1(d_3+\alpha_2)}\left[\alpha - \frac{\beta d_3}{d_2+m_1}\right]$$

and thus,

$$\frac{\partial R_0}{\partial p} > 0 \text{ if } \alpha > \frac{\beta d_3}{d_2+m_1}$$

Development of carriers has great impact on the basic reproduction number under the above condition. There will be an increase in the reproduction number as the probability of carriers increases.

We can also analyze the effect of diagnosis rate α_2 on R_0.

Analyze the effect of α_2 on R_0.

$\frac{\partial R_0}{\partial \alpha_2} = -\frac{1}{(d_3+\alpha_2)^2}\left[-\alpha + \frac{\beta d_3}{d_2+m_1}\right]$
and thus,

$\frac{\partial R_0}{\partial \alpha_2} < 0 \text{ if } \alpha > \frac{\beta d_3}{d_2+m_1}$

Hence, from the above analysis we observed that the parameter p and α_2 have opposite effects on R_0, i.e., they are inversely proportional to each other, the higher probability p of carrier increases R_0, a higher diagnosis rate α_2 of carriage decreases R_0.

24.4.1 Local Stability

Model I:

The Jacobian corresponding to Model I is:

$$J(S^*, I^*, R^*, C^*) = \begin{pmatrix} \alpha_1 - \alpha C - d_1 & 0 & \delta & -\alpha S \\ \alpha C & -d_1 & 0 & \alpha S \\ 0 & \beta & -\delta - d_1 & 0 \\ 0 & \gamma & 0 & -d_2 \end{pmatrix} + e^{-\mu T}\begin{pmatrix} 0 & 0 & 0 & 0 \\ 0 & -\beta & 0 & 0 \\ 0 & 0 & 0 & 0 \\ 0 & 0 & 0 & 0 \end{pmatrix}$$

$$= \begin{pmatrix} \alpha_1 - \alpha C - d_1 & 0 & \delta & -\alpha S \\ \alpha C & d_1 - \beta e^{-\mu T} & 0 & \alpha S \\ 0 & \beta & -\delta - d_1 & 0 \\ 0 & \gamma & 0 & -d_2 \end{pmatrix}$$

$$= \begin{pmatrix} \alpha_1 - \alpha C - d_1 - \mu & 0 & \delta & -\alpha S \\ \alpha C & d_1 - \beta e^{-\mu T} - \mu & 0 & \alpha S \\ 0 & \beta & -\delta - d_1 - \mu & 0 \\ 0 & \gamma & 0 & -d_2 - \mu \end{pmatrix}$$

The characteristic equations corresponding to trivial equilibrium point is:

$$(\alpha_1 - d_1 - \mu)(d_1 - \beta e^{-\mu T} - \mu)(-\delta - d_1 - \mu)(-d_2 - \mu) = 0 \tag{24.11}$$

Case 1: $T = 0$

$$(\alpha_1 - d_1 - \mu)(d_1 - \beta - \mu)(-\delta - d_1 - \mu)(-d_2 - \mu) = 0 \tag{24.12}$$

$$\mu_1 = d_1 - \alpha_1, \mu_2 = \beta - d_1, \mu_3 = \delta + d_1, \mu_4 = d_2 \tag{24.13}$$

Since μ_3 and μ_4 will remain positive even if μ_1 and μ_2 are negative. Hence, the trivial equilibrium will always remain unstable.

Case 2: $T > 0$

$$(\alpha_1 - d_1 - \mu)(d_1 - \beta e^{-\mu T} - \mu)(-\delta - d_1 - \mu)(-d_2 - \mu) = 0 \tag{24.14}$$

$$\mu^4 + a_3\mu^3 + a_2\mu^2 + a_1\mu + a_0 + e^{-\mu T}(y_3\mu^3 + y_2\mu^2 + y_1\mu + y_0) = 0 \quad (24.15)$$

where
$a_3 = d_1 + d_2 - \alpha_1 + 2d_1 + \delta$
$a_2 = d_2d_1 - (d_1 + d_2)(\alpha - 2d_1 - \delta) - \alpha_1\delta - \alpha_1 d_1 + d_1\delta + d_1^2$
$a_1 = (d_2d_1)(\alpha_1 - 2d_1 - \delta) + (\alpha_1 S + \alpha_1 d_1 - d_1 S - d_1^2)(d_1 + d_2)$
$a_0 = -(\alpha_1 S + \alpha_1 d_1 - d_S - d_1^2)(d_1 + d_2)$
$y_3 = \beta,\ y_2 = \beta(\mu + 2d_1 + d_2 + \delta - \alpha_1)$
$y_1 = \beta(d_1d_2 - \alpha_1(\delta + d_1 + d_2))$
$y_0 = d_1d_2(d_1 - \alpha_1)$
Let $\mu = i\eta$

$$\eta^4 - a_3 i\eta^3 - a_2\eta^2 + a_1 i\eta + a_0 + e^{-i\eta T}(-iy_3\eta^3 - y_2\eta^2 + iy_1\eta + y_0) \quad (24.16)$$

Separating real and imaginary part of (24.16), then by squaring and adding both the equations, we get,

$$\eta^8 + (-2a_2 + a_3^2 + y_3^2)\eta^6 + (a_2^2 + 2a_0 + y_2^2 - 2a_1a_3 - 2y_1y_3)\eta^4$$
$$+(-2a_2a_0 - 2y_2 2y_0 + a_1 + y_1^2)\eta^2 + (a_0^2 + y_0^2) = 0$$

Let $s = \eta^2$, then above equation becomes:

$$s^4 + (-2a_2 + a_3^2 + y_3^2)s^3 + (a_2^2 + 2a_0 + y_2^2 - 2a_1a_3 - 2y_1y_3)s^2$$
$$+(-2a_2a_0 - 2y_2 2y_0 + a_1 + y_1^2)s + (a_0^2 + y_0^2) = 0$$

By Routh Hurwitz criteria, above equation has roots with negative real parts if $a > 0, b > 0, c > 0, ab - c > 0, ad - c > 0$, where $a = -2a_2 + a_3^2 + y_3^2$, $b = a_2^2 + 2a_0 + y_2^2 - 2a_1a_3 - 2y_1y_3$, $c = a_2^2 + 2a_0 + y_2^2 - 2a_1a_3 - 2y_1y_3$, $d = a_0^2 + y_0^2$.

But $s = \eta^2 > 0$, this proves that the assumption $\mu = i\eta$ is wrong and thus the equation has no positive roots and the real part of all eigenvalues are negative for all $T > 0$. Therefore, system is stable for $T > 0$.

Endemic Equilibrium Point:

Case 1: T = 0

The Jacobian corresponding to endemic equilibrium point is:

$$= \begin{pmatrix} \alpha_1 - \alpha C^* - d_1 & 0 & \delta & -\alpha S^* \\ \alpha C^* & d_1 - \beta e^{-\mu T} & 0 & \alpha S^* \\ 0 & \beta & -\delta - d_1 & 0 \\ 0 & \gamma & 0 & -d_2 \end{pmatrix}$$

$$\begin{pmatrix} \alpha_1 - \alpha C^* - d_1 - \mu & 0 & \delta & -\alpha S^* \\ \alpha C^* & d_1 - \beta e^{-\mu T} - \mu & 0 & \alpha S^* \\ 0 & \beta & -\delta - d_1 - \mu & 0 \\ 0 & \gamma & 0 & -d_2 - \mu \end{pmatrix}$$

Characteristic equation corresponding to endemic equilibrium point is:

$$\lambda^4 - (A + B + C + D)\lambda^3 + (AB + BC + CD + DA)\lambda^2 \\ - (ABC + BCD + CDA)\lambda + ABCD = 0 \tag{24.17}$$

where, $A = \alpha_1 - \alpha C - d_1,\ B = d_1 - \beta,\ C = -\delta - d_1,\ d = -d_2$.

Eigenvalues corresponding to above characteristic equation are negative if, $A + B + C + D$ and $ABC + BCD + CDA$ are negative.

Case 2: The local stability of endemic equilibrium for $T > 0$ can be obtained in the same way as has been done for trivial equilibrium.

Model II: To examine the local stability of the disease-free equilibrium, we calculate Jacobian matrix at $E_0(\frac{\alpha_1}{d_1}, 0, 0, 0)$, which is:

$$E_0 = \begin{bmatrix} \alpha_1 - d_1 & -\beta\left(\frac{\alpha_1}{d_1}\right) & -\alpha_2\left(\frac{\alpha_1}{d_1}\right) & 0 \\ 0 & (1-p)\beta\left(\frac{\alpha_1}{d_1}\right) - (d_2 + m_1) & (1-p)\alpha_2\left(\frac{\alpha_1}{d_1}\right) + \alpha_2 & 0 \\ 0 & p\beta\left(\frac{\alpha_1}{d_1}\right) & p\alpha_2\left(\frac{\alpha_1}{d_1}\right) - (d_3 + \alpha_2) & 0 \\ 0 & m_1 & 0 & -(d_4 + \delta) \end{bmatrix} \tag{24.18}$$

One of the eigenvalues is $\lambda_1 = -(d_1 - \alpha_1) < 0$. The others eigenvalues $\lambda_2, \lambda_3, \lambda_4$ are obtained by 3×3 matrix.

$$\begin{bmatrix} (1-p)\beta\left(\frac{\alpha_1}{d_1}\right) - (d_2 + m_1) & (1-p)\alpha_2\left(\frac{\alpha_1}{d_1}\right) + \alpha_2 & 0 \\ p\alpha_2\left(\frac{\alpha_1}{d_1}\right) & p\alpha_2\left(\frac{\alpha_1}{d_1}\right) - (d_3 + \alpha_2) & 0 \\ m_1 & 0 & -(d_4 + \delta) \end{bmatrix} \tag{24.19}$$

Second eigenvalue is $\lambda_2 = -(d_4 + \delta) < 0$. Now reduce the above matrix M 2×2

$$M = \begin{bmatrix} (1-p)\beta\left(\frac{\alpha_1}{d_1}\right) - (d_2 + m_1) & (1-p)\alpha_2\left(\frac{\alpha_1}{d_1}\right) + \alpha_2 \\ p\alpha_2\left(\frac{\alpha_1}{d_1}\right) & p\alpha_2\left(\frac{\alpha_1}{d_1}\right) - (d_3 + \alpha_2) \end{bmatrix} \tag{24.20}$$

In order to prove that the other two eigenvalues are negative, we will prove that $tr(M) < 0$ and $det(M) > 0$ when $R_0 < 1$.

Consider $tr(M)$

$$tr(M) = (1-p)\beta\left(\frac{\alpha_1}{d_1}\right) - (d_2+m_1) + p\alpha_2\left(\frac{\alpha_1}{d_1}\right) - (d_3+\alpha_2)$$
$$= (d_2+m_1)\left(\frac{(1-p)\beta\left(\frac{\alpha_1}{d_1}\right)}{(d_2+m_1)} - 1\right) + (d_3+\alpha_2)\left(\frac{p\alpha_2\left(\frac{\alpha_1}{d_1}\right)}{(d_3+\alpha_2)} - 1\right)$$

If, $R_0 = [\frac{(1-p)\beta}{(d_2+m_1)} + p(\frac{m_1}{(d_3+\alpha_2)} + \frac{\alpha_2\beta}{(d_3+\alpha_2)}\frac{1}{(d_2+m_1)})]\frac{\alpha_1}{d_1} < 1$ then, $\frac{\beta(1-p)\frac{\alpha_1}{d_1}}{(d_2+m_1)} < 1$ and $\frac{\alpha_2 p\frac{\alpha_1}{d_1}}{(d_3+\alpha_2)} < 1$. Thus, $tr(M) < 0$.

Consider $det(M)$;

$$det(M) = \left[\left((1-p)\beta\left(\frac{\alpha_1}{d_1}\right) - (d_2+m_1)\right)\left(p\alpha_2\left(\frac{\alpha_1}{d_1}\right) - (d_3+\alpha_2)\right)\right] - \left[\left(p\alpha_2\left(\frac{\alpha_1}{d_1}\right)\right)\left((1-p)\alpha_2\left(\frac{\alpha_1}{d_1}\right) + \alpha_2\right)\right]$$

$$= (1-p)\alpha_2\beta p\left(\frac{\alpha_1}{d_1}\right)^2 - (1-p)\beta\left(\frac{\alpha_1}{d_1}\right)(d_3+\alpha_2) - \alpha_2 p\left(\frac{\alpha_1}{d_1}\right)(d_2+m_1) + (d_2+m_1)(d_3+\alpha_2) - \alpha_2^2 p(1-p)\left(\frac{\alpha_1}{d_1}\right)^2 - \alpha_2^2 p\left(\frac{\alpha_1}{d_1}\right)$$

$$= (d_2+m_1)(d_3+\alpha_2) - \left[(d_2+m_1)\alpha_2 p\left(\frac{\alpha_1}{d_1}\right) - (d_3+\alpha_2)(1-p)\beta\left(\frac{\alpha_1}{d_1}\right) - \alpha_2^2 p(1-p)\left(\frac{\alpha_1}{d_1}\right)^2 - \alpha_2^2 p\left(\frac{\alpha_1}{d_1}\right)\right]$$

$$= (d_2+m_1)(d_3+\alpha_2)[1-R_0]$$

Therefore $det(M) > 0$ if and only if $R_0 < 1$.

Proposition 24.3 *E_0 is locally asymptotically stable if $R_0 < 1$ and is unstable if $R_0 > 1$.*

24.4.2 Global Stability

Model I:

Proposition 24.4 *E^* is globally asymptotically stable if it satisfies the following conditions:*

(i) $6a_{12}^2 < a_{11}a_{22}$
(ii) $4a_{14}^2 < a_{11}a_{44}$
(iii) $3a_{23}^2 < a_{22}a_{33}$
(iv) $6a_{24}^2 < a_{22}a_{44}$

where, $a_{11} = -\alpha_1 + d_1$, $a_{44} = (\delta + d_4)$, $a_{33} = d_2$, $a_{12} = -\alpha C^*$, $a_{14} = -\delta$, $a_{23} = -\alpha S_{max} - \gamma$, $a_{24} = -\beta$, $a_{22} = (\beta + d_1)$.

Proof To study the global stability we will make use of Lyapunov function V(S, I, C, R) of the form:

$V(S, I, C, R) = \frac{1}{2}(S-S^*)^2 + \frac{1}{2}(I-I^*)^2 + \frac{1}{2}(C-C^*)^2 + \frac{1}{2}(R-R^*)^2$

We will show the derivative of Lyapunov function as negative definite.

Let $z_1 = S - S^*, z_2 = I - I^*, z_3 = C - C^*, z_4 = R - R^*$, then we get, which implies

$$\dot{V} = z_1[\alpha_1 S - \alpha SC + \delta R - d_1 S] + z_2[\alpha SC - \beta I - d_1 I] + z_3[\gamma I - d_2 C] + z_4[\beta I - \delta R - d_1 R]$$

which on further simplification gives the following:

$$\dot{V} \leq -\Big[\frac{a_{11}}{2}z_1^2 + a_{12}z_1z_2 + \frac{a_{22}}{3}z_2^2 + \frac{a_{22}}{3}z_1^2 + a_{23}z_1z_3 + a_{33}z_3^2 + \frac{a_{11}}{2}z_1^2 + a_{14}z_1z_4 + \tfrac{a_{44}}{2}z_4^2 + \tfrac{a_{22}}{3}z_2^2 + a_{24}z_2z_4 + \tfrac{a_{44}}{2}z_3^2\Big]$$

Thus, $\dot{V} \leq 0$.

Hence E^* is global asymptotically stable under the conditions stated in the theorem.

Model II:

Proposition 24.5 *E_0 is globally asymptotically stable if $R_0 \leq 1$.*

Proof Here, we will use the method of Lyapunov functions

$$L = \left[\frac{\beta}{d_3 + \alpha_2} + \frac{\gamma\alpha_2}{(d_3 + \alpha_2)(d_2 + m_1)}\right] C + \left(\tfrac{\gamma}{d_2 + m_1}\right) I$$

$$\frac{dL}{dt} = \left[\frac{\beta}{d_3 + \alpha_2} + \frac{\gamma\alpha_2}{(d_3 + \alpha_2)(d_2 + m_1)}\right] C' + \left(\frac{\gamma}{d_2 + m_1}\right) I$$

$$\left[\frac{\beta}{d_3 + \alpha_2} + \frac{\gamma\alpha_2}{(d_3 + \alpha_2)(d_2 + m_1)}\right] p\alpha_2 SC + p\beta SI - (d_3 + \alpha_2)C +$$

$$\left(\tfrac{\gamma}{d_2 + m_1}\right)(1 - p)\alpha_2 SC + (1 - p)\beta SI + \alpha_2 C - (d_2 + m_1)I$$

$$= \left[\frac{p\beta}{d_3 + \alpha_2} + \frac{p\gamma\alpha_2}{(d_3 + \alpha_2)(d_2 + m_1)} + \frac{(1 - p)\gamma}{(d_2 + m_1)}\right] S(\alpha_2 C + \gamma I) - (\alpha_2 C + \gamma I)$$

$$= \left[\frac{d_1}{\alpha_1} R_0 S - 1\right] [\alpha_2 C + \gamma I] \text{ by } S \geq \frac{\alpha_1}{d_1}$$

$$\frac{dC}{dt} \leq (R_0 - 1)(\alpha_2 C + \gamma I) \leq 0$$

so $\dfrac{dL}{dt} \leq 0$ if $R_0 \leq 1$

further $\dfrac{dL}{dt} = 0$ (=) $C = I = 0$ *or* $R_0 = 1$ and $S = \frac{\alpha_1}{d_1}$.

Hence, by LaSalle's Invariance Principle, E_0 is globally asymptotically stable.

Next, we will prove the global stability of endemic equilibrium point.

Proposition 24.6 *E^* is globally stable if it satisfies the following conditions:*

(i) $6a_{12}^2 < a_{11}a_{22}$
(ii) $9a_{13}^2 < a_{11}a_{33}$
(iii) $6a_{14}^2 < a_{11}a_{44}$
(iv) $9a_{23}^2 < a_{22}a_{33}$
(v) $6a_{34}^2 < a_{33}a_{44}$

where, $a_{11} = -\alpha_1 + d_1 - \beta I^* + \alpha C^*$, $a_{22} = -(1-p)\beta S^* + \pi + d_2$, $a_{33} = -p\alpha S^*$, $a_{44} = \delta + d_4$, $a_{12} = -(1-p)\beta I^* + (1-p)\alpha C^*$, $a_{13} = p\alpha C^* + p\beta I^*$, $a_{23} = -(1-p)\alpha S^* - \alpha_2 - p\beta S^*$, $a_{14} = -\delta$, $a_{34} = -\pi$.

Proof To study the global stability we will make use of Lyapunov function V(S,I,C,R) of the form:

$$V(S, I, C, R) = \frac{1}{2}(S - S^*)^2 + \frac{1}{2}(I - I^*)^2 + \frac{1}{2}(C - C^*)^2 + \frac{1}{2}(R - R^*)^2$$

We will show that the derivative of Lyapunov function as negative definite.

Let $z_1 = S - S^*$, $z_2 = I - I^*$, $z_3 = C - C^*$, $z_4 = R - R^*$, then, we get,

$$\dot{V} = z_1[\alpha_1 S - \alpha SC - \beta SI - d_1 S + \delta R] + z_2[(1-p)\alpha SC + (1-P)\beta SI + \alpha_2 C - m_1 I - d_2 I] + z_3[p\alpha SC + p\beta SI - d_3 C - \alpha_2 C] + z_4[m_1 I - \delta R - d_4 R]$$

which on further simplification gives the following:

$$\dot{V} \leq z_1^2[\alpha_1 - d_1 + \beta I^* - \alpha C^*] + z_2^2[(1-p)\beta S^* - \pi - d_2] + p\alpha S^* z_3^2 - (\delta + d_4)z_4^2 + z_1 z_2[(1-p)\beta I^* - (1-p)\alpha C^*] + z_1 z_3[-p\alpha C^* - p\beta I^*] + z_2 z_3[(1-p)\alpha S^* + \alpha_2 + p\beta S^*] + \delta z_1 z_4 + \pi z_3 z_4$$

Then,

$$\dot{V} \leq -\left[\frac{a_{11}}{3}z_1^2 + a_{12}z_1 z_2 + \frac{a_{22}}{2}z_2^2 + \frac{a_{11}}{3}z_1^2 + a_{13}z_1 z_3 + \frac{a_{33}}{3}z_3^2 + \frac{a_{11}}{3}z_1^2 + a_{14}z_1 z_4 + \frac{a_{44}}{2}z_4^2 + \frac{a_{22}}{3}z_2^2 + a_{12}z_2 z_3 + \frac{a_{33}}{3}z_3^2 + \frac{a_{33}}{3}z_3^2 + a_{34}z_3 z_4 + \frac{a_{44}}{2}z_4^2\right]$$

Thus, $\frac{dV}{dt} \leq 0$.

Hence E^* is global asymptotically stable under the conditions stated in the theorem.

24.5 Conclusion

In this paper, we have studied two models where population is divided into four classes: susceptible, carrier, infected and recovered. In Model I we have incorporated delay in time to come to recovered class from infected class. In Model II we have introduced a probability factor and we have also incorporated the interaction between susceptible and infective and the diagnosis rate of carrier. We have obtained boundedness, existence of equilibrium points for both the models. In the case of Model I, the disease-free equilibrium point does not exist. Further, we have also obtained the local and global stability of endemic equilibrium point. In the case of Model II, we have obtained basic reproduction number and obtained the local and global stability analysis of disease-free equilibrium point depending on the basic reproduction number. We have also proved that as the probability of carrier increases, the basic reproduction number increases and as diagnosis rate of carrier increases, basic reproduction number decreases. Thus, carrier can be taken as a control parameter for disease but with Model II, we also have interaction between susceptible and infective; hence to control number of infective, we should go for various vaccination policies in addition to the control of disease.

References

1. Shim, E.: An epidemic model with immigration of infectives and vaccination. University of British Columbia (2004)
2. Crump, J.A., Luby, S.P., Mintz, E.D.: The global burden of typhoid fever. Bull. World Health Organ. **82**, 346–353 (2004)
3. Black, R.E., Morris, S., Bryce, J.: Where and why are 10 million children dying every year?. **361**(9376), 2226–2234 (2003)
4. Riggs, M.M., Sethi, A.K., Zabarsky, T.F., Eckstein, E.C., Jump, R.L., Donskey, C.J.: Asymptomatic carriers are a potential source for transmission of epidemic and non epidemic Clostridium difficile strains among long term care facility residents. **45**, 992–998 (2007)
5. Guo, H., Li, M.Y.: Global dynamics of a staged progression model for infectious diseases. Math. Quart. **13**, 313–323 (2005)
6. Naresh, R., Pandey, S., Misra, A.K.: Analysis of a vaccination model for carrier dependent infectious diseases with environmental effects. Nonlinear Anal. Model. Control **13**, 331–350 (2008)
7. Viral Hepatitis and Emerging Bloodborne Pathogens in Canada.: CCDR, 27S3, Public Health Agency of Canada (PHAC) (2001)
8. WHO.: Fact Sheet on Hepatitis B. http://www.who.int/mediacentre/factsheets/fs204/en/index.html (2008)
9. Goldstein, S.T., Zhou, F., Hadler, S.C., Bell, B.P., Mast, E.E., Margolis, H.S.: A mathematical model to estimate global hepatits B disease burden and vaccination impact. Int. J. Epidemiol. **34**, 1329–1339 (2005)

Chapter 25
Resume of Some Invited and Contributed Talks

Pooja

Abstract In this chapter we present a summary of those talks that could not find a place in the volume due to its limited extent.

Keywords Resume

In this chapter we present a summary of those talks that could not find a place in the volume due to its limited extent.

Professor Leon Chua, University of California, Berkeley delivered a long lecture entitled, "Principle of local activity and the Edge of Chaos". His lecture provided the mathematical foundation for the currently very active research area named as "Complexity", which is based on anecdotes and computer simulations. In particular it will provide a definitive foundation for Turning's theory of the morphogenesis and a resolution of hitherto unsolved Smale's paradox.

Professor Goetz Pfander, Jacobs University, Germany, delivered a talk on "The BalianLow Theorem for subspaces in higher dimensions" based on his joint work with Carlos Cabrelli and Ursula Molter. An abstract of his talk is given below.

The classical BalianLow Theorem states that time and frequency well-localized window functions cannot give rise to a Gabor orthonormal basis for the space of square integrable functions on Euclidean space. Certainly, well-localized orthonormal bases of Gabor type are possible for subspaces. We establish BalianLow type theorems for subspaces, i.e., for spaces generated by a discrete set of time frequency-shifted copies of a single window function. We show that whenever the generating system forms an Riesz bases for its closed linear span, and if the closed linear space has a nontrivial shift invariance, then the generating window function is again not well localized in time or in frequency.

Pooja (✉)
Guru Nanak Dev University, Amritsar, India
e-mail: poojarai_44@yahoo.com

P. Manchanda et al. (eds.), *Industrial Mathematics and Complex Systems*,
Industrial and Applied Mathematics, DOI 10.1007/978-981-10-3758-0_25

The abstract of talk by Prof. Graeme Fairweather, Mathematical Reviews, American Mathematical Society, Ann Arbor, Michigan U.S.A. entitled, "Sixty Years of Alternating Direction Implicit (ADI) Methods" is given below.

It is sixty years since the publication of the landmark paper by D. W. Peaceman and H. H. Rachford in which ADI methods were first introduced, in the context of finite difference methods, for parabolic and elliptic problems in two space variables. The attraction of the ADI approach is that it replaces the solution of a multidimensional problem by sequences of one-dimensional problems in the coordinate directions. ADI methods in conjunction with various types of spatial discretizations continue to be studied extensively today, especially for the numerical solution of time-dependent problems. In this talk, we present a brief overview of the history of ADI methods, followed by a discussion of such methods developed recently for parabolic partial integro-differential equations, Schrodinger systems, and nonlinear reaction-diffusion systems.

One of his collaborators, Santosh Kumar Bhal, presented a joint research paper with him entitled, "Some Observations of Orthogonal Spline Collocation Methods for the two-dimensional Helmholtz equation with discontinuous coefficients". An abstract of this paper is given below.

In this paper, we use orthogonal spline collocation method (OSCM) for two-dimensional Helmholtz equation with discontinuous coefficients. Monomial cubic basis functions are used in X direction and piecewise cubic Hermite basis functions are used in Y direction to approximate the solution. We use the matrix decomposition algorithms (MDA) to find the approximate solution effectively with minimum operations count. Finally, we perform several numerical experiments with different wave numbers and using grid refinement analysis, we compute the order of convergence of the numerical scheme.

Professor Govindan Rangarajan, IISc, Bangalore, presented a talk on "Applications of Granger Causality to Neuroscience". An abstract of his talk is given below.

Detecting connectivity patterns in a brain network is crucial to the subsequent analysis of the network structure. Once these connectivity patterns are detected, there is also tremendous interest in determining how these patterns change with time. This is important since changes in connectivity patterns can serve as functional biomarkers for the onset of diseases or can be used to detect changes in the underlying states. Granger causality is a tool that can be used to detect and quantify connectivity patterns. We propose extensions of Granger causality that enable it to be applied to a much wider variety of complex systems. We also demonstrate how changes in connectivity patterns can be measured using these extensions. If time permits, we will consider block coherence, a new tool that we have proposed to study connectivity patterns.

Professor A. Adimurthi of TIFR CAM, Bangalore, presented his contribution on the Hyperbolic conservation law one space dimension with discontinuous flux. Under suitable hypothesis on the flux, the existence of an optimal control given at any time $T > 0$ is proved and a numerical scheme proposed.

Dr. Venky Krishnan, TIFR CAM, Bangalore, presented his work in an emerging area namely, "Inversion of restricted ray transforms of symmetric rank m tensor fields in n-dimensional Euclidean space", which is joint work with Rohit Kumar Mishra. An abstract of his talk is given below.

We consider the integral geometry problem of recovering rank m symmetric tensor fields from its Euclidean ray transform, that is, from its integrals along lines in n dimensional space. We focus on ray transforms restricted to lines passing through a fixed smooth curve. Under such conditions on the curve, we will present microlocal inversion results for the recovery of a component of the symmetric tensor field from its ray transform.

Professor Samares Pal, University of Kalyani, Kalyani, discussed his research work on the topic "Effects of macroalgal toxicity and overfishing on the resilience of coral reefs". An abstract of this paper is given below:

Macroalgae and corals compete for the available space in coral reef ecosystem. While herbivorous reef-fish play a beneficial role in decreasing the growth of macroalgae in coral reef ecosystem. Abundance of macroalgae changes the community structure towards macroalgae dominated reef ecosystem. We investigate coral-macroalgal phase shifts by means of a continuous time model in a food chain. It is observed that in presence of macroalgal toxicity and overfishing the system exhibits hysteresis through saddle-node bifurcation and transcritical bifurcation. We also examine the effects of time lags in the liberation of toxin by macroalgae and recovery of algal turf in response to grazing of herbivores on macroalgae.

Dr. L. M. Saha, Shiv Nadar University, presented a paper entitled, "Nonlinear Dynamics, Chaos And Complexities" based on joint work with Prof. M. K. Das and Prof. Rashmi Bhardwaj (GGSIP University, New Delhi). An abstract of this paper is given below.

Almost all evolving real systems emerging around us are nonlinear in nature and their dynamics are not as simple as in cases of linear systems. Principle of superposition is no more applicable to real system and, to study them, one must apply the recent rules and methodology suggested in nonlinear dynamics. Because of nonlinearity in nature, real systems show complexities in behavior while evolving and chaos is one such complexity. Principles of nonlinear dynamics can only help to understand complex and chaotic behaviors observed in any nonlinear system. Various tools which have been discovered due to growing researches in this area are helpful to understand phenomena of evolutions in real systems. In addition to the basic tools (e.g., time series and phase plane graphs), some tools have been suggested recently which help to understand better the complexities and chaotic motion in a dynamical system, such as Lyapunov exponents (LCEs), topological entropies, correlation dimension, Poincaré map, etc. To distinguish regular and chaotic motion, some recent tools have to be discussed with their working limitations.

The present talk aims to explain evolutionary dynamics of nonlinear systems and to explain about complexities observed during that processes. Some specific models proposed for real systems would be discussed and calculations of LCEs, topological entropies, correlation dimensions have been obtained as a part of complexity measure. Numerically calculated results are displayed graphically with complete interpretation.

Vikram Sharma, D.A.V. College, Amritsar, presented a paper entitled, "Nonuniform wave packet frames in $L^2(R)$" based on joint work with Prof. Pammy Manchanda. An abstract of this paper reads as.

"Wave packet systems are generated by the combined action of translations, modulations, and dilations on a finite family of functions. We construct wave packet frames in $L^2(R)$ over a translation set Λ, which is not necessarily a group. We call it nonuniform wave packet frame and present necessary and sufficient condition for the wave packet system $\{D_{(2N)^j} T_\lambda E_{c_m} \psi\}_{j,m \in \mathbb{Z}, \lambda \in \Lambda}$ to be a frame for $L^2(\mathbb{R})$".

Pooja, GNDU, Amritsar, presented a paper on "Solving Variational Problems using Haar Wavelet" based on joint work with Prof. Pammy Manchanda. An abstract of this paper is given below.

Haar wavelets are used for approximating solution of variational problem of calculus of variations. The variational problem is converted into differential equation using Euler Lagrange's equation and then Haar collocation method is applied to solve this differential equation. We present the numerical solution to some of the problems of calculus of variation.

Mamta Rani, GNDU, Amritsar, presented a paper on "Non Uniform Haar Wavelet Matrix Method for Numerical Solution of Ordinary Differential Equations" based on joint work with Prof. Pammy Manchanda. An abstract of this paper is presented below.

Wavelet based algorithms have become an important tool in numerical analysis for solving differential equations. Uniform Haar wavelet matrix methods have been used to find the numerical solution of ordinary differential equations. We have applied the non uniform Haar wavelet matrix method to find the numerical solution of the ordinary differential equations and presented error estimation comparison for uniform and non uniform Haar wavelet matrix methods.

Abdullah, Zakir Husain College, New Delhi, presented his result on "Characterization of Scaling Functions Associated With Nonuniform Multiresolution Analysis". Prof. Renu Chugh and Mandeep Kumari from MDU, Rohtak presented a research paper. In this paper, common zeros of a finite family of m-accretive operators based on modified proximal point algorithm in Hadamard manifolds are investigated and applications are studied. Santosh Kumar, AMU, Aligarh, presented a paper entitled, "An efficient PDE-based model for image restoration" based on joint work with Prof. M. K. Ahmad. An abstract of this paper is given below.

"In this paper, we propose new time dependent model for solving total variation (TV) minimization problem in image restoration. The main idea is to apply a priori smoothness on the solution image. The total variation of the image is minimized subject to constraints involving the point spread function (PSF) of the blurring process and the statistics of the noise. The blurring operator provides useful information in restoration. The constraints are implemented using Lagrange's multipliers. The solution is obtained using the gradient-projection method of Rosen. Proof of the existence, uniqueness and stability of the viscosity solution of our model. The results of our model using explicit numerical schemes are compared with other known image restoration models".

Deepti Gupta, Jamia Millia Islamia, New Delhi, presented a paper entitled, "Support Vector Machine: A Classification Technique". Puneet Kaur investigated the Rayleigh–Bénard convection under sinusoidally varying temperatures of the horizontal rigid-planes bounding a laterally infinite fluid layer for the bicritical states. The coexistence of both the harmonic and subharmonic behavior in response to the excitation applied in some parametric space is called as bicritical state. The problem is analogous to the well-studied Faraday-instability and Rayleigh-Bénard convection under gravity modulation. Under modulation, the neutral instability curve is found to alternate between the conventional harmonic and subharmonic tongues in the space of the dimensionless wave number of disturbance and the control parameter. The transition between harmonic and subharmonic critical instability responses is found to occur via a bicritical state, where the two instability responses coexist with different wave numbers. These bicritical states are found to depend upon the modulation parameters and the Prandtl number.

Mazibar Rahman presented a paper entitled, "Fluctuating free convective flow and heat Transfer along an infinite vertical porous plate". The free convective flow and heat transfer along an infinite vertical porous plate are investigated when a transverse sinusoidal suction velocity distribution fluctuating with time is applied. Due to this transverse velocity the flow of the fluid is three-dimensional. A series expansion method is applied to get the solution of the governing equations and the expressions for velocity and temperature fields are obtained. The skin friction and the rate of heat transfer at the plate are discussed in detail.

Javed Miya, UTU, Dehradun, presented a paper entitled, "Threshold Based Segmentation and Analysis Of Medical Image Compression" based on joint work with Dr. M. A. Ansari. An abstract of this paper states.

Medical environment is moving toward computerisation, digitization and centralization, resulting in prohibitive amounts of digital medical image data. Compression techniques are, therefore, essential in archival and communication of medical image. Although lossy compression has much higher compression rates, the medical community has relied on lossless compression for legal and clinical reasons. In this paper, we have done a region-based segmentation and image analysis with application to medical image. Image segmentation is important for object and its boundary detections. It is also important for a variety of image analysis and visualization tasks both inside and outside the medical image domain. With the help of Sobel filter we consider different edges of the medical image, which is useful for image compression algorithm for effective compression of medical image.

Ashok Kumar Sahoo, Sharda University, Greater Noida, presented a paper entitled, "Computer Recognition of Isolated Numeral Signs of Indian Sign Language" based on joint work with Gouri Sankar Mishra. An abstract of this paper reads as.

Sign language recognition is helpful in communication to exchange ideas and information among hearing impaired community and with people with normal speaking capability. In India, the official sign language is known as Indian Sign Language (or ISL). For the purpose of automating the process of ISL recognition, a system is developed. A comprehensive set of ISL signs are captured by the help of a digital camera and are used in experiments. In this research, a system for computer–human interface for ISL recognition is proposed. The paper describes a system for automatic recognition of ISL of static numeral signs. The system is capable of recognizing isolated numeral signs. A set of 5000 images are created for ISL numeral signs (0–9). Structural, pixel and histogram features are extracted from these ISL signs and are used as training and testing samples the recognition system. After feature extraction phase; k-Nearest Neighbor (kNN), Naïve Bayes, decision tree and neural network classifiers are used to test the performance of the proposed system. The recognition results obtained are with maximum accuracy rate of 97.17%.

Dr. Shelly Arora and Amandeep Kaur, Punjabi University, Patiala, presented a paper entitled, "Solution of Burger's Equation Using Orthogonal Collocation Technique With Lagrangian Basis". An abstract of this paper is presented below.

Burger equation is a stiff nonlinear equation of parabolic type. It is the class of few nonlinear problems possessing the analytic solution forming a base for the comparison of numerical results obtained. Technique of orthogonal collocation with Lagrangian basis have been proposed to solve the Burger's equation numerically. Zeros of Chebyshev polynomials have been taken as collocation points. The number of interior collocation points varied from 3 to 12. Numerical values have been presented graphically in the form of 2D and 3D graphs. Comparison of numerical and analytic values is presented in tabular form.

Dr. S. Prabhakaran and Prof. L. Jones T. Doss, Anna University, Chennai, presented a paper entitled, "Total variation diminishing scheme for multi dimensional multi-species transport with first order reaction network". An abstract of this paper is given below.

A total variation diminishing (TVD) scheme for multi-species transport with first order reaction network in multidimensional space is discussed in this article. The partial differential equations which describe this multi-species transport with chain reactions scenario are in a form of a coupled system. This system is solved by total variation diminishing scheme with various flux limiters. The numerical diffusion controlled by the flux limiters is explained in detail both theoretically and numerically. An attempt is made to use radial basis functions as flux limiters. The stability and consistency conditions of the TVD scheme are also derived. The explicit relation between the flux limiters and mesh parameters is obtained from this work.

Vivek Kumar and Prof. Bhola Ishwar, B. R. A. Bihar University, presented a paper entitled, "Normalization of Hamiltonian in photogravitational elliptic restricted three body problem with Poynting-Robertson drag". An abstract of this paper is given below.

In this paper, we have taken bigger primary as radiating and smaller as an oblate spheroid. We include Poynting–Robertson drag also. We have performed second order normalization in our problem. We have used Birkhoff's normalization of the Hamiltonian. For this, we have utilized Henrard's method and expanded the coordinates of the infinitesimal mass in double D'Alembert series. Finally we obtained the third order part H_3 of the Hamiltonian in term of $l_1^{1/2}$, $l_1^{1/2}$. We conclude that H_3 is zero.

Noor e Zahra, Sharda University, Greater Noida, presented a paper entitled, "Tumor Detection Using Wavelet and its Variants", based on joint work with Aakarshna. An abstract of this paper is given below.

A tumor, or neoplasm, is referred to an abnormal growth of tissue that may be solid, or fluid-filled. There are various kinds of tumors and their names reflect their shape and the type of tissue they appear in. To put it simply, it is a kind of swelling or lump. If a patient is suffering from brain tumor, his/her cells grow and multiply uncontrollably. If the growth becomes more than 50%, then the patient is not able to recover. Hence, the detection of brain tumor needs to be very fast and accurate. The paper objective is to provide an efficient algorithm to detect the edges of a brain tumor. The first step starts with the acquisition of the MRI image of the brain and then imaging techniques, wavelet and its variants are applied to find out the exact location of the tumor.

Ruchira Aneja presented a paper entitled, "Image Compression Using Alpha-Molecules" based on joint work with Prof. A.H Siddiqi. The paper describes a simple and efficient method of image compression using alpha molecules and Huffman coding technique. Image compression using wavelet transform, curvelet transform and contourlet transform followed by Huffman coding exists in the literature. In this paper we propose it using Shearlet transform which has not been done yet. Various biomedical images of MRI of different parts of body such as lungs, shoulder, cardiac and brain are compressed using transform-based coding. Wavelet and its variants such as curvelet, contourlet and Shearlet are used for transformation. The approximation and detail coefficients being extracted are processed for coding followed by decoding and inverse transformation. A comparative analysis is drawn between various transforms on the basis of various quantitative measures such as peak signal to noise ratio, mean square error, average difference, normalized absolute error, etc.

Mijanur Rehman of AMU, Aligarh, presented a paper "Existence Results for Strong Mixed Vector Equilibrium Problem for Multivalve Mappings". His results generalize, improve, extend, and unify some existence results proved earlier.

Nagma Irfan presented a paper based on joint work with her supervisor Prof. A.H. Siddiqi on application of sine-cosine wavelets to Radon transform and atmospheric tomography. Nitendra Kumar presented his results based on joint work with his supervisor Dr. Khursheed and co-supervisor Prof. A.H. Siddiqi, devoted to applications of wavelet method to classification of EEG signal, denoising of Raman spectroscopy and classification of EEG applying ANN and SVM. Padmesh Tripathi presented a paper based on his joint work with his supervisor, Prof. A.H. Siddiqi related to Noise removal in EEG signal.

Krishan Kumar and Dr. M.A. Ansari, Gautam Buddha University, Greater Noida presented a paper on "Distributed Energy Resources and Microgrid Systems challenges and Scope".

MIX
Papier aus verantwortungsvollen Quellen
Paper from responsible sources
FSC® C105338

If you have any concerns about our products,
you can contact us on
ProductSafety@springernature.com

In case Publisher is established outside the EU,
the EU authorized representative is:
Springer Nature Customer Service Center GmbH
Europaplatz 3, 69115 Heidelberg, Germany

Printed by Libri Plureos GmbH
in Hamburg, Germany